Hydrometry

Hydrometry

Principles and Practices

Edited by

R. W. Herschy

Department of the Environment
Water Data Unit, Reading

A Wiley–Interscience Publication

JOHN WILEY & SONS

Chichester · New York · Brisbane · Toronto

Library of Congress Cataloging in Publication Data:

Main entry under title:

Hydrometry: principles and practices.

  'A Wiley–Interscience publication.'
  1. Stream measurements. 2. Hydraulic
measurements.
I. Herschy, R. W.
TC175.H98   627'.12'028   78-4101

ISBN 0 471 99649 1

Printed in Great Britain by
J. W. Arrowsmith Ltd., Bristol BS3 2NT

Acknowledgements

The editor and contributors to *Hydrometry:Principles and Practices* wish to thank the following editors, publishers and individuals for permission to reproduce the material specified.

Chapter 1
    Figs. 1.1 and 1.2   Scottish Development Department, Edinburgh, UK
    Fig. 1.3   British Standards Institution, London, UK
    Figs. 1.4 to 1.10   Scottish Development Department, Edinburgh, UK
    Figs. 1.11 and 1.12   British Standards Institution, London, UK

Chapter 3
    Figs. 3.1 to 3.4, 3.6, 3.8   John Wiley and Sons Ltd.
    Figs. 3.5, 3.7 and 3.9   Hydraulics Research Station, Wallingford, Oxfordshire, UK. Crown copyright; reproduced with the permission of the Controller of Her Majesty's Stationery Office

Chapter 4
    Figs. 4.1 to 4.9   Water Research Centre, UK

Chapter 5
    Figs. 5.1 to 5.10   United States Geological Survey, Reston, Virginia, USA

Chapter 6
    Figs. 6.1 to 6.3   United States Geological Survey, Reston, Virginia, USA
    Figs. 6.4 to 6.17   Thames Water Authority, Reading, UK

Chapter 7
    Figs. 7.7 to 7.10   The Plessey Company Ltd, UK
    Figs. 7.11 to 7.20   Water Research Centre, UK

Chapter 8
    Figs. 8.1(a) and 8.1(b)   The British Standards Institution, London, UK
    Figs. 8.1(c)   Scottish Development Department, Edinburgh, UK
    Fig. 8.1(d)   United States Geological Survey, Reston, Virginia, USA
    Fig. 8.2 to 8.4   United States Geological Survey, Reston, Virginia, USA

v

Figs. 8.5 to 8.7 Leupold & Stevens Instruments Inc., USA

Fig. 8.8 Fischer & Porter Company, USA

Fig. 8.9 Leupold & Stevens Instruments Inc., USA

Figs. 8.10 to 8.16 United States Geological Survey, Reston, Virginia, USA

Fig. 8.17 Valeport Developments Ltd, UK

Fig. 8.21 to 8.27 United States Geological Survey, Reston, Virginia, USA

Fig. 8.28 Reproduced by permission of the Minister of Supply and Services Canada

Fig. 8.29 Leupold & Stevens Instruments Inc., USA

Fig. 8.30 United States Geological Survey, Reston, Virginia, USA

Figs. 8.31 to 8.32 Reproduced by permission of the Minister of Supply and Services Canada

Figs. 8.33 to 8.44 United States Geological Survey, Reston, Virginia, USA

Chapter 9

Figs. 9.1 to 9.22 Anglian Water Authority, UK

Chapter 10

Fig. 10.1 International Organization for Standardization, Geneva

Chapter 11

Fig. 11.10 Institute of Hydrology, UK

Chapter 12

Figs. 12.1 to 12.4 World Meteorological Organization, Geneva

Fig. 12.5 Reproduced by permission of the Minister of Supply and Services Canada

Figs. 12.6 United States National Environmental Satellite Service

Fig. 12.7 US Army, Corps of Engineers

Fig. 12.8 Reproduced by permission of the Minister of Supply and Services Canada

Chapter 13

Figs. 13.1 to 13.9 Orstom, Paris, France

Chapter 14

Figs. 14.1 and 14.2 World Meteorological Organization, Geneva

Contributing Authors

H. H. Barnes	United States Geological Survey, Reston, Virginia, USA
F. G. Charlton	Hydraulics Research Station, Wallingford, Oxfordshire, UK
J. Davidian	United States Geological Survey, Reston, Virginia, USA
K. K. Framji	International Commission on Irrigation and Drainage, 48 Nyaya Marg, Chanakyapuri, New Delhi, India
M. J. Green	Water Research Centre, Medmenham, Buckinghamshire, UK
R. A. Halliday	Environment Canada, Inland Waters Directorate, Water Resources Branch, Ottowa, Ontario, Canada
R. W. Herschy	Department of the Environment, Water Data Unit, Reading, UK
V. V. Kuprianov	Hydrometric Service of the USSR, State Hydrological Institute, Moscow D-376, USSR
J. C. Lambie	Scottish Development Department, Edinburgh, UK
R. J. Mander	Thames Water Authority, Reading, UK
T. J. Marsh	Department of the Environment, Water Data Unit, Reading, UK
J. Němec	World Meteorological Organization, Geneva, Switzerland
K. E. C. Powell	Anglian Water Authority, Boston, Lincolnshire, UK
M. Roche	Office de la Recherche Scientifique et Technique Outre-Mer, 75008 Paris, France
J. Rodier	Office de la Recherche Scientifique et Technique Outre-Mer, 75008 Paris, France
G. F. Smoot	United States Geological Survey, Reston, Virginia, USA
K. E. White	Water Research Centre, Stevenage, UK
W. R. White	Hydraulics Research Station, Wallingford, Oxfordshire, UK

Contents

Preface

Hydrometry: Principles and Practices consists of a collection of interrelated chapters on the measurement of flow in open channels.

These chapters are written by international experts who are practising engineers and scientists but who are also concerned with the formulation and implementation of international standards and with research.

The book deals with both traditional methods and new methods of flow measurement and is aimed at all those concerned with the management of water.

CHAPTER 1

Measurement of Flow— Velocity-area Methods

J. C. LAMBIE

1.1   INTRODUCTION

Streamflow is caused by the runoff of precipitation and, on a natural undeveloped river system, the fluctuations in stage and discharge result from variations in the duration, frequency, intensity and areal cover of precipitation and in the catchment characteristics which control the rate and amount of runoff.

Over any given region the quantity of water flowing in streams can vary widely in both time and space, and knowledge of these variations and of the total quantities involved is essential for the proper development of man's activities. Records of floods are required for the design of river structures such as bridges, dams and floodbanks and for the operation of flood warning systems; records of low flows are needed for the evaluation of drought conditions, the control of abstractions and the design of water conservation measures; and records of longer term mean flows are necessary for the design of power production and water resource systems. It is universally recognized that the problems created as a result of the continually increasing demand for water for domestic, industrial and agricultural use can only be solved by the adoption of extensive water conservation measures. The assessment, management and control of water resources can only be effective if there is access to accurate and continuous information on streamflow, and this information can only be obtained satisfactorily from a network of river gauging stations usually, but not necessarily, sited and operated as part of a more general hydrologic network.

A gauging station is a site on a river which has been selected, equipped and operated to provide the basic data from which systematic records of water level and discharge may be derived. Essentially it consists of a natural or artificial river cross-section where a continuous record of stage can be obtained and where a relation between stage and discharge can be determined. The proper selection of each site is of major importance, as the quality of the site and particularly the hydraulic characteristics of the channel affect the type of equipment and the method of operation. adopted for the station and ultimately the accuracy of the record produced.

The general requirement is to produce a reliable, accurate and continuous record which will stand up to close and critical inspection, but before this can be ensured there are many problems to be overcome in the siting, equipping and operation of the station. Because of this it is advisable, when designing a gauging station network, to have a clear picture of the need for gauging at any particular site and to identify the various uses to which the data will be put. Some aspects of river management may justify a high refinement in flow measurement, whereas others do not; for example, the determination of acceptable irrigation allocations or abstractions is much more sensitive to the accuracy of flow measurement data than the yield of a catchment for water supply purposes. It may well be that in certain cases the uses do not demand a high level of accuracy in the basic data and therefore a lower grade station would be theoretically acceptable, but it is important to realize that needs can change very quickly and a catchment of little material importance today may well be of interest tomorrow.

In the early days of river gauging, when the network of gauging stations throughout the United Kingdom was sparse and when methods in use for obtaining and analysing the data were, by today's standards, rather suspect, there was little opportunity of being able to correlate results from one station to another and consequently no great incentive towards striving for high accuracy or even for showing that accuracy was being obtained. Nowadays, however, with the need to ensure efficient management of river basins, each river system is being closely covered by measuring stations and accuracy of a high order is essential to ensure that data from one station will correlate with its neighbour. This makes it essential that each station, particularly those for national network purposes where it is hoped that some standardization will be effected, should be sited, constructed and equipped to a high standard and operated in a way that will take into consideration any peculiarities of the site.

1.2 METHODS FOR MEASURING RIVER FLOW

1.2.1 General

There are several methods for measuring river flow—the longer established velocity-area, structural and dilution techniques and the more recent adaptations involving ultrasonic, electromagnetic and moving boat methods—each method having advantages and limitations which makes its choice dependent on site conditions and on the equipment and resources available to the gauging authority. In this chapter the velocity-area method only is considered.

1.2.2 Velocity-area method

The primary object of a gauging station is to provide a record of the discharge of the open channel on which it is sited. At a gauging station calibrated by the velocity-area method this is achieved by producing a continuous measurement of water level (stage) and transforming this to discharge by means of a stage–discharge relation which correlates discharges to either the water level at a single section of the channel or to the water levels at each end of a reach. In the former case a single-gauge station is employed but in the latter a twin-gauge station is necessary and a different procedure for establishing the stage–discharge relation is used.

The stage–discharge relation is determined from field measurements of stages and corresponding discharges and the calibration so established holds good only so long as no significant alteration takes place in the characteristics of the reach. Since a river channel is continuously in course of development, it is inherently unstable and its characteristics are subject to changes which may affect the calibration. These changes may take place gradually as a result of slow processes of erosion or siltation or they may occur suddenly as a consequence of alterations in the channel. In addition, temporary changes may be caused by the growth and decay of aquatic weed, by the formation and break-up of ice cover, or by the deposition of debris. It is therefore, essential to keep an established stage–discharge relation continuously under review in order to be assured that its validity is maintained and to redetermine the relation when it is shown to have been significantly altered by the changes which have occurred.

In its simplest form gauging by the velocity-area method involves the measurement of the area of a river cross-section and the mean velocity of flow through it, discharge being the product of the two quantities. The area of cross-section is determined by means of soundings at a number of verticals on the cross-section, together with the measurement of the distance of these verticals from a reference point on the bank, and velocities are determined mainly by current meter observations. The accuracy obtained during any one gauging depends on the integrity of the current meter rating and on the sufficiency and competency of the observations of velocity and depth. In general, the characteristics of the site, together with a knowledge of the accuracy required, will determine the detail necessary in the gauging procedure and in the operation of the station.

Velocities may also be determined by floats or by measuring the surface slope of the flowing water, but it is generally accepted that the current meter has a decided superiority in accuracy over these methods and its use is virtually universal. The other two methods still have a place in any broad river gauging programme and recourse to them may be necessary when the use of current meters is impracticable and where the need for accuracy is not so great.

1.3 SELECTION OF SITE

1.3.1 General

The overall accuracy of the record from a gauging station is governed to a large extent by the physical and hydraulic characteristics of the channel in which it is placed and particularly those which control the stability and sensitivity of the stage–discharge relation, and the precise location of a station is decided by the availability of a site where the desired characteristics are present.

At an ideal site with a permanent control, discharge would be a unique function of stage so that during rising and falling flood cycles a given stage would always represent the same discharge. It is not often possible to fulfil the requirements of an ideal site and there can be cases where, because of the many special purposes which the stations are required to serve, the choice may be limited by the need to locate a station for some specific river management purpose—for example, at a dam site, at a key point for drainage outlets, at a critical flooding section, or on a tidal reach. In other cases there may be a lack of suitable sites throughout an entire river basin and location in poor conditions may have to be accepted. Within these limitations the ultimate requirements of an effective national network of gauging stations should be kept constantly in mind and every effort made to select locations which will fit into and form part of such a network. In general, however, the site which will give the greatest continuity and accuracy of record should be chosen.

1.3.2 Controls

Controls may be natural or artificial but in all cases their function is to stabilize and sensitize the relation between stage and discharge at the measuring section (see Figures 1.1 and 1.2). Stability is important to reduce the extent of recalibration required and sensitivity is an indication of the extent of the response to an increase in discharge by an increase in stage. Where a small increase in discharge produces a relatively large increase in stage the relation is said to be sensitive; where a relatively large increase in discharge produces a small increase in stage the relation is said to be non-sensitive. In a broad river gauging programme where a wide variety of stations are operated it is likely that variations in sensitivity between the two extremes will be found.

The relation between rate of change of stage and rate of change of discharge is governed by those physical characteristics of the river channel at, or near, the gauging site which form the control of the station. It should be noted that in this connection the control can be either a section control, caused by a sharply defined physical feature such as a weir, ledge of rocks, boulder-covered riffle; a channel control where the relationship between stage and

Figure 1.1 Typical river reach showing pooled section and downstream natural control

Figure 1.2 Typical artificial control

discharge is governed only by the dimensions, slope and roughness of the river bed and banks; or a combination of both. From the point of view of sensitivity the extremes can best be illustrated by reference on the one hand to a V-shaped control section which causes a large increase in head for a small increase in discharge, and on the other hand to a control section formed by a long weir of horizontal crest on which a large increase in discharge, especially at low flow, will cause only a small increase in stage. The degree of sensitivity affects the record of the gauging station during the process of calibration, but it must also be realized that its effect is felt at all stages in the collection of the streamflow record and is reflected in the accuracy of the final record of discharge. It is clear that a sensitive record of stage can be converted more accurately into a record of discharge than a non-sensitive one. For example, at a sensitive station it may be sufficient to be able to read water level to an accuracy of 5 mm, but at a less sensitive station the accuracy required will have to be 1 mm to obtain the equivalent accuracy in discharge. This will influence the choice and design of the reference gauge and recording instrument and also the quality of control and maintenance required.

As far as can be ascertained it appears that there is no standard method of defining sensitivity in a way which would allow a comparison of this quality to be made, station to station. At any single station sensitivity varies with discharge, in most cases tending to decrease as discharge decreases and, for a true comparison to be made, it should be calculated for each station at the same level of flow such as annual average, annual minimum, or some intermediate state such as an agreed percentile of flow. As the effect is greatest at low flow, it is desirable that some point in the lower range should be used, but to use absolute minimum is possibly too rigorous and it is suggested that the 95 percentile be adopted. For this purpose the definition of sensitivity has been taken as 'the increase in stage in millimetres caused by an increase in discharge of 1% at the stage corresponding to the long term 95 percentile level of flow'. This means that the sensitivity of the control and therefore of the stage–discharge relation will be better than the adopted value for 95 per cent of the time (347 days in a year on average) but worse for 5 per cent of the time (18 days in a year on average).

In practice, the value of sensitivity determines the degree of accuracy required in taking off records of stage. At a station with a sensitivity of 5, an error of 5 mm in reading stage would give an error in the computed discharge of 1 per cent, which could be regarded as satisfactory. At a station with a sensitivity of 0.5, an error of 5 mm in stage would give an error in discharge of about 10 per cent, which is unsatisfactory—in this case the station should be equipped to a standard that would ensure a more accurate reading of stage. It is essential therefore that due consideration be given to sensitivity when establishing a gauging station and in some cases it may be advisable or even mandatory to build an artificial control shaped to provide a suitable sensitivity value.

1.3.3   Requirements

It is particularly important to check that the site is free from continuous or seasonal weed growth which could affect the stage–discharge relation or prevent the full use of current meters, and that no abnormal conditions occur during rising and falling stages. In addition to hydraulic considerations it is necessary to take into account such practical points as cost, accessibility, the availability of local observers to provide attendance, the availability of power lines, the feasibility of building artificial controls, etc. The site selected should be such that it is possible to measure the whole range and all types of flow which may be encountered or which it is required to measure. All measurements should be referred to the water level at one reference gauge only, but may be made at a single section for all flows or at two or more sections for different ranges of flow. Either single- or twin-gauge stations may be employed depending upon conditions, but the former should be preferred.

In general the most favourable conditions for recording stage, for taking accurate single measurements of discharge (gaugings) and for establishing a stable stage–discharge relation will be obtained when certain requirements are met and these should be given full consideration:

(1) The channel should be straight and of uniform cross-section and slope to ensure parallel and non-turbulent flow and to reduce the chance of abnormal velocity distribution. Ideally the length of straight should be at least three times the channel width with the measuring section mid-way, but where this is not possible the measuring section should be within the downstream half of the reach. It should, however, be remote from any natural or artificial obstructions on the banks or in the channel likely to cause disturbance, distortion or reversal of flow.

(2) The depth and velocity of water at minimum flow and the velocity and turbulence at maximum flow should be within the limits imposed by the type of measuring equipment to be used.

(3) The physical characteristics of the channel should ensure a substantially consistent and stable relation between stage and discharge. The channel itself should be stable and there should be no variable backwater such as from tidal influences, downstream tributaries, locks, sluices, off-takes and other structures.

(4) The channel and especially the control should be free from weed in all seasons.

(5) Flows at all stages should be confined to a well-defined channel or channels or within an unobstructed floodway having stable boundaries.

(6) The site should be accessible at all times and at all stages of flow.

(7) The orientation of the reach should be normal to the prevailing wind, particularly where the reach is long and straight and has a flat surface slope.

(8) The site should be sensitive, so that a small increase in discharge will produce a relatively large increase in stage.
(9) The field of view of the measuring section and the upstream reach should be clear and unobstructed.
(10) A local observer should be available to provide routine attendance.

It is often difficult or impossible to find a site in the required reach of a river which completely satisfies all these requirements, and experience has shown that the practical considerations are of prime importance. A site which is quite ideal from the hydraulic standpoint is useless if arrangements cannot be made for regular attendance or if it is impossible or excessively dangerous to approach the station, with the necessary apparatus, to make velocity measurements at all stages of the river and by night as well as by day. For that reason it may sometimes be necessary to accept sites which are not in every respect ideal hydraulically. Thus while parallel flow and adequate control are indispensable some irregularity of cross-section and some over-bank flow may be accepted provided neither is excessive. Some degree of instability is almost inevitable in any natural river channel.

The necessity to have sufficient depth of water to immerse a current meter satisfactorily and at the same time to have velocity within the range of accurate registration of the meter is nearly always a difficulty at low stages of flow and generally it becomes necessary to make such measurements at a different site from that selected for measuring medium and high discharges. In a few cases it may be possible to locate a gauging station where a stream is crossed by a bridge, so that velocities may be observed by suspending a current meter over the bridge rail, but in general a bridge site seldom satisfies completely the most important hydraulic requirements. On the other hand a very effective control is often found at a bridge and a good station site may exist a short distance upstream from it.

Where some of the desirable features are not naturally present, consideration should be given to the possible improvement of these conditions by remedial works (see Section 1.5).

1.3.4 Preliminary survey

In all cases a preliminary survey should be made to check the physical and hydraulic features of a site and to demonstrate the extent to which the desired characteristics are present. Where possible this should be done well in advance of deciding on the permanent location to allow opportunity to survey under different flow and seasonal conditions. However, it is seldom that circumstances allow the opportunity for such an extended study and locations are generally made on the basis of a reconnaissance by an experienced engineer followed by a survey to pick up the topographical and hydraulic details of the site. This should include a plan of the site, indicating the width of the water surface at a stated stage, the edges of the natural banks of the channel, the line of any definite discontinuity of the slope of these banks, and the toe and crest of any artificial floodbank. All obstructions in the channel or channels or floodway should be indicated. A longitudinal section of the channel should be drawn from below the control where this exists to the upstream limit of the reach, showing the level of the deepest part of the bed and water surface gradients at low and high stages.

At least five cross-sections should be surveyed in the measuring reach, two cross-sections upstream from the measuring section and two downstream covering a minimum of one bank-full width of the channel in each direction. The control should be defined either by one or more cross-sections or by a close grid of spot levels over the area. The survey of the reach should be extended through the floodway to an elevation well above the highest anti-cipated stage of flood. The spacing of levels or soundings should be close enough to reveal any abrupt change in the contour of the channel, and the bed of the reach should be examined carefully for the presence of rocks or boulders, particularly in the vicinity of the measuring section.

Where velocities are to be measured by current meter, exploratory current meter runs should be made in the proposed measuring section and in the cross-sections immediately upstream and downstream using, where possible, the method of velocity distribution (Subsection 1.7.2).

When floats have to be used for velocity measurements, trial runs of floats should be closely spread across the width of the channel to ensure that any abnormality of flow will be detected. In both cases the velocity measurements should be repeated at more than one stage.

1.3.5   Definitive survey

After the construction of the station a final survey should be made which should include the accurate determination of the elevations and relative positions of all the station installations and any other key points or significant features of the site. Of particular importance are the elevations of the station bench mark, the zero of the reference gauge, the inverts of intake pipes, the level of the control, and any datum marks within the instrument house for setting and checking the recorder. The elevations should be connected with ordnance survey datum through the station bench mark.

The bed profile of the channel should be carefully determined along the line of each cross-section, preferably when the flow is at a low stage and greater accuracy is possible. Measurements of depth should be made at intervals close enough to define the cross-section accurately and at least two depth measurements should be taken at each point, when sounding by rod or weighted line, and the mean value adopted. In general the intervals should not be greater than one-fifteenth of the width, in the case of regular bed profiles, and not greater than one-twentieth of the width, in the case of irregular bed profiles. In channels of sufficient depth, or where the velocity is high, an echo-sounder can be used but only if it is regularly and frequently calibrated under the same conditions of water salinity and temperature as those in which it is employed. Where possible, the depth at a point should be taken as the average of several readings but where this is not practicable individual depths may be taken from a single reading of the instrument. It is essential, of course, that when depths are determined by soundings referred to the water surface, frequent readings of water level should be made on the reference gauge to ensure that all measurements can be corrected to the same plane.

For current meter stations a standard profile on the measuring cross-section should be drawn, indicating the position of the cross-section markers, the positions selected for the measuring verticals, and the depth of water at each vertical when water level is at the zero of the reference gauge. The profile should be checked at regular intervals and a copy kept in the instrument house for reference and use during gauging.

1.3.6   Inaccuracies in sounding

Inaccuracies in sounding by rod, line or echo-sounder are dependent on the bed conditions and the type of equipment being used but are most likely to occur due to the following:

(1) The departure from the vertical of the sounding rod or line, particularly in deep water. A sounding line may deviate from the vertical owing to the force exerted by the stream on the line itself and on the sounding weight, but the amount of deviation may be minimized by using fine wire line (2.5 mm diameter or less) and a streamline weight. Where necessary, corrections to the indicated depth should be applied to compensate for such deviation.
(2) The penetration of the bed by the sounding weight or rod. This difficulty may be alleviated by fitting a base plate.
(3) The presence of boulders or large stones. The influence of these may be reduced by taking a greater number of soundings.
(4) The presence of soft deposits on the bed which may give rise to a double echo when an echo-sounder is used.

1.4   SELECTION OF EQUIPMENT

1.4.1   General

The choice of equipment to be used will depend on the characteristics of the site, particularly those affecting the stability and sensitivity of the stage–discharge relation, the type and quality of data required, the manpower and back-up facilities available to the gauging authority, the initial capital and future maintenance costs, and the availability of local attention. Essentially, however, the apparatus will consist of the equipment described in the following subsections.

1.4.2   Non-recording gauges to measure stage (see also Chapter 8)

These may be vertical or inclined gauges placed in the stream and from which readings of water level can be taken directly, or hook, point, chain, wire-weight or float gauges with which water level is measured indirectly by reference to a fixed point above the water surface. Non-recording gauges are cheap and simple to install and maintain and on certain locations, such as lakes or large rivers which rise and fall slowly, daily or twice daily readings may suffice for the station stage record. They have, however, certain disadvantages—regular and systematic reading by an observer is essential but not always possible and there is a loss of accuracy when trying to estimate the fluctuations which have taken place between readings. Their principal use, therefore, is as a reference gauge from which readings can be taken to check and adjust the operation of a continuous recorder.

1.4.3   Recorders to provide a continuous record of stage (see also Chapter 8)

The accuracy of the record of discharge is governed principally by the accuracy of the record of stage, and it is essential that the instrumentation used can sense and record water level efficiently and with an accuracy sufficient for the purposes for which the measurements are required. The record of stage is commonly produced by a water level recorder actuated by a float and counterweight or tensator spring system working within a stilling well, the movement of the float being used to operate a recording mechanism such as a pen or punching head which can produce either an analogue record on a chart or a digital record on punched tape. Essentially the recorder consists of a time element and a water height element which operate together and produce on the chart or tape a record of the rise and fall of the water surface with respect to time. The time element consists of a clock, actuated by a spring, weight or electrical mechanism, driving in the case of an autographic recorder either a chart drum or a recorder pen, or in the case of a digital recorder rotating a cam which initiates the punching cycle. The water height element consists of a float, combined with a counterweight or tensator device, and some form of mechanical linkage to connect either directly or through reduction gearing to the recording device.

   In the case of digital recorders an encoding or digitizing device is used to convert the analogue measurement of water level to a digital output. The recording device can be a pen or pencil, in the case of an autographic record, or a punching head, in the case of a digital record. The essential requirement of such a recorder is that it is reliable in performance with respect to both stage and time, but in any such system there are possible sources of error which can cause a difference in the actual and recorded level—these must be eliminated or taken into account.

Stage may also be sensed by pressure bulb or gas purge (bubbler) devices which sense pressure changes caused in water depth. The first type comprises a pressure bulb connected by a small-bore pipe to a pressure indicator reading in terms of water level, the calibration being adjusted for the density of the liquid. In the second type, a small quantity of air or inert gas is allowed to feed through a tube and bubble into the liquid to be measured. A gauge measures the pressure of the gas required to displace the liquid in the pipe and this pressure is directly proportional to the head above the end of the tube and can be measured in terms of liquid level. It is generally accepted that the performance of float-operated recorders is more reliable than that of pressure-actuated ones, although the accuracy of data produced by the two systems is about the same when operated and maintained to a high standard. Pressure gauges, therefore, are used mainly for proving sites, for short-term stations, or for long-term stations where site conditions are not suitable for the installation of a stilling well.

1.4.4 Stilling well and instrument house to protect the recorder

To protect the float and to eliminate, or at least reduce, the effect of surface waves and short period surging in the natural channel it is customary to provide a stilling well, usually set back in the bank of the river and connected to it by one or more intake pipes, but sometimes placed directly in the stream. The accuracy of the recorded water level will then depend partly on the sensitivity of the instrument but mainly on any difference between the water level inside the stilling well and that in the river. The stilling well and intake pipe assembly must therefore be designed to ensure that these levels correspond within acceptable limits. The instrument house is required to protect the instrument from adverse environmental conditions and vandalism and may vary in size, type and sophistication depending on its location.

1.4.5 Current meters for measuring velocities

These may be vertical axis, cup-type or horizontal axis, propeller-type meters both of which give results of sufficient accuracy when used in the proper manner and under the proper conditions. They are described in detail in Chapter 2.

1.4.6   Suspension equipment (see also Chapter 8)

When taking discharge measurements it is necessary to measure accurately the area of cross-section and the mean velocity of the flowing water. For this purpose a suspension system is required which provides means of measuring accurately the depth of water and which is suitable for mounting the current meter and placing and holding it at any predetermined part of the cross-section. This can involve the use of wading rods with tag lines, cableway systems with cored suspension cables and weights, or longer rods, handline or crane equipment for use from boats or bridges. In general, gaugings are taken either by wading at low levels with the meter mounted on a wading rod, from boats or bridges with the meter mounted on long rods or suspended on cables, or from slack line cableways at higher levels with the meter suspended from a thin cable and controlled from the bank or from a cable car. Only in exceptional cases are bridges used, as experience has shown that bridge sites do not often provide the hydraulic conditions necessary for a station where a stage–discharge relation is to be established.

1.5   DESIGN AND INSTALLATION

1.5.1   General

Basically the station will consist of one or more cross-sections with facilities for measuring width, depth and velocity of the water, one or more reference gauges to indicate stage, one or more recorders to produce a continuous record of stage, and any other miscellaneous structures such as a stilling well, recorder house, station bench mark which may be required. The design and layout of the works should be based on the features disclosed by the topographical survey, and the choice of equipment and instrumentation to be installed should be made from considerations of station stability and sensitivity, the type of data to be produced, and its required accuracy.

At a gauging station, the measurement of stage is the most important of all the measurements made in the course of flow data collection, and is the elevation of the plane of the water surface at the site by reference to a fixed datum normally taken as the level of the zero mark of the station reference gauge. This is the base to which all other measurements are related and to which the setting and checking of the water level recorder is referred. To avoid the possibility of negative gauge readings, the level of the zero mark should be selected to be well below the elevation of zero flow on the control. A permanent datum should be maintained so that one only is used for the stage record throughout the life of the station. Where an arbitrary datum plane is used, therefore, it should be referred to one or more station bench marks located at the site but independent of the gauge structure. These bench marks should in turn be referred to a bench mark of known elevation above sea level so that the precise datum may be recovered if the gauge and reference marks are destroyed. It is essential that the relation between the gauge zero and the bench marks should be checked at least annually to ensure that no change has taken place.

The station bench mark or marks should be set in a position offering maximum security against disturbance or damage, but to facilitate accurate levelling between them and gauge zero they should be so located that the transfer of level may be carried out by reciprocal levelling or with equally balanced foresights and backsights.

The section control which regulates the stage at low flows at the gauging section is situated at the downstream end of the reach, and the measuring section should be sufficiently remote from it to be clear of any distortion of flow which might occur in that vicinity. It should be close enough, however, to reduce or eliminate the possibility of a variable stage–discharge relation being introduced through the effect of wind or weed growth in the channel. At higher stages, when the low flow control is drowned out, flow may be regulated by other section controls further downstream but more often is controlled by the general characteristics of the channel. Where possible, these should be identified and evaluated for use during the calibration stage.

The reference gauge and recorder should be located as closely as possible to the measuring section in the case of a current meter station or near the mid-point of the measuring reach in the case of any other method of measurement.

Trees obstructing a clear view of the measuring section or measuring reach should be trimmed or removed. To reduce the danger of damage to the meter assembly the field of view should extend far enough upstream to enable floating debris to be seen in sufficient time to permit the removal of the meter from the water.

Where not already existing, a suitable access to the site should be constructed to provide safe passage, at all stages of flow and in all weathers, for personnel and for any vehicles used for the conveyance of instruments and equipment.

All key points of the site should be permanently marked on the ground by markers sunk far enough below the surface to secure them against movement. Cross-section markers should be set on the line of the cross-section to facilitate the repetition of levels or soundings when the section is checked.

Where the main requirements for a suitable gauging site are not present conditions may be improved by remedial works. For example, the loss of water by spillage from the main channel can be avoided by constructing floodbanks to confine the flow in a defined floodway. Minor irregularities in the bank or bed causing local eddies may be eliminated by trimming the bank to a regular line and a stable slope and by removing from the bed any large stones or boulders. It is advisable to protect unstable banks and such protection should, in the case of a current meter station, extend upstream and downstream of a measuring section for a distance equal to at least one-quarter the bank-full width of the channel in each direction. In the case of float measurements the whole length of the measuring reach should be protected.

Instability of the bed may be improved by introducing an artificial control which can also be designed to improve the stability and sensitivity of the stage–discharge relation, to create conditions in the measuring section for instruments to be effectively used and, where necessary, to reduce or eliminate variable backwater effect. Most controls of this type are relatively simple structures designed in accordance with the conditions at the site to function effectively at the low and possibly medium ranges of stage only—it is seldom that the effects of downstream conditions can be eliminated over the higher ranges. Although laboratory calibrations can be obtained for some types of control, such as the flat-V weir, the station calibration should in all cases be established by the current meter method. Four major points should be considered when preparing the design of controls:

(1) The structure should be stable and permanent.
(2) The shape of the structure should permit the passage of water and debris without creating undesirable disturbances of the channel regime above or below the control.
(3) The height of the structure should be sufficient to eliminate the effects of variable conditions downstream, particularly at low flows.
(4) The profile of the crest should be designed to be sensitive, so that a small change in discharge at low stages will cause a measureable change in stage.

Where, at low stages, there is insufficient depth or velocity in the primary measuring section, it may be possible to gauge in the same general reach of the stream at another section which is more suitable under low flow conditions.

1.5.2   Reference gauges (see also Chapter 8)

The non-recording reference gauge is the basic instrument for the measurement of stage whether at a regular flow measuring station or at a site where only casual observations are made. It can be sited as an outside gauge to allow a direct reading of water level in the stream or as an inside gauge to indicate the level in a stilling well, and is used for setting and checking the water level recorder, for indicating the stage at which discharge measurements are taken, and for emergency readings when the recorder is out of action. There are various forms, the choice being decided by the site conditions and the specific use to which it will be put.

Vertical and inclined gauges

A vertical (staff) gauge consists of a scale marked or securely attached to a suitable vertical surface and can be installed as a single gauge covering the whole range of stage or as a series of stepped gauges normal to the direction of flow. An inclined (ramp) gauge consists of a scale marked on or securely attached to a suitably inclined surface and can be set on one continuous slope or on a compound of two or more slopes on a cross-section normal to the direction of flow. In all cases it is essential that the gauges extend above and below the highest and lowest levels expected at the site, that they are accurate and clearly marked, and that they are simple to install and use and durable and easy to maintain. In particular the material used for the construction should be durable in alternating wet and dry conditions, particularly in respect of resistance to wear or fading of the markings.

Where possible the gauge should be placed at the edge of the stream, but where this is impractical because of adverse site conditions such as excessive turbulence, wind effect or inaccessibility, it can be placed in a suitably protected stilling bay in which the wave actions are damped and the level of the water surface closely follows the fluctuations of the water level in the mainstream. It should be located as close to the measuring section as possible without affecting the flow conditions at this point, but should not be placed where the water is disturbed by turbulence or where there is danger of damage by drift—bridge abutments or piers are generally unsuitable. It must be readily and conveniently accessible so that the observer can make readings as nearly as possible at eye level; at the majority of sites this requires the construction of a flight of steps.

For vertical gauges a suitable backing is provided by the surface of a wall having a vertical or nearly vertical face parallel to the direction of flow, the gauge board being securely fastened to the surface so as to present a truly vertical face to receive the gauge plates. Where a pile is used as the backing post it must be driven firmly into the river bed or bank or set in concrete so as to be secure against damage or displacement, and any anchorage must extend below the ground surface to a level free from disturbance by frost. In order to avoid velocity effects which prevent accurate reading, a pile should be shaped to present streamline cut-waters upstream and downstream or alternatively situated in a bay where it will not be exposed to the force of the current.

Gauge plates are normally manufactured in convenient lengths of enamelled iron plate or similar material with subdivisions clearly and accurately marked for easy identification. A typical gauge plate is shown in Chapter 8 (Figure 8.1a). For inclined gauges it is usual to construct a special backing structure to carry the graduated scale. It should be securely anchored to the bank and founded at a depth below frost influence or any slip plane and it should be sufficiently removed from and have no connection with any heavy structure liable to cause subsidence. The graduated scale should be fixed to the backing but provision should be made for removing it when required for maintenance or adjustment. Except where use is made of manufactured gauge plates designed to be set to a specified slope, the gauge should be calibrated *in situ* by careful levelling from the station bench mark. A typical example of an inclined gauge installation is shown in Chapter 8 (Figure 8.1b).

Electrical gauges

There are a number of electrical devices used to detect and indicate water surface level. The electrical point gauge is a needle gauge provided with an electrical device to indicate the instant when the point touches the water surface. The electrical tape gauge consists essentially of a graduated steel tape fastened to a cylindrical weight and linked to an electrical circuit which is completed when the weight touches the water surface. Both types are used

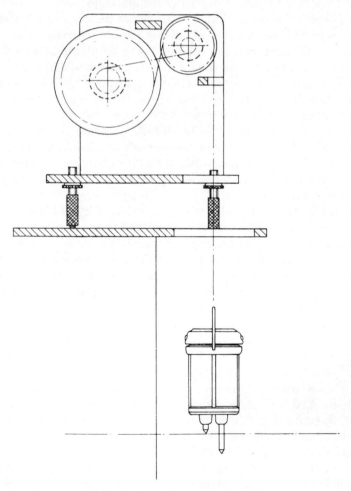

Figure 1.3   Electrical depth gauge

mainly as reference gauges inside stilling wells but can be used outside where conditions are suitable.

A typical electrical point gauge is shown in Figure 1.3. Other types of gauge in common use in different parts of the world are described in Chapter 8.

1.5.3 Stilling wells and intakes

The function of a stilling well is to accommodate the water level recorder and protect the float system, to provide within the well an accurate representation of the mean water level in the river and to damp out natural oscillations of the water surface. The function of the intakes is to allow water to enter or leave the stilling well so that the water in the well is maintained at the same elevation as that in the stream under all conditions of flow, and to permit some form of control with which to limit lag and oscillating effects within the well.

The well itself can be constructed from any suitable building material such as concrete blocks, bricks or stone, or fabricated from sections of concrete, steel, fibre glass, galvanized iron or similar type of pipe. It can be set into the bank of the stream or directly in the stream when attached to a rigid support such as a bridge pier or abutment (see Figure 1.4). When placed in the bank it should have a sealed bottom to prevent seepage into or leakage out of the chamber and should be connected to the stream by one or more intake pipes; when placed directly in the stream the intakes may take the form of holes or slots cut in the well itself. It is essential that the well remains stable at all times and it must therefore be firmly founded when placed in the bank and firmly anchored when standing in the stream. Its dimensions should allow unrestricted operation of all equipment installed in it and it is recommended that where a single float is used within the well the clearance between walls and float should be at least 75 mm, and where two or more floats are used clearance between them should be at least 150 mm. In silt-laden rivers it is an advantage to have the well large enough to be entered and cleaned.

The well and all construction joints of well and intake pipes should be watertight so that water can enter or leave only by the intake itself. It should be vertical within acceptable limits and have sufficient height and depth to allow the float to travel freely the full range of water levels. The bottom of the well should be at least 300 mm below the invert of the lowest intake pipe to provide space for sediment storage and to avoid the danger of the float grounding at times of low flow. The well itself should not interfere with the flow pattern in the approach channel and if set behind a control should be located far enough upstream to be outside the area of draw-down to the control.

The intakes may take the form of one or more pipes connecting the well to the river, when the well is set back into the bank, or a series of holes or slots cut into the well itself, when it is set directly into the river. In rivers with a

Figure 1.4   Stilling well installed in river showing vertical gauge
and instrument house

permanent high silt content a well set in the stream may have a hopper-
shaped bottom to serve as an intake and also as a means of self-cleansing.

The dimensions of the intakes should be large enough to allow the water
level in the well to follow the rise and fall of river stage without appreciable
delay but must also be small enough to damp out oscillations caused by wave
action or surges. These requirements are opposed and a suitable balance

should be achieved. For example, to effectively eliminate surging it may be necessary to restrict the cross-sectional area of the intakes to 0.1 per cent of the cross-sectional area of the well, whereas to reduce lag effect to acceptable limits the ratio may have to be at least 1 per cent. This will depend on site conditions, the type and length of intakes and the surface area of the well. Because of this, no firm rule can be laid down for determining the best size of intake but it is advisable to make the connection too large rather than too small, as a restriction may be added if found necessary. As a general guide, the total cross-sectional area of the intakes should not be less than 1 per cent of the cross-sectional area of the well.

To ensure continued operation of the system if the lowest intake pipe becomes blocked it is often advisable to install two or more intakes, one vertically above the other. The lowest intake should be at least 150 mm below the lowest anticipated stage. In cold climates this intake should be below the frost line. Where practicable the intake pipe should be laid level and straight on a suitable foundation which will not subside or at a constant gradient with the highest point at the stilling well. The pipe should enter the stream normal to the direction of flow, and where it terminates in a head wall it should be set flush with the wall.

Intake pipes more than 20 m in length should be provided with an intermediate manhole fitted with internal baffles to act as a silt trap and provide access for cleaning. Means of cleaning the intakes should be provided either by a flushing system where water under several metres of head can be applied to the stilling well end of the intake, by pumping water through the intake, or by hand cleaning with collapsible draining rods. Where velocity past the river end of the intake is high, draw-down of the water level in the well may occur but this can be reduced by attaching a capped and perforated static tube to the river end of the intake and extending it horizontally downstream.

In cold climates the well should be protected from the formation of ice which could restrict or prevent the free operation of the float system. This may be done by the use of well covers, sub-floors to act as a frost barrier, heaters or oil on the water surface. When oil is used the oil surface will stand higher than the water level in the stream and a correction must be applied when setting the recorder. When discharge in the river increases, the water level in the channel rises and water flows from the channel into the stilling well. If this flow is appreciable, head losses in the connecting pipe including any control fittings used (valves, outlet tees, static tubes, etc) can cause a difference between water level in the river and that in the well. This is due to stilling well lag and for any given combination of well size, intake size and control fittings the magnitude will be a function of the rate of change of stage. For a given rate of change of stage the amount of lag can be determined by the relationship

$$Sh_1 = \frac{W}{2g}\left(\frac{A_w}{A_p}\right)^2\left(\frac{dh}{dt}\right)^2$$

where Sh_1 is the amount of lag (m); g the acceleration of gravity (m s^{-1} s^{-1}); A_w the horizontal cross-sectional area of stilling (m^2); A_p the cross-sectional area of intake pipe (m^2); dh/dt the rate of change of stage in the river (m s^{-1}); and W is the coefficient of head losses in the intake pipe and fittings.

Note: For a single straight intake pipe with no control fittings

$$
\left.
\begin{aligned}
\text{entry loss} &= 0.5\frac{v^2}{2g} \\[2mm]
\text{exit loss} &= 1.0\frac{v^2}{2g} \\[2mm]
\text{friction loss} &= \frac{4FL}{D}\frac{v^2}{2g}
\end{aligned}
\right\} \text{i.e.} \qquad W = 1.5 + \frac{4FL}{D}
$$

where F is the Darcy–Weisback coefficient of total drag; L the length of intake pipe (m); and D is the diameter of intake pipe (m).

1.5.4   Recorder housing

Some form of housing is essential to protect the recording and measuring equipment from the elements or from unauthorized attention or vandalism, and also to give shelter to the servicing and gauging staff who operate the station. The size and quality of the housing adopted will depend on conditions at the site and on the type and range of equipment to be installed in it, but in general the minimum required is a well-ventilated, weatherproof and lockfast hut set on a stable foundation and of such dimensions as will permit normal servicing of the recorders and, where relevant, the operation of the cableway traversing gear.

There are many locations where timber huts over a concrete pipewell are adequate and may indeed have certain advantages—for example, in times of hard frost—but various forms of construction such as brick, stone or reinforced concrete are used to suit local conditions. There should be provision for heating the hut in winter conditions, for excluding vermin and insects, and for restricting the entry of water vapour from the pipewell. One form of timber hut is shown in Figure 1.5.

The recorder should normally be mounted on a rigid table firmly fixed to the foundations of the hut but independent of the framework of the building. It should be placed so that the float and counterweight have ample clearance from the sides of the stilling well walls throughout their full range of travel, and in a position which allows convenient access to the service staff for chart or tape changing and general maintenance. Where a cover is placed over the well, the holes or slots for the various cables and tapes must be located accurately to eliminate any risk of rubbing or fraying. An arrangement for two recorders is shown in Figure 1.6.

Figure 1.5   Typical instrument house with ramp gauge, steps and cableway

Figure 1.6   Arrangement of analogue and digital recorders

1.5.5 Suspension equipment

To provide a means of accurately measuring the depth of water and of mounting the current meter and placing and holding it at any predetermined part of the cross-section, some form of suspension system is required at or near the gauging station site. Where site conditions are suitable the measurement can be done by wading with the meter mounted on a short, rigid, graduated wading rod hand-held by the operator, or from a point above water level such as a cableway, boat or bridge, with the meter mounted on a longer, rigid, graduated rod which may be hand-held or mechanically operated. In deep or fast-flowing rivers where rigid rod suspension is not feasible, some form of flexible cable suspension is required.

Where depth and velocity of flow are suitable, low water measurements are almost invariably made by wading with the cross-section and observation points delineated by a metallic tape or, more usually, a graduated tag line of steel wire. These measurements are frequently made at a cross-section near to but not at the actual gauging station site where low flow conditions may be better. For medium and high water measurements, where the depth and velocity make wading impracticable, the current meter can be suspended from a slack-line cableway spanning the stream with the meter placement and velocity recording controlled from one bank or from a cable-car running on the cableway. Such cableways require towers and anchorages structurally adequate to suit the full load for which the cableway is designed, a supporting cable to carry the cable-car or meter travelling block, traversing cables to operate the cable-car or block, a hanger bar and connector assembly to support the meter and weight and connect it to the suspension cable, a streamline weight or weights to hold the meter steady in the current and keep the suspension cable as vertical as possible under the point of support, single-drum mechanical winches to handle efficiently the heavy weight where a cable-car is used, and double-drum mechanical winches to traverse the meter travelling block where control is from the bank.

Where a bridge is used, and velocity and depth are not great, suspension can be by long rigid rods hand-held or mechanically operated or by flexible handline equipment. Where depth and velocity require heavy weights to hold the meter in position, a gauging reel mounted on a portable crane is essential. Suspension equipment is described in detail in Chapter 8.

1.6 OPERATION OF STATION

1.6.1 General

The successful operation of a gauging station involves the production of a continuous, accurate and reliable record of both stage and discharge.

Depending on the quality and stability of the site, the relation between stage and discharge can be sensitive or insensitive, stable or unstable, accurate or inaccurate and the method of equipping and operating a station will depend to a great extent on these qualities. An acceptable record can only be obtained if the station is maintained in full operating order at all times, which requires efficient attendance on the station, its equipment and its calibration.

1.6.2 Recording of stage

The accuracy of the computed discharge is dependent on the accuracy of the record of stage produced at the station, and a prime requirement is that the instrumentation used for this purpose operates efficiently and within the tolerance limits of its specification. An inspection and maintenance routine for the servicing staff should be established and any installation, operation and maintenance instructions provided by the manufacturer and the gauging authority must be followed carefully. Field personnel responsible for the operation of recorders should be trained to recognize recording errors, and be able to identify the source of such errors and carry out any necessary adjustments. It should be impressed upon them that the final accuracy of the discharge record will rest primarily on the standard of operation of the station and the recognition by the operator of the many problems which have to be overcome.

In order to maintain accuracy and continuity of record it is important that errors originating from sources other than the recorder should be anticipated and prevented. Where the recorder is set in a stilling well, for example, it is necessary to ensure that there is no significant difference between the water level being recorded inside the stilling well and that in the river. Certain aspects of the station fabric, therefore, which could affect a recorder must be checked regularly:

(a) the stilling well and housing must be maintained in good physical condition;
(b) the intake pipes and well must be kept clear of silt and other debris;
(c) the well must be protected against ice;
(d) the housing must be protected against extreme humidity changes;
(e) the reference gauges must be kept at the correct level relative to the station datum.

In any recording system using a float to sense water level there are possible sources of error which can result in a difference between actual and recorded water levels. Some of these errors can be minimized by careful design of the instrumentation and by proper operational procedures, but it is unlikely that all can be eliminated and precautions must be taken to reduce such errors to acceptable limits. Errors caused by the float system can result from various combinations of possible faults:

(a) a change, from its initial setting, in the depth of submergence of the float in the water;
(b) submergence of counterweight and float line;
(c) backlash in gearing;
(d) friction in mechanism;
(e) inadequate diameter of float or badly matched float and counterweight;
(f) overriding or displacement of wires on float or counterweight pulleys;
(g) overriding or displacement of wires on pen carriage movement;
(h) kinks in float suspension cables;
(i) slipping of the perforated float tape on the sprockets of the float pulley;
(j) build-up of silt on the float pulley affecting the fit of the float tape perforations on the sprockets.

A number of these faults can cause a lag both in the response of the float to changing water levels and in the response of the recording head to changes in float position. The result of this lag is to produce a difference between the indicated height on rising and falling stages, the sign of the error being dependent on the conditions of stage at the time of the original setting of the recorder. The error can be systematic and is of particular importance during a period of recession to low summer levels when comparatively small errors in water level can produce large relative errors in discharge. Because of this, there is some advantage in setting the recorder on a falling stage.

   Backlash is caused by badly cut or badly paired gears and produces an error on either the rising or falling limb of the hydrograph. Friction is a function of the design of the mechanism and, more particularly, the quality of the maintenance given, and has to be overcome by the driving force developed by the float. The driving force depends on the buoyancy of the float and, on a float of regular diameter, varies directly with the depth of flotation and with the square of the diameter at the water surface. With changing water levels, the depth of flotation is affected by change of weight induced by shift of suspension line from one side of the float pulley to the other or by submergence of part of the line and its counterweight. To some extent the friction factor increases with the weight on the pulley shaft which varies directly with the mass of the counterweight. For the best results, therefore, the instrument should be maintained to a high standard, the float should be as large as practicable and the counterweight as light as possible, consistent with being able to fully tension the suspension cables or tapes, which should be light and flexible.

There are many other sources of error which must be recognized and taken account of. On analogue recorders, for example, errors in both level and time can be caused by the simple operation of setting the chart on the chart drum:

(a) improper setting of the chart on the recorder drum;
(b) improper joining of the chart edges;
(c) distortion and/or movement of the chart paper;
(d) improper setting of the pen on the chart;
(e) distortion or misalignment of the chart drum;
(f) faulty operation of the pen system;
(g) clock inaccuracy.

From the point of view of accuracy and consistency most analogue and digital recorders, if selected after careful consideration of the conditions in which they have to operate, can be expected to sense and record a changing water level with sufficient accuracy for most river gauging purposes. It must be realized, however, that this will be so only with a high standard of maintenance and constant checking. The performance of all machines deteriorates with age and water level recorders are no exception.

Figure 1.7 shows a result from an analogue recorder of standard make after many years in the field. This particular recorder had a float of diameter 152 mm and weight 1985 g, a counterweight of 2353 g and a recording scale set to 1 : 8. On fast and slow tests, total lag was 24 and 26.5 mm, respectively, 11 mm of this being due to backlash in gearing and an average of 14 mm due to float lag. As the buoyancy of a float this size is 18.25 g mm^{-1}, the average driving force necessary to work the mechanism was 242 g and to have provided this with an immersion of only 1 mm would require a float of 555 mm diameter. This level of performance is unacceptable and to keep errors to a minimum and ensure satisfactory operation regular and careful attention is necessary. On each visit the field operator should check that the recorder is functioning properly, the clock is at the correct time and the record shown on the chart or tape is legible and accurate. The reference gauge or gauges should be read, the readings compared with that shown by the recorder and the information entered on a check sheet with the relevant times. The use of a portable stilling box will increase the accuracy of reading if surface waves are present.

On a chart recorder the gauge readings can be related exactly to the chart trace by raising and lowering the float wire to mark the chart with a vertical line. If the readings do not agree within an acceptable tolerance further investigation must be carried out until the source of the error has been found. This should be done systematically by reference to the errors listed above, but particular attention should be paid to the intake pipe and stilling well to ensure that they are not obstructed and to the float to check that it is not stuck or damaged or has debris lodged on it.

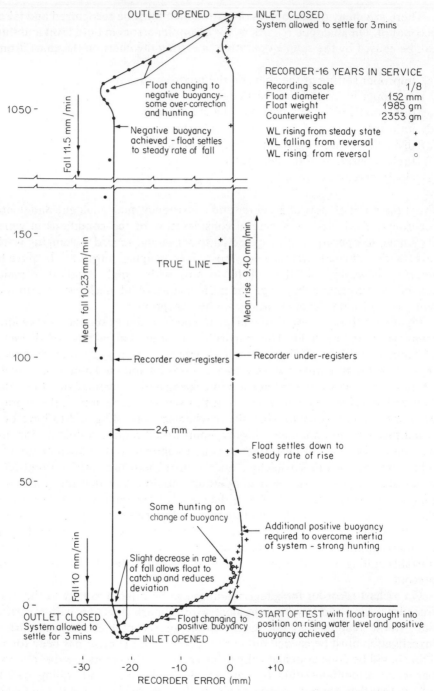

Figure 1.7   Result of test on analogue recorder

1.6.3   Procedure for routine chart and tape changing

After installation of the recorder a routine for changing charts or tapes, taking check readings of stage from the reference gauge and providing regular maintenance must be set up. This may be done by an engineer from the gauging authority or by a local observer, and the procedures adopted will depend on the ability and experience of the personnel employed.

To ensure efficient and continuous operation of the recorder it is advisable to furnish each observer with printed instructions detailing a specific routine which has to be followed. The exact nature of the routine will depend on the type of instrumentation but consists essentially of checking the condition of the recorder and its installation, of recording on the check sheet details of reference gauge readings, and of chart or tape changing. It is imperative that this is done on a regular basis so that a breakdown would not last long enough to cause a significant gap in the water level record. Where possible, additional visits should be made between chart-changing periods to verify and record that the recorder is operating satisfactorily. The following routines are designed for specific types of analogue and digital recorders but the same principles could apply generally.

Analogue recorders

(1) Read the outside reference gauge, cleaning if necessary.
(2) Read the inside reference gauge, if one is installed.
(3) Read the indicated level on the chart and compare with the above. If the readings do not agree find the cause and remedy it.
(4) Check that the clock is running and read the time indicated by the pen on the chart.
(5) Enter all readings of water level, recorder time and clock time on the check form and/or the chart. For this purpose the operator should be provided with a reliable watch checked to local time.
(6) Mark chart with a short vertical line by raising the float wire.
(7) Remove the stylus from the chart.
(8) Remove the chart drum from the recorder.
(9) Remove the chart from the drum by cutting cleanly with a sharp blade. Do not cut at the joint as it is essential to be able to examine the joint to determine any error.
(10) Place new chart on drum, making sure that it fits properly on the rim and that it matches at the joining edges.
(11) Rewind the clock.
(12) Check the stylus assembly to ensure that it is working properly and recharge with ink if necessary.
(13) Replace the chart drum on the recorder.
(14) Check the float and counterweight assembly and clean float if necessary.

(15) Clean and oil the recorder mechanism according to manufacturer's instructions.
(16) Rotate the drum anti-clockwise to eliminate backlash.
(17) Reset the stylus on the chart at the correct time and level.
(18) Enter the readings of water level and time on the new check form and/or on the new chart.
(19) Before leaving the station check that the instrument is working properly.

Digital recorders

(1) Check that the machine is operating normally and at the correct punching time, in case any special action is required.
(2) Check gear train and shafts on float tape pulley.
(3) Check for wave action or slow surge.
(4) Read outside reference gauge shortly before a punch is due.
(5) Immediately punch registers, press test switch and hold until seven extra punches are made.
(6) Turn up by the ratchet wheel 12 inches (0.3 m) of extra blank tape then part the tape. Do not advance tape by pulling the paper through.
(7) Draw off the used tape and lay it aside. Make no attempt to roll it at this stage.
(8) Enter all readings of water level, recorder time and clock time on the check form and on the end of the old tape.
(9) Adjust machine for time and height as required, checking carefully for surge. Where wave action is present estimate carefully the mean reading.
(10) Thread new end of tape or fit a new roll and, again utilizing the ratchet, feed out the minimum 12 inches (0.3 m) plus sufficient to bring up to the punching positon a point on the time scale seven or more spaces before the 'ON' time.
(11) Enter the readings of water level and time on the end of the new tape.
(12) Enter the readings of water level and time on the new check form.
(13) Tighten the passed tape on the take-up spool, and adjust the spring band.
(14) Press test switch and allow to punch up to and including the last tape 'OFF' time, i.e. the punch time before the new 'ON' time. If a new tape roll has been fitted adjust punching position to ensure the holes cut a line.
(15) Wind up the old tape to leave the 'OFF' details on the outside end.
(16) When the machine makes the first punch read off the figure punched and enter up the 'punched reading' and 'error' columns on the new check form.
(17) Check that the spring band had been set on to both pulleys.
(18) Check that the punch tray has been replaced.

(19) Check again that no slow surge has altered the reading or, if wave action is present, that the extremes still mean out.
(20) Before leaving the station check that the instrument is working properly. The operator should await one punch sequence.

1.7   CURRENT METER MEASUREMENTS

1.7.1   General

In calibrating a gauging station a relation is established at all levels of flow between stage and discharge by taking measurements of discharge (gaugings) at different increments of stage. Essentially each gauging involves measurements of depth and velocity at a number of verticals spaced transversely along a cross-section, and measurements of width across the stream to each vertical from a reference point on the bank. Velocity is measured by current meter, depths are measured by hand-held or mechanically operated rods, by a weight suspended from a cable which again can be hand-held or operated from a gauging reel, or by an echo-sounder, and widths are measured by steel tape, tag line, cableways fitted with distance counters or taken from permanent marks on bridges.

The cross-section is divided into segments by selecting verticals at a sufficient number of points to ensure an adequate sample of bed profile and the velocity distribution. The spacing of the verticals can vary from site to site and depends on the width, the character and unevenness of the bed and the variations in velocity distribution throughout the cross-section. In general the intervals should not be greater than one-fifteenth of the width, in the case of regular bed profiles, or one-twentieth the width, in the case of irregular profiles, but in special cases such as small lined channels with a regular geometric profile the number of verticals can be reduced. Tests should be made at each vertical to prove their suitability as permanent observation locations—if necessary the spacing should be adjusted to ensure the difference between mean velocities on adjacent verticals does not exceed 20 per cent or that the discharge through any one segment does not exceed 10 per cent of the total discharge. Increasing the number of verticals increases the time required to complete the gauging and, where stage is fluctuating, a balance must be struck between the need for accuracy in the gauging operation and the need for an accurate determination of mean stage.

Velocity observations are usually made at the same time as measurements of depth and this is mandatory when gauging where the channel bed is unstable. At each vertical, depth is measured first and the value used to compute the depth setting of the current meter. The mean velocity in each vertical can be determined by a number of different methods, each method involving a velocity observation at one or more points in the vertical. The method chosen may vary from site to site and depends on a number of factors whose effects have to be balanced to give the best overall result: the time available, the width and depth of the water, the bed conditions in the measuring section and the approach channel, the rate of change of stage, and the accuracy required. There are three principal methods and a number of others which, although not recommended for general use, can be applied where site conditions make normal methods inadvisable or impossible.

1.7.2 Principal methods

The three principal methods are as follows:

The velocity-distribution method

In this method, also known as the multiple-point or the vertical-velocity curve, velocity observations are made in each vertical at a sufficient number of points distributed between the water surface and bed to define effectively the vertical-velocity curve, the mean velocity being obtained by dividing the area between the curve and the plotting axis by the depth. The number of points required depends on the degree of curvature, particularly in the lower part of the curve, and usually varies between 6 and 10. Observations should always be made at 0.2, 0.6 and 0.8 of the depth from the surface, so that the results from the vertical-velocity curve can be compared with various combinations of these reduced points methods, and the highest and lowest points should be located as near to the water surface and bed as possible without contravening the requirements laid down for the type of current meter being used.

   This method is the most accurate if done under ideal, steady-stage conditions but is not considered suitable for routine gauging due to the length of time required for the field observations and for the ensuing computation. It is used mainly for checking velocity distribution when the station is first established and for checking the accuracy of the reduced points methods.

The 0.6 depth method

In this method, velocity is observed at a single point at 0.6 of the depth from the surface and the value obtained is accepted as the mean for the vertical. This assumption is based both on theory and on results of analysis of many vertical-velocity curves, which showed that in the majority of cases the 0.6 method produced results of acceptable accuracy. The value of the method is its essential reliability, the ease and speed of setting the meter at a single point and the reduced time necessary for completion of a gauging.

The 0.2 and 0.8 depth method

In this method, velocity is observed at two points at 0.2 and 0.8 of the depth from the surface and the average of the two readings is taken as the mean for the vertical. Here again this assumption is based on theory and on the study of vertical-velocity curves, and experience has confirmed its essential accuracy.

More time is required than in the 0.6 depth method but it is considered to give slightly better results when used in the right conditions. The 0.6 depth method, however, is to be preferred if stage is changing rapidly and speed is essential, if depth is below 750 mm or if surface debris or floating ice make observation at 0.2 depth inadvisable or impossible.

1.7.3 Other methods

There are several other methods which can be used in conditions where the normal methods are not applicable such as extremely high velocities, abnormal vertical-velocity distribution or where sounding is not possible and depth cannot be reliably estimated.

In the integration method a mean velocity is obtained by raising and lowering the meter through the vertical at a uniform speed. This is feasible with propeller-type meters, but cup-type meters such as the Watts will overregister and the method is seldom used except in estuarial reaches.

Where velocities are high and it is not possible to measure depth accurately or place the meter at 0.6 or 0.8, a single observation at 0.2 of the depth or some other similar distance below the surface can be used and a coefficient applied to convert the observed value to a mean for the vertical.

Where the vertical-velocity distribution is abnormal, observations can be made at 0.2, 0.6 and 0.8 of the depth, the mean being obtained by taking the arithmetical mean of all three or, more accurately, by taking the mean of the 0.2 and 0.8 values and averaging this result with the 0.6 value. This method is not suitable for depths below 750 mm.

Other possibilities are the five-point method using values below the surface, above the bed and at 0.2, 0.6 and 0.8; the six-point method where 0.4 is added; or the one-point method using velocity at 0.5 of the depth with an appropriate coefficient. There is no particular merit in these, except in special circumstances, and they are not in common use.

1.7.4 Procedures

General

When gauging by wading, from a boat or from an unmarked bridge, a measuring tape or graduated tag line is stretched across the river at right angles to the direction of flow and the positions of the successive verticals used for depth and velocity are located by horizontal measurements from a reference marker on the bank, usually defined by a pin or monument. When gauging by unmanned cableway controlled from the bank, the verticals are located by a distance counter set to register zero when the meter and weight assembly is at the reference marker; when gauging by manned cableway, the cable itself is graduated for distance. The gauging starts at the water edge of the near bank, where depth and velocity may or may not be zero. At each chosen vertical, the depth is measured and the value used to compute the setting or settings for the current meter depending on the method to be used—usually 0.6 or 0.2 and 0.8 depth. After the meter is in position the rotor should be allowed to adjust to the stream velocity before the count of rotor revolution is started—this should take only a few seconds where velocities are over about 0.3 m s^{-1} but a longer period is necessary at lower velocities or if cable suspension is being used. A revolution count should then be taken at each selected point for a minimum of 60 s but where it is known that the velocity is subject to short period variations or pulsations it is advisable to continue the observation for at least 3 min, noting the revolution count at the end of each one-minute period. The velocity at the point should then be taken as the average of all the separate readings unless it is apparent that the difference is due to some cause other than pulsation of the flow.

The phenomenon of pulsation has an effect on the measurement of point velocities, and therefore on current meter gauging in general, which is not always fully appreciated. The velocity at any point in the river, even when the discharge is constant and when the surface is apparently smooth and free from surges and eddies, is continuously fluctuating with time. This pulsation is caused by secondary currents developed by hydraulic conditions upstream of the gauging site; for example, by obstructions in the approach channel, by surging produced at riffles or rapids being continued through pooled reaches, or by the acceleration of the water at bends. Generally, the velocity at any point changes in cycles which can vary from a few seconds to possibly more than 1 h, and the extent of these pulsations—both short term and long term—should be known before deciding on the duration of exposure of the meter at each velocity point. For reliable results it is necessary to observe for a time sufficient to average the effects of the pulsation or to determine the velocity limits of the cycle. This is specially important when trying to determine the relationship between mean velocity as determined by different reduced point methods and that obtained by drawing a vertical-velocity curve.

The extent of the error caused by pulsation and its effects on overall accuracy have been evaluated by tests carried out on a variety of river types and sizes, and it was found that the magnitude of the errors vary from site to site, with stage and velocity, with the meter depth setting, and particularly with the length of observation. Recommended values for the uncertainty in velocity due to pulsations are given in Chapter 10.

It is accepted that the error in a complete discharge measurement due to velocity pulsation will be randomized due to the number of verticals at which velocity is measured (see Chapter 10). To obtain the most efficient performance in gauging, the optimum duration of observation should be that which will give the minimum error consistent with spending the least amount of time, and this has led in some cases to the necessity of taking a three-minute reading at each vertical. This can, however, be modified at sites where there is detailed knowledge of the pulsation effect and in many cases a reduction in observation time is acceptable.

1.7.5 Wading measurements

Where possible, gauging should be done by wading, with the current meter supported on a graduated wading rod which rests on the bed of the stream. Wading measurements should be more accurate than those from cableways or bridges, as the operator has more control over the general gauging procedure, particularly in the selection of the cross-section which may not be the normal station measuring section, in the selection of verticals and in the measurement of depth. In the lower ranges of flow greater accuracy is called for, as errors in the discharge measurement procedure, which would be regarded as minor at

Figure 1.8   Typical wading gauging—preparing to position the meter
and wading rod

higher stages, can have a much greater effect. Limitations on wading are imposed by the combination of depth and velocity and by the quality of footing on the bed, and the advisability or otherwise of wading has to be judged by the operator at each individual site.

The position of the operator is important because he must ensure that his body does not affect the flow pattern at or approaching the current meter. The best position is to stand facing one or other of the banks, slightly downstream of the meter and an arm's length from it. The rod should be kept vertical throughout the measurement with the meter parallel to the direction of flow. Rotor revolutions can be registered on an electric counter, but in clear water when using a cup-type meter a visual count can be made if one of the cups is painted a distinctive colour. A typical wading gauging is shown in Figures 1.8 and 1.9.

Figure 1.9   Typical wading gauging—showing position of the operator
and wading rod

1.7.6   Cableway measurements

When using an unmanned cableway a meter and weight assembly is suspended from a pulley block which can be traversed across the river by means of a winch on one bank. The positioning of the meter at each vertical and at any depth setting in the vertical is achieved by distance and depth counters on the winch, and meter revolutions are returned through a suspension cable with an electrical core and registered on a revolution counter. A heavy weight, set at a known distance below the meter, is required when sounding or when setting the meter at the depth for velocity observation, to ensure that the suspension is as near vertical as possible and the meter is held steady in the current. The size of weight required depends on the velocity and depth of water and can be decided by experience. To obtain the depth, the weight is lowered until the bottom just touches the water surface, the depth counter is zeroed, the weight is then lowered to the bed and the depth read off the counter. Where drift of the meter downstream is not eliminated the angle of drift should be measured or estimated and both dry- and wet-line corrections applied to obtain the actual vertical depth. A short calculation, taking account of the distance from the bottom of the weight to the centreline of the meter, is then made and the assembly raised until the meter is at the velocity observation point. A typical cableway gauging is shown in Figure 1.10.

1.7.7   Bridge measurements

Although bridges do not often offer the right conditions for gauging, measurements from them may be necessary where suitable sites for wading or for a cableway are not present. There are advantages and disadvantages in the use of either the upstream or downstream faces, and the decision on which to use should be made for each individual bridge after due consideration of all the factors involved. A handline can be used where the weight of the meter assembly does not exceed about 20 kg, but where heavier assemblies are required a winch or reel mounted on some form of portable crane is necessary (see Chapter 8). When sounding by handline the weight is positioned on the water surface and then lowered to the bed while measuring the amount of line paid out. When sounding by a winch fitted with a depth counter the same procedure as for a cableway gauging is followed.

1.7.8   Current meter errors

Irrespective of the method being used, attention must be paid to the condition of the current meter to ensure that it does not become affected by weed, leaves, grass, ice or other floating debris, and where this is suspected the meter should be removed from the water for cleaning and servicing. Where, however, the air temperature is below the freezing level the meter should be

Figure 1.10   Cableway gauging

kept in the water when traversing from one vertical to the next. Current meter errors will be introduced in the following circumstances:

(1) If the meter is used to measure velocities below or above its established rating. Where such use is unavoidable it should be noted for consideration and possible adjustment at the computation stage.
(2) If the flow is turbulent and the meter is not held steady in the current at the exact observation point. In such conditions the meter can yaw, drift and move vertically, causing over-registration by a cup-type meter and under-registration by a propeller-type meter.

(3) If the flowlines are not parallel to the propeller-type meter, causing under-registration.
(4) If the flowlines are oblique to the line of the cross-section and no corrections are made. The extent of this error will vary with the type of suspension, the type of meter and the stream direction. If oblique flow is unavoidable, the angle of the direction of flow to the perpendicular to the cross-section must be measured and the measured velocity corrected. Special instruments have been developed for measuring the angle and velocity at a point simultaneously, but where these are not available and there is insignificant wind, the angle of flow throughout the vertical may be taken to be the same as that observed on the surface. The correct velocity is obtained by multiplying the measured velocity by the cosine of the angle formed by the direction of flow and the perpendicular to the cross-section.
(5) If there is significant disturbance of the surface by wind.

1.8   COMPUTATION OF DISCHARGE

1.8.1 General

From a current meter gauging the discharge can be computed either arithmetically by the mid-section or mean-section methods or graphically by the depth–velocity integration or velocity-contour methods. The choice of computation method depends upon the equipment and observational method used during the gauging, the flow condition at the time of gauging, the type of stream and the accuracy required. For the computation of routine gauging at standard stations arithmetical methods are preferred as they give sufficient accuracy, are simpler and quicker to perform and can be programmed for computer processing. Graphical methods are only appropriate when multiple-point velocities have been taken and are used for checking on flow characteristics when setting up a station or for obtaining a reliable base to which the accuracy of reduced point methods can be related.

1.8.2 Arithmetical

Mean-section method
In this method the cross-section is regarded as being made up of a number of segments, each bounded by two adjacent verticals. The discharge through each segment is taken as

$$q = \left(\frac{\bar{v}_1 + \bar{v}_2}{2}\right) \cdot \left(\frac{d_1 + d_2}{2}\right) \cdot b$$

where q is the discharge through segment; $\bar{v}_1$ the mean velocity in first vertical; $\bar{v}_2$ the mean velocity in second vertical; d_1 the depth in first vertical; d_2 the depth in second vertical; and b is the width of segment.

This is repeated for each segment, the values for each segment being summated to give the total discharge. The mean velocity in each of the two segments adjacent to the banks may be calculated by assuming zero depth and velocity at the water's edge. In the extreme case, where the segments are exceptionally wide due to a gradually shelving beach, the discharge in these segments is obtained by multiplying the calculated mean velocity by the area of the corresponding segment. The total discharge is obtained by adding together the discharges from all the segments.

Mid-section method

In this method the $\bar{v}d$ product at each vertical is multiplied by the water surface width taken as the sum of half the distances to adjacent verticals. In this case the segment discharge is taken as

$$q = \bar{v}d\left(\frac{b_1 + b_2}{2}\right)$$

where $\bar{v}$ is the mean velocity in vertical; d the depth in vertical; b_1 the width of segment on one side of vertical; and b_2 is the width of segment on other side of vertical. In most cases the value of $\bar{v}d$ in the end sections next to the banks is negligible and can be estimated.

1.8.3   Graphical methods

Depth–velocity integration

In this method the depth–velocity curve for each vertical is drawn by plotting the velocity observations against the corresponding depth and drawing a smooth curve through the points. The areas contained by each curve are measured by planimeter and their values (equivalent to $\bar{v}d$) are plotted over the water surface line and a smooth curve drawn through the points. The area enclosed between this curve and the water surface line represents the discharge. The graphical integration method is shown in Figure 1.11.

Velocity-contour method

Based on the velocity distribution curves of the verticals, a velocity distribution diagram for the cross-section is prepared showing contours of equal velocity. The areas enclosed by the velocity contours and the water surface line, if this is cut, are measured by planimeter and plotted in a second diagram in which the ordinate represents velocities, and the abscissa the areas enclosed by the corresponding velocity contours. The area enclosed by the resulting velocity area curve represents the discharge. An example of the velocity-contour method is shown in Figure 1.12.

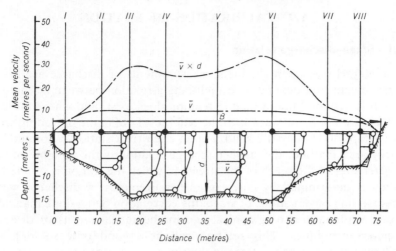

$$Q = \sum_{0}^{B} \bar{v}d \, \Delta B$$

Figure 1.11   Graphical integration method

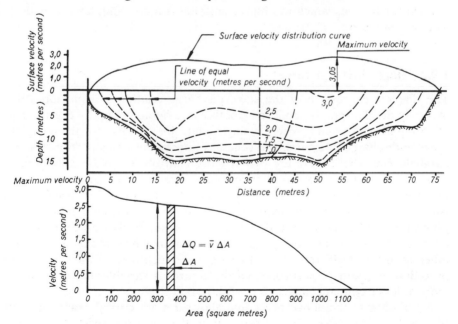

$$Q = \sum_{0}^{A} \bar{v} \Delta A$$

Figure 1.12   Velocity-contour method

1.9 CALIBRATION OF STATION

1.9.1 Stage–discharge relation

In the majority of open channels, measurements of discharge are made by indirect means and depend on establishing a relation between discharge and stage. It is assumed that a unique relation does exist, and where the station has been sited in accordance with the criteria set out in the preceding sections this assumption is normally valid. For an open river station calibrated by the velocity-area method, the stage–discharge relation is the epitome of all the characteristics of the reach of the stream in which the station is situated. This relation is determined by correlating measurements of discharge with the corresponding observations of stage, the correlation being done manually by various combinations of graphical and mathematical means or directly by computer techniques. This relation will hold good only so long as no significant alteration takes place in the characteristics of the reach, and since most stream channels are continuously in the course of development and are inherently unstable, it is essential to keep such a relation continuously under review to ensure either that its validity is maintained or that it is redetermined when it is shown to have been altered. The treatment of the stage–discharge relation for reaches which are highly unstable is more complex and is dealt with in Chapter 6.

1.9.2 Stage–discharge curve

The simplest expression of the stage–discharge relation is a plot on arithmetically divided graph paper, a smooth curve being drawn by eye through the array of data points. Where a single control is effective for the complete range of discharge, the curve will generally approximate to a portion of some parabola, but where there are two or more controls it may be composed of parts of different parabolas with inflexions and reversals in curvature indicating changes from the influence of one control to another. The drawing of a curve by eye through an array of data points is inevitably a subjective procedure and however carefully it is done the result is likely to contain some element of bias and, at some gauging stations, it may be that the inaccuracies inherent in the measurement of discharges are such that no more elaborate procedure is justified, particularly if the control is variable and necessitates more or less frequent changes of the calibration curve.

A variation of the above procedure allows a more critical examination of the results and is based on the proposition that over a restricted range the curve of a polynomial function may be closely approximated by an exponential function which plots as a straight line on logarithmically divided graph paper. Over a more extended range, therefore, a polynomial curve may be represented by one, two or more straight lines on a logarithmic graph. In practice, it is generally found that stage–discharge data plotted on logarithmic paper do, in fact, line up very well on successive straight lines and this is a great aid to obtaining a best line, although it does not eliminate the subjectiveness of the operation. The discontinuation in the straight line usually represents changes in control and this should be kept in mind when settling on the weak points between successive straights. Nevertheless, a single control may operate over a considerable range, and discontinuity could also occur owing to the need to have more than one exponential expression to represent a relative polynomial. In such a case the inflexion is likely to be small.

In plotting stage discharge data, particularly when using logarithmic paper, it is necessary to adjust the value of stage observations so that they are referred to a zero which coincides with the level at which flow past the station ceases but which may not coincide with the zero of the reference gauge. The value of this adjustment (the 'a' value in the equation $Q = C(h + a)^\beta$ may be obtained by graphical methods or by mathematical iteration to produce the best correlation coefficient, but these can only give an approximation to the true value of a and wherever possible it should be determined on the site. The $(h + a)$ value is actually the reading which the reference gauge would give when flow is on the point of ceasing. In the case of a section control such as rock ledge or gravel shoal it is the elevation of the point at which the least flow would cross the control, adjusted if necessary for surface slope between station and the control and referred to the zero on the reference gauge.

Great care must be taken in making the measurement, which is often best done by sounding over the control with a rod, either wading or from a boat, when the river is at a very low stage. If the control tends to be gullied the effective point may lie in a gully at some distance downstream from the lip of the control. Probably the best indication of a change in the a value is by examining the lower ranges of the plot. If the sample points indicate that a convex or concave curve would give a better fit, the a value should be checked immediately.

A typical curve on log paper is shown in Figure 1.13. When plotting on log paper it is sometimes possible to obtain an apparent advantage if the divisions on standard graph paper are assigned values which are constant multiples of their logarithmic values. This device permits the standard graph paper to be adjusted to the data to be plotted, opening up the scale and increasing the general slope of the line. While there may on occasion be reason to adopt this course, it is generally preferable to plot all data to the same standard scaling in order to allow comparison to be made between stations. The slope of the graph is an index of the sensitivity of the station, and the relativity is lost if any adjustment is made to the scale. The curve should be examined for hysteresis.

In general, the sites chosen for gauging stations have channel characteristics (a gradient sufficiently steep and a downstream channel of sufficient capacity) which ensure that, except for occasional changes in control which may affect calibration, the relationship between stage and discharge is substantially

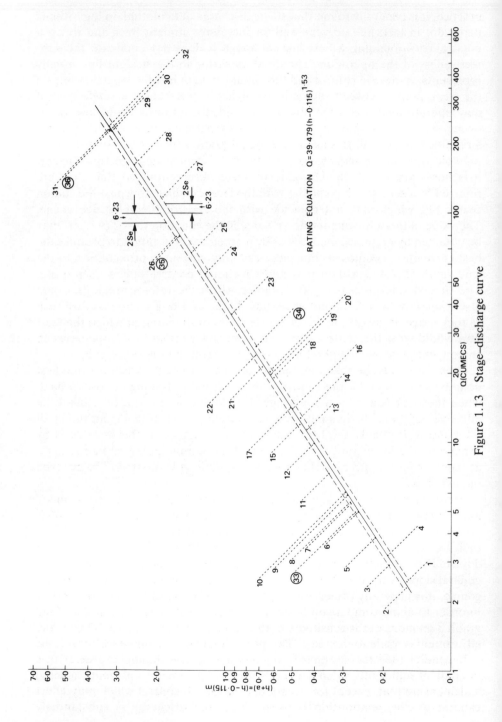

Figure 1.13 Stage–discharge curve

consistent, i.e. one particular gauge height will indicate one corresponding discharge. Under certain conditions (flatter gradients and constricted channels) the phenomenon known as hysteresis (the effect on the stage–discharge relation at a gauging station subject to variable slope where, for the same gauge height, the discharge on a rising stage differs from that on the falling stage) can occur where a looped stage–discharge curve is obtained for floods with differing stage–discharge relations for rising and falling levels. The shape of loop rating curves can vary from station to station, and also at the same station with the height of flood, but generally the curve for the rising stage will plot to the right of that for the falling stage, indicating a higher discharge for the same water level. This is usually due to the fact that during a rising flood the slope of the flood wavefront is significantly steeper than the steady-state hydraulic gradient of the river, the reverse applying during the recession. Difference in discharge caused by this effect can be significant.

If discharge measurements are made equally on rising and falling stages, an average rating curve falling between the two is obtained and this in most cases is usually of sufficient accuracy. In practice, however, there is a tendency for flood gaugings to be made on the falling stage only, especially on rivers which rise quickly and carry quantities of debris on the rising flood. On stations, therefore, where channel conditions are favourable for hysteresis, precautions should be taken to check the extent of the effect before a decision is made on whether to use an average rating curve or a series of looped curves.

1.9.3 Rating table

For some purposes a graphical expression of the calibration may be adequate, but when a station is being operated to obtain a continuous record of discharge it is more convenient to have a rating table in which discharges are tabulated against the corresponding stages at intervals usually of 1 or 5 mm of stage. If the record is to be processed by a computer, it is essential to have the calibration either in tabular form or in the form of equations.

1.9.4 Stage–discharge equation

While it is possible to prepare a rating table by taking readings at intervals from an arithmetic or logarithmic graph and to interpolate between these readings for intermediate values, the operation is extremely laborious and requires great care to ensure that successive differences increase or decrease in uniform progression and in the proper sense. Because of this it is considered much more appropriate to determine the equations to the curve and to compute the values to be tabulated from these. This should be done mathematically by the least-squares procedure, either manually using a calculator or by a computer direct from the gauging data, although it is still advisable to plot the gaugings for preliminary examination to determine if they should be split into ranges for separate treatment.

This method is subjective only in so far as the selection of the several straights for which best lines are to be computed is at the option of the operator. If a sufficient number of well-distributed data points is available there is usually little difficulty in making the selection. Where a group of data points is available within close proximity to an expected break point it is permissible to use all the points in the group in the determination of both the upper and lower straights. The actual break point is determined by simultaneous solution of both the relevant equations.

In the majority of cases the stage–discharge relation at a station can be expressed by an equation of the form $Q = C(h + a)^{\beta}$ (where Q is the discharge, h the gauge height, C and β are coefficients, and a is the difference in height between the zero level of the reference gauge and the true level of zero discharge) over the whole range of discharges, or more often by two or more similar equations each relating to a portion of the range. The constant a for each equation must first be fixed either from consideration of the levels or by logarithmic plotting, and the values of C and β can then be obtained from the equations

$$\Sigma(Y) - N(\log C) - \beta\Sigma(X) = 0$$

$$\Sigma(XY) - \Sigma(X)(\log C) - \beta\Sigma(X^2) = 0$$

where $\Sigma(Y)$ is the sum of all values of $\log Q$
      $\Sigma(X)$ is the sum of all values of $\log(h + a)$
      $\Sigma(X^2)$ is the sum of all values of the square of X
      $\Sigma(XY)$ is the sum of all values of the product of X and H
      N is the number of observations.

The preparation of the data and the solution of the equation is simplified by employing a tabulation, as shown in Table 1.1, or by computer, a typical print-out of which is shown in Table 1.2. The coresponding tabulation for determining plotting points or a rating table is shown in Table 1.3.

1.9.5   Extrapolation of stage–discharge relation

Stage–discharge relation curves are primarily intended for interpolation, and their extrapolation beyond the highest recorded high water or the lowest recorded low water may be subject to error. Physical factors like over-bank spills at high stages, shifts in controls at very low and very high stages, changes in rugosity coefficients at different stages, etc., materially affect the nature of the relationship at the extreme ends and must be taken into account. Extrapolation should be avoided as far as possible, but where this is desirable the results obtained should be checked by more than one method. The

Table 1.1   Computation of stage–discharge relation

Observation Ref No.	Q $(m^3 \, s^{-1})$	Stage 'h' (m)	$(h+a)$ where $a=-0.115$ (m)	$\text{Log } Q = Y$	$\text{Log }(h+a)=X$	XY	X^2
1	2.463	0.272	0.157	0.391 5	−0.804 1	−0.314 8	0.646 6
2	2.325	0.273	0.158	0.366 4	−0.801 3	−0.293 6	0.642 1
3	2.923	0.303	0.188	0.465 8	−0.725 8	−0.338 1	0.526 8
4	3.242	0.307	0.192	0.510 8	−0.716 7	−0.366 1	0.513 7
5	3.841	0.334	0.219	0.584 4	−0.659 6	−0.385 5	0.435 1
6	4.995	0.374	0.259	0.698 5	−0.586 7	−0.409 8	0.344 2
7	5.410	0.393	0.278	0.733 2	−0.556 0	−0.407 7	0.309 1
8	5.422	0.394	0.279	0.734 2	−0.554 4	−0.407 0	0.307 4
9	5.883	0.402	0.287	0.769 6	−0.542 1	−0.417 2	0.293 9
10	6.154	0.410	0.295	0.789 2	−0.530 2	−0.418 4	0.281 1
11	7.376	0.463	0.348	0.867 8	−0.458 4	−0.397 8	0.210 1
12	9.832	0.520	0.405	0.992 6	−0.392 5	−0.389 6	0.154 1
13	11.321	0.548	0.433	1.053 9	−0.363 5	−0.383 1	0.132 1
14	12.372	0.576	0.461	1.092 4	−0.336 3	−0.367 4	0.113 1
15	11.825	0.580	0.465	1.072 8	−0.332 5	−0.356 7	0.110 6
16	13.826	0.616	0.501	1.140 7	−0.300 2	−0.342 4	0.090 1
17	14.102	0.626	0.511	1.149 3	−0.291 6	−0.335 1	0.085 0
18	19.020	0.721	0.606	1.279 2	−0.217 5	−0.278 2	0.047 3
19	19.790	0.739	0.624	1.296 4	−0.204 8	−0.265 5	0.041 9
20	20.280	0.747	0.632	1.307 1	−0.199 3	−0.260 5	0.039 7
21	21.204	0.796	0.681	1.326 4	−0.166 9	−0.221 4	0.027 9
22	23.996	0.846	0.731	1.380 1	−1.136 1	−0.187 8	0.018 5
23	36.242	1.041	0.926	1.559 2	−0.033.4	−0.052 1	0.001 1
24	54.591	1.340	1.225	1.737 1	0.088 1	0.153 0	0.007 8
25	67.327	1.526	1.411	1.828 2	0.149 5	0.273 3	0.022 4
26	79.050	1.761	1.646	1.897 9	0.216 4	0.410 7	0.046 8
27	110.783	2.010	1.895	2.044 5	0.277 6	0.567 6	0.077 1
28	162.814	2.632	2.517	2.211 7	0.400 9	0.886 7	0.160 7
29	227.600	3.265	3.150	2.357 2	0.498 3	1.174 6	0.248 3
30	228.800	3.280	3.165	2.359 5	0.500 4	1.180 7	0 250 4
31	228.500	3.306	3.191	2.358 9	0.503 9	1.188 6	0.253 9
32	236.600	3.340	3.225	2.374 0	0.508 5	1.207 2	0.258 6
			Σ	40.730 5	−6.766 3	−0.553 4	6.697 5

For a least-squares regression where $Q = C(h+a)^\beta$:

$$\Sigma(Y)-N\,(\log C)-\beta\Sigma(X)=0$$

(where N is the number of observations)

$$\Sigma(XY)-\Sigma(X)\,(\log C)-\beta\Sigma(X^2)=0$$

$$\text{Then } 40.730\,5 - 32 \log C + 6.766\,3 = 0 \qquad \text{(i)}$$

$$-0.553\,4 + 6.766\,3 \log C - 6.697\,5 = 0 \qquad \text{(ii)}$$

Multiplying equation (i) by $\dfrac{6.766\,3}{32}$ and adding gives:

$$\beta = 1.530\,1$$

$$C = 39.47\,9$$

$$\text{then } Q = 39.479\,(h-0.115)^{1.530\,1}$$

Table 1.2 Computer print-out for the stage–discharge relation
(observed and predicted flows in ascending order of stage)

STAGE	GAUGED FLOW	FLOW FROM RATING EQUATION	ACTUAL DEVIATION	PERCENTAGE DEVIATION
0.272	2.463	2.323	−0.140	−6.0
0.273	2.325	2.345	0.020	0.9
0.303	2.923	3.060	0.137	4.5
0.307	3.242	3.160	−0.082	−2.6
0.334	3.841	3.865	0.024	0.6
0.374	4.995	4.996	0.001	0.0
0.393	5.410	5.568	0.158	2.8
0.394	5.422	5.598	0.176	3.2
0.402	5.883	5.846	−0.037	−0.6
0.410	6.154	6.097	−0.057	−0.9
0.463	7.376	7.851	0.475	6.0
0.520	9.832	9.902	0.070	0.7
0.548	11.321	10.968	−0.353	−3.2
0.576	12.372	12.072	−0.300	−2.5
0.580	11.825	12.233	0.408	3.3
0.616	13.826	13.711	−0.115	−0.8
0.626	14.102	14.132	0.030	0.2
0.721	19.020	18.345	−0.675	−3.7
0.739	19.790	19.185	−0.605	−3.2
0.747	20.280	19.563	−0.717	−3.7
0.796	21.204	21.931	0.727	3.3
0.846	23.996	24.442	0.446	1.8
1.041	36.242	35.098	−1.144	−3.3
1.340	54.591	53.855	−0.736	−1.4
1.526	67.327	66.859	−0.468	−0.7
1.761	79.050	84.631	5.581	6.6
2.010	110.783	104.989	−5.794	−5.5
2.632	162.814	162.095	−0.719	−0.4
3.265	227.600	228.478	0.878	0.4
3.280	228.800	230.145	1.345	0.6
3.306	228.500	233.044	4.544	1.9
3.340	236.600	236.854	0.254	0.1

SUMMARY

STAGE–DISCHARGE EQUATION DERIVED FROM 32 GAUGINGS

$$(Q = 39.479\ 0(H + (-0.115))^{1.5301}$$

EQUATION VALID FOR RANGE $0.272 < H < 3.340$

Table 1.3   Rating table

h (m)	(h − 0.115) (m)	Q (m³ s⁻¹)	dQ (m³ s⁻¹)	dh (m)	Increment for 2 mm (m³ s⁻¹)
0.300	0.185	2.986			
0.400	0.285	5.784	2.798	0.1	0.056
0.500	0.385	9.164	3.380	0.1	0.068
0.600	0.485	13.048	3.884	0.1	0.078
0.700	0.585	17.382	4.334	0.1	0.086
0.800	0.685	22.129	4.747	0.1	0.094
0.900	0.785	27.259	5.130	0.1	0.102
1.000	0.885	32.748	5.489	0.1	0.110
1.100	0.985	38.576	5.828	0.1	0.116
1.200	1.085	44.727	6.151	0.1	0.122
1.300	1.185	51.186	6.459	0.1	0.128
1.400	1.285	57.941	6.755	0.1	0.134
1.500	1.385	64.980	7.039	0.1	0.140
1.600	1.485	72.294	7.314	0.1	0.146
1.700	1.585	79.875	7.581	0.1	0.150
1.800	1.685	87.713	7.838	0.1	0.156
1.900	1.785	95.801	8.088	0.1	0.160
2.000	1.885	104.133	8.332	0.1	0.166
2.100	1.985	112.692	8.559	0.1	0.170
2.200	2.085	121.506	8.814	0.1	0.176
2.300	2.185	130.534	9.028	0.1	0.180
2.400	2.285	139.785	9.251	0.1	0.184
2.500	2.385	149.253	9.468	0.1	0.188
2.600	2.485	158.933	9.680	0.1	0.194
2.700	2.585	168.822	9.889	0.1	0.198
2.800	2.685	178.916	10.094	0.1	0.202
2.900	2.785	189.212	10.296	0.1	0.204
3.000	2.885	199.705	10.493	0.1	0.208

physical conditions of the channel, i.e. whether the channel has defined banks over the entire range, or only up to a stage and over-bank spill above that stage, as well as whether the channel has fixed or shifting controls, should govern the methods to be used in the extrapolation. There are a number of methods but those given below are suitable for a channel with defined banks and fixed controls, as well as for a channel with spill:

(1) If the control does not change beyond a particular stage it may be possible to fit a mathematical curve, as indicated in Section 1.9.2, and obtain the values in the range at the upper or lower end of the stage discharge curve to be extrapolated.

(2) Another simple method is to consider separate extensions of the stage area curve and the stage–mean velocity curve. The latter has little curvature under normal conditions and can be extended without significant

error. The product of the corresponding values of A and $\bar{v}$ may be used for extending the discharge (Q) curve.

(3) The mean velocity curve may also be extended by adopting the following procedure. The hydraulic mean radius (R_h) can be found for all stages from the cross-section. A logarithmic plot of $(\bar{v})$ against (R_h) generally shows a linear relationship for the higher measurements and the values of $\bar{v}$ for the range to be extrapolated may be obtained therefrom.

(4) A variation of the last method is by the use of Manning's formula:

$$Q = A\bar{v} = \frac{AR_h^{2/3}S^{1/2}}{n} \text{ metric units}$$

Assuming $S^{1/2}/n$ remains constant and substituting $\bar{D}$ for R_h a curve can be prepared for Q against $A\bar{D}^{2/3}$. After the bank-full stage, the discharge of the spill portion is worked out separately by assuming an appropriate value of n. If accurate gauges do not exist for computing the slope (S), it may be roughly estimated from the flood marks.

(5) Chezy's formula can be used in the same way in place of the Manning's formula. In this case the slope of the $\log \bar{v} - \log R_h$ line is taken as $1/2$.

(6) Yet another method that may be used, when another discharge site exists on the same stream, upstream or downstream, is to assess the relation between the rating curves of the two stations and find out the discharge at the required station. This presumes that the corresponding discharge at the other station is either known or can be worked out correctly taking additions, withdrawals, channel and valley storage into consideration.

1.10   ACCURACY OF THE VELOCITY-AREA METHOD

When considering the accuracy being obtained by the velocity–area method, it is necessary to distinguish between the accuracy of a single gauging and that of the relation between stage and discharge determined from a series of discharge measurements.

The uncertainty in a single determination of discharge cannot be calculated or predicted exactly, but a statistical estimate of the likely tolerance within which the true value will lie can be made by analysing the individual component measurements which are made during the gauging. There are uncertainties in the measurement of width, depth and point velocities, all of which are dependent on the quality of the instruments and methods being used, and in the velocity-area method itself because of the limitations imposed on the number of verticals used, the number of points tested in each vertical and the time allowed for velocity observation at each point. Methods for determining the uncertainty of a single gauging are described in detail in Chapter 10.

The determination of the uncertainty in the stage–discharge relation involves two separate but related issues, the first the uncertainty in each gauging and the second the calculated relation derived from them. The accuracy of the mean relation is better than that of the individual measurements in inverse ratio to the square root of the number of measurements from which the mean is determined and, although it would be possible to assess the standard error of the mean from the errors of the individual measurements, it is preferable to use a mathematical procedure. The scatter or deviation of the plotted points about the mean line of relation is measured by the standard error of estimate (S_e) and the uncertainty in the relation itself is measured by the standard-error of the mean relation (S_{mr}). For methods of determining S_e and S_{mr} see Chapter 10.

A high value indicates a high uncertainty in the relation and may result from a poor gauging site, faulty instrumentation or gauging techniques, insufficient number of gaugings or combinations of all three. The errors found in this manner are independent for each section of the curve and where the curve consists of two or more lines of different gradient, the S_e and S_{mr} values should be calculated for each range allowing two degrees of freedom for each range in the calculationof S_e. At least 20 gaugings should be available in each range before a statistically acceptable esimation of S_e and S_{mr} can be made.

After the stage–discharge relation has been determined it is essential that check gaugings in each range should be made from time to time. It is often difficult to assess the true significance of the deviations of the check gaugings when plotted on the log curve and it is useful to prepare a 'deviation diagram', in which the percentage deviations of the observed discharges from the corresponding discharges given by the stage–discharge equations are plotted as abscissae around a central ordinate representing the latter discharge. To an experienced operator a change in control necessitating a change in the stage–discharge relation may appear obvious from a set of check gauging data but there can be cases when check gaugings fall within the prescribed tolerance limits but are biased to one side of the mean—in such cases it is often advisable to check if the amount of bias is significant, and the Student's 't' test is a convenient way of doing this. The efficiency of this test depends upon the number of check measurements available, but the indication given by the test should be applied with some discretion and should be related to other factors which might be relevent. A typical calculation is shown in Chapter 10.

The final accuracy of the discharge record will rest primarily on the standard of operation of the station and the recognition by the operator of the many problems which have to be overcome. It is evident that there still exists a considerable amount of disagreement on the comparative accuracy of various methods of river flow measurement and in some quarters gauging by current meter is regarded as suspect mainly because of a lack of understanding of the principles involved and of the analytical methods necessary for treating the field data. It should be remembered that more sophisticated methods of handling velocity-area stations are now in use, in instrumentation, in gauging procedures and in the handling and analyses of the gaugings themselves. Because of this, the velocity-area method, with velocity measured by current meter, can well stand comparison with any of the other methods of flow measurement. The accuracy of the data from any particular station depends on the accuracy of recording stage and of converting stage to discharge by means of a rating equation. A useful measure of this is obtained by linking the S_{mr} value, which assesses the accuracy of the stage–discharge relation, with the sensitivity value which indicates the ease or otherwise of recording stage (see Chapter 10).

BIBLIOGRAPHY

BS 3680: Part 3, 1964. Methods of measurement of liquid flow in open channels—velocity area methods, British Standards Institution, London.

ISO 748: 1973. Liquid flow measurement in open channels by velocity area methods, International Organization for Standardization, Geneva.

ISO 1100: 1973. Liquid flow measurement in open channels—establishment and operation of a gauging station and determination of the stage–discharge relation, International Organization for Standardization, Geneva.

ISO 3454: 1975. Liquid flow measurement in open channels—sounding and suspension equipment, International Organization for Standardization, Geneva.

Lambie, J. C., 1975. Some specific problems in the operation of a gauging station, Proceedings of International Seminar on Modern Developments in Hydrometry, vol. 11 (September 8–13, Padova, Italy), WMO No. 427, World Meteorological Organization, Geneva.

CHAPTER 2

Current Meters

F. G. CHARLTON

2.1 HISTORY OF MEASUREMENT

2.1.1 Velocity and discharge

Although many of the ancient civilizations depended for their wealth and power on the conservation and control of water for irrigation, it is remarkable that flow measurements were based solely on measurements of water level and duration of flow. Discharge was assumed to be a function of the cross-sectional area of flow only; the influence of speed on flow was not considered. Even the Romans, excellent practical engineers, continued in this severely limited approach, and it was not until about A.D. 100 that Hero of Alexandria suggested that discharge in a channel was the product of both area and speed of flow. Surprisingly this basic concept was lost and remained unknown for a further 1500 years until it was derived independently by Benedetto Castelli in 1628.

2.1.2 Development of velocity measuring instruments

The speed of flow in channels had been measured prior to Castelli's work even though its importance was overlooked. Towards the end of the fifteenth century, Leonardo da Vinci's work with floats is the first recorded serious attempt, and included an investigation of the variation of speed across a channel. About 1610, Sartorio constructed the first current meter in the form of a hinged vane which was displaced by the thrust of flowing water on it. This, however, only gave qualitative results. The essential requirement of any quantitative measuring instrument, i.e. calibration, was not possible with the limited hydraulic knowledge of the day.

About 1663, Robert Hooke invented an instrument, the forerunner of the propeller-type meter, to measure the distance run by a ship. This was followed in 1681 by the paddle-type instrument invented by Marsigli. Later, Guglielmi developed the ball-and-quadrant instrument in which the flowing water deflected the ball and the angle of inclination of the suspension cable was a measure of the speed. The Pitot tube was invented, in about 1732, by Henry de Pitot but it was not until about 1784 that Woltman made the first propeller meter intended specifically for measuring the speed of flow in a channel. This was followed by many similar designs in which the propellers were made of flat plates mounted on spokes.

In 1870, Revy attempted to reduce the frictional resistance by forming the propeller of a hollow sphere with vanes attached such that the assembly was weightless when immersed in water. This principle has been revived recently, using helical-bladed propellers formed of neutrally buoyant plastic material.

A great improvement in meters was made in 1868, when Henry designed an instrument formed of several conical cups mounted on a vertical axis. This was developed by Price, from 1882 onwards, and became the most popular meter in the USA and continues in wide use today.

In 1868, Haskel developed a meter using a method invented by Ritchie for the remote reading of a magnetic compass and was thus able to measure both speed and direction of flow. This principle has been developed for use in estuaries and offshore work where wide ranges of flow direction occur.

A practical current meter requires a device to sense and display or record its operations. Many early devices followed Hooke's original design, in which the propeller drove a graduated circular scale by means of a worm gear. This was unsatisfactory due not only to the resistance of the gear train, but also because the indicator was attached to the meter and was therefore remote from the observer. It was thus necessary to arrange a mechanical method of starting and stopping the counter in order to define the observation period and to bring the meter to the surface after each observation, to read the counter. Acoustic indicators were later developed which allowed the observer to use a stethoscope tube from above the water surface to count the blows of a small hammer driven by the submerged propeller.

The greatest advance in this field was the introduction by Henry, in 1868, of an electric make-and-break switch mechanism, driven by the propeller, with a low voltage supply from portable batteries. The electrical indicator is now used exclusively, graduating from the early designs using a galvanometer, earphone, lamp bulb or buzzer to the digital counters and improved switches of present practice. With all of these the observer notes either the number of revolutions in a chosen period of time or the time required for a chosen number of revolutions. Modifications of this indicating method are to integrate the series of electrical pulses into a smooth, variable current or to replace the switch and battery by a propeller-driven electric generator with a similar output. This makes the instrument capable of nearly instantaneous direct indications of flow speed, depending on the inertia of the moving part. These are, however, not generally used in open channel measurements.

Recently some miniature current meters were developed for use in laboratories, and these use the principle of the variation in electrical impedance, caused by the propeller blades passing a nearby electrode, to generate a signal. This avoids the problem of switch operation, or indeed of any linkage to the propeller, but there is no known meter of this type suitable for measuring flow in rivers.

Continuous development of the current meter over 300 years has resulted in the general adoption, for river flow measurements, of an instrument with an underwater unit in which a rotating element operates an electrical make-and-break switch. In British and European practice this rotating element has helical blades mounted on a horizontal shaft, and the latest models generate a signal using a reed switch, sealed in a glass capsule, which responds to small magnets rotating with the propeller. A simple electrical lead runs to the indicator unit above the water surface. The unit contains an on–off switch, digital computer, low voltage electrical battery and sometimes a timer, otherwise the observer uses a hand-held stopwatch (see Chapter 1, Figure 1.9). The underwater unit is supported either on a rod held by the observer wading in the channel, or by a cable suspended from a point above the channel. In the latter case a streamlined sinker weight and tail fin unit, to ensure alignment of the water, are both additionally necessary (see Chapter 1, Figure 1.10).

2.2 TYPES OF METER

2.2.1 Methods of measuring discharge

There is a variety of methods of measuring discharge. These methods may be listed under two categories—continuous measurements and occasional and calibration methods.

Continuous measurements

(a) Stage–discharge method;
(b) fall discharge or slope stage–discharge method;
(c) weirs;
(d) flumes;
(e) ultrasonic method;
(f) electromagnetic method;
(g) use of existing facilities.

Occasional and calibration methods

(a) Velocity-area method;
(b) slope-area method;
(c) dilution methods;
(d) use of existing facilities.

Some of the continuous methods of measurement necessitate calibrating the channel structure or equipment. The velocity-area method is the one most frequently used. It requires little equipment but is expensive in manpower and time. It is based on the concept of measuring the speed of flow and the area of cross-section of the channel from which the discharge may be deduced. The instruments for measuring the velocity of flow in an open channel are enumerated below.

2.2.2 Types of velocity measuring instrument

(a) Horizontal axis propeller-type current meters;
(b) vertical axis cup-type meters;
(c) floats;
(d) pendulum or drag meters;
(e) thrust meters;
(f) Pitot tubes;
(g) velocity head rods;
(h) electromagnetic meters;
(i) Thrupp's wake method;
(j) optical current meters;
(k) deep water isotopic current meter;
(l) heated element (thermo)velocity meters;
(m) eddy shedding current meter;
(n) acoustic velocity meter;
(o) Doppler shift velocity meter.

2.2.3 Rotating element current meters

Rotating element current meters may be considered under two broad headings:

(1) Meters with a propeller and a horizontal axis (see Chapter 8, Figures 8.17 and 8.18).
(2) Meters with a series of conical cups rotating about a vertical axis (see Chapter 8, Figure 8.14).

Types and patterns of the former are more widely varied than the latter and are now extensively used in the UK and Europe, but vertical axis meters remain predominant in North America. Each type includes instruments intended for gauging most rivers, and smaller versions for use in narrow and shallow rivers and canals. The diameter of the common propeller meter is about 100 mm, while the smaller versions have diameters about half this value or less. Although the small meters were often intended for use in the laboratory or in lined channels, experience has shown that with care they can be used successfully in natural streams. Some horizontal axis meters, both large and small, also provide a range of propellers of different pitch and diameter, to be selected according to the prevailing flow speeds.

Propellers of horizontal axial flow meters are usually made of metal but more recently some manufacturers have supplied propellers of plastic which are neutrally buoyant in water. They are cheaper to produce but their main advantage is that they respond more quickly to changes in velocity. Since flow in natural channels fluctuates such propellers improve the accuracy of the measurement of turbulent flow, particularly at low velocities.

A few types of horizontal axis meter have the additional facility of indicating direction of flow as well as speed. This ability is rarely required in unidirectional river measurement, but is of great value in estuary and offshore use where it may be compared with further sensors of water temperature, salinity, pH value, etc., to produce a multi-function instrument. Some meters now in production can be coupled to a recording system, either punched or magnetic tape, to obviate continuous attendance for measurements of long duration. It should be remembered, however, that an undetected instrument failure could result in the loss of a part of the record.

2.2.4   Other velocity measuring devices

Of the instruments without a rotating element the simplest is the float. An elementary float, consisting perhaps of an empty bottle or an orange, indicates the speed and direction of flow of the surface layer only and is affected by wind. A long cylinder weighted to float upright with no more than the upper end showing is more accurate as it integrates the average flow conditions over the depth of immersion of the cylinder, while a subsurface drogue may be connected to a small floating marker to show conditions at a chosen depth. It is often difficult to position a float at the required distance from the river bank, particularly on wide rivers, and subsurface floats can ground on shoals; however, they are cheap, can be quickly fabricated and indicate direction as well as speed.

Thrupp's wake method employs a simple device for indicating surface velocity, which consists of two pins projecting a few inches apart from a flat supporting surface. When placed so that the pins touch the surface of flowing water, the distance downstream of the intersection of the wake patterns gives a measure of flow speed.

A major group of non-rotating meters is that which makes use of the conversion of the velocity energy of flowing water to potential energy, which may be seen as a displacement or static head. The simplest form is the velocity rod, a staff which is held in the stream and indicates speed of flow past it by the height of the 'bow-wave' against graduations on the rod. A more complex instrument is the modern version of the late-seventeenth century ball-and-quadrant meter, improved by using a range of carefully designed submerged bodies, thin steel suspension wire, mathematical analysis which allows compensation for drag on the wire, and an indicator capable of showing direction as well as speed. This instrument is rarely used in river gauging, but it would best be mounted on a rigid bridge structure across the channel. Of the same class, the Pitot tube consists essentially of a horizontal open-ended tube facing into the flow connecting either to a vertical limb of the same tube or a distant manometer. Velocity energy in the filament of water approaching the tube orifice is converted to a static head which may be observed in the vertical column.

Continuous detail improvement has resulted in the Bentzel velocity tube of the 1930s and other contemporaneous standard designs. Although it is very simple, and can be made small in order not to disturb flow conditions, its greatest disadvantage is a very small indication at low flow speeds. It is widely used in the laboratory and sometimes in small lined channels, but not in natural rivers due to the problems of operating a manometer and obstruction of the Pitot by solids.

There are more complex non-rotating current meters, such as the electromagnetic, ultrasonic and laser designs. The first two make use of the distortion caused by the flow of water, respectively, on an induced electromagnetic field or sound signal. The laser velocimeter uses the scatter of light from the small particles which abound in all natural water. The equipment is expensive and sophisticated, and it is not generally likely to show a worthwhile advantage over simpler instruments.

The remaining types of velocity device in Section 2.2.2 are dealt with in Chapter 8.

2.3 TYPES OF SIGNALLING DEVICE

Apart from floats, all other forms of current meter require a device to indicate the operation of the meter itself. The indicator for most rotating element meters is a self-contained unit incorporating an electromagnetic digital counter, low voltage battery, on–off switch, and sometimes a range selection switch and an inbuilt electronic timer. As the action of most meters is simply to operate a switch, the indicating units may be interchangeable between different manufacturers of meter. Battery life is long, of the order of 100 h, and is from a commonly available inexpensive dry battery.

Most electromagnetic pulse counters are unable to deal reliably with signal rates greater than about 10 Hz, which for many meters limit flow measurements to speeds not greater than about 2.5 m/s. At the same time, at low speeds the signal rate is so low that a long observation period is advisable for accuracy. There are several methods of overcoming the problem of generating a suitable signal frequency; in the current meter either variable gearing or a range of propellers of different pitch, or in the indicating unit some range-selecting device. All give rise to the possibility of misinterpreting observations, and perhaps the most promising recent advance is the introduction of wholly electric digital counters. These are able to cover a much wider range of signal frequency, and so operate without adjustment in conjunction with a single impeller of well-chosen pitch.

All meters generating discrete pulses require a means of timing the number of pulses in a given period, which may either be contained in the indicator unit as a built-in electronic unit or be a mechanical linkage from the electrical on–off switch to a normal stopwatch. Otherwise, the observer uses a hand-held stopwatch.

In those meters which are designed to give a direct reading analogous to flow speed, rather than a series of pulses, the indicator unit and frequently the connecting cable also are peculiar not only to that type of instrument but usually to the particular example. Indication may be visual on a milliammeter suitably scaled in speed units, or recorded on a chart, punched tape or magnetic tape.

2.4 CALIBRATION OF METERS

2.4.1 Reasons for calibrating meters

Manufacturers design current meters to possess specific qualities, but in order to establish the characteristics of a given meter it is necessary to calibrate it. In addition, although every effort is made to ensure uniformity of production the characteristics vary from meter to meter. Since the cost of calibration is high and forms a substantial proportion of the basic cost of a meter, two types of calibration are available: (a) individual calibrations, and (b) group calibrations.

The first type is necessary if improved accuracy is required, or to establish the characteristics after a period of use, if wear is appreciable, the meter has been damaged, or the standard of maintenance is low.

The second type of calibration is adequate for most engineering purposes if a high standard of uniformity can be achieved in manufacture.

2.4.2 Methods of calibration

Meters are usually calibrated by suspending them from a rod or a cable attached to a trolley which tows them through still water. The tanks are long rectangular channels about 100 m in length by 2 m wide and 2 m deep. Sometimes, however, they are circular and the meter is suspended from the end of a rotating beam. A less satisfactory and less accurate method is to suspend the meter in a channel of flowing water and compare its behaviour with that of a standard meter of known characteristics.

2.4.3 Calibration equations

The number of equations used to express the calibration of a meter depends on the shape of the graph relating speed and rate of revolution of the rotor, and is peculiar to the different types and makes of instrument.

Three linear equations are generally adequate to describe the calibration of cup-type meters. The calibration of many propeller-type meters is expressed by a curve at lower velocities, and above about 0.25 m/s the relation is, for practical purposes, a straight line. The calibration of propeller meters with a small pitch may, however, be expressed by two or even one linear equations.

The linear equation is of the form

$$V = A + Bn \tag{2.1}$$

where V is the speed of meter through still water, n the rate of revolution of the meter, and A and B are constants.

2.4.4 Group calibration

In view of the cost of individual calibrations there is an increasing incentive to adopt group calibrations. When manufacturing tolerances are tightened, group calibrations become more acceptable. In the USA this control of manufacture is practised and the conclusion has been reached that for the Price cup-type meter the uncertainty of calibration of a number of meters of the same type and of the same batch is no greater than the uncertainty of a number of repeat calibrations of a single meter.

Provided that sufficient meters are calibrated to establish a group calibration, this procedure provides an acceptable result. International standards recommend that a minimum of 10 meters should be used, but it is suggested that this number should be increased to at least 20. Users are, however, warned against adopting for one meter an individual calibration intended for another, as this can lead to appreciable errors. The use of a nominal propeller pitch over a range of observations may also lead to serious error.

In view of the increasing tendency to accept group calibrations, for economic reasons, some form of check calibration which is cheaper than the normal calibration based on about 16 measurements, but is at the same time an adequate indication of the characteristics of the meter after a period of use, is desirable. Attention is being given to checking the rating curve of a meter without a previous overhaul by observing its response at no more than about six points. This procedure can result in a substantial reduction in the calibration charges (about half the normal cost), and users of meters adopting group calibrations or even individual calibrations are recommended to accept this as a suitable and cheaper alternative.

2.4.5 Repeatability of measurements

The accuracy of a calibration depends on (a) the repeatability of the calibrated measurements; (b) the closeness of fit of the calibration equations; and (c) the spread of the results when group calibrations are made using a number of meters of a similar pattern.

The term 'repeatability' of a measurement' has three meanings:

(1) It can refer to the extent of agreement between more than one measurement at a given speed.

(2) It may refer to the agreement between a pair of measurements, the second measurement being made after the meter has been dismantled, cleaned and reassembled.
(3) It also refers sometimes to the agreement between calibrations made at stages in the working life of a meter.

Items (1) and (2) reflect the precision of the calibration measurement and depend on the quality of the measuring equipment. Item (3) is dependent on the durability of the meter, the use to which it has been subject, and the efficiency of maintanance.

2.4.6 Epper effect

When towing a meter through still water in a tank, a wave may develop ahead of the meter. The speed at which this wave occurs is related to the depth of water in the tank and is given by

$$V = (gD)^{0.5} \tag{2.2}$$

where V is the speed of progression of the wave, g the acceleration due to gravity, and D the depth of water.

Measurements in the vicinity of this speed should be avoided, as the wave affects the rotation of the meter causing appreciable deviations (often about 1.5 per cent) from the anticipated values.

2.4.7 Comparison of calibrations in different tanks

Tests have been made to compare calibrations made in different rating tanks. These have shown that the spread of results at speeds in excess of 0.5 m/s is less than about 1.5 per cent and may be about 0.7 per cent.

2.4.8 Change of calibration with use

The trend in calibration characteristics of a meter is greatly affected by the conditions in which it is used and by the standard of maintenance, but after some use the effective pitch tends to become slightly shorter; thereafter, variation in characteristics does not exceed the accuracy of the equipment used in the measurements. Meters in general, therefore, appear to maintain their characteristics within acceptable limits, but care is necessary in the operation of some direct-reading instruments where alteration of the controls can introduce errors of the order of 40 per cent.

In general, meters which are easiest to maintain and to calibrate maintain their calibrations better than those which require more difficult adjustments.

2.4.9 Effect of servicing on calibration

Efficient maintenance of a current meter is the most important factor govern-ing the accuracy of measurements. This is particularly the case for measure-ments of low speeds.

Barring accidental damage, the characteristics of a current meter show only small variations with use, particularly at the higher velocities.

It is most important that meters are carefully serviced and checked, in accordance with the maker's instructions, immediately after every occasion on which they are used.

Meters which have been carefully serviced and maintained show little variation in characteristics when recalibrated. Meters which have not been properly maintained show large differences between the calibration charac-teristics before and after overhaul and repair. Poor maintanance can even result in a meter failing to operate at all.

2.5 PERFORMANCE OF METERS

2.5.1 Minimum speed of reponse

The minimum speed of response of a meter, sometimes referred to as the threshold speed, is the lowest speed at which the rotor develops a continuous and steady motion. The performance of a meter depends on its type and make, but generally propeller-type meters have a threshold speed of between about 0.025 m/s and 0.035 m/s. The spread of results near this value may be about 20 per cent. At higher speeds the spread of results is reduced, being approximately 1 per cent at speeds in excess of 0.5 m/s. In general, the spread of results at the higher speeds is less for propeller-type meters than for cup meters.

Although meters will rotate steadily at speeds near the threshold it is not recommended that they be used for field measurements in this range. Inter-national standards, in fact, recommend that the minimum operational speeds should be not less than about 0.15 m/s.

2.5.2 Method of suspension

Current meters may be rated and subsequently used suspended either from a wading rod or from a cable. When used with a cable the amount by which the meter is swept backwards is reduced by adding a heavy weight. These take various forms; the weight may have the meter mounted on the nose with a tail tube and fins to the rear, or it may be mounted below the meter.

The weights used vary in shape. The commonest form is the bomb shape, in which the downward force on the meter is due to the submerged weight of the mass only. Some weights are of a similar shape in profile but are elliptical in cross-section. The downward force is then a combination of the submerged weight and the hydrodynamic force caused by the flow of water past the mass (a variation of the lift caused by an aircraft wing). A variation of this type depends almost entirely on the downward hydrodynamic pull, and the weight is designed to produce the maximum pull. These have proved less successful as they tend to oscillate from side to side, thus affecting the stability of the meter.

Meters should be calibrated suspended in the same manner in which they are to be used in the field, as the method of suspension and sinker weight affect the calibration characteristics. Tests show that the difference in performance of an axial flowmeter on rod and cable suspension with a small sinker weight may be about 1 per cent, the cable-mounted meter tending to rotate faster than when mounted on a rod. The difference is also small when the axial meter is mounted on the nose of the sinker weight.

When a large sinker is suspended at a distance of about 200 mm or less below the axial flowmeter, the propeller may turn about 2 per cent faster than when mounted on a rod. When mounted on the nose of a large sinker weight, the propeller may rotate about 0.5 per cent more slowly.

Some propellers are mounted behind the body of the meter and the suspension unit. In such cases the performance may differ from that of the same meter with a forward-mounted propeller, and the performance may be erratic, between 5 per cent and 10 per cent lower, reaching about 25 per cent at speeds of about 0.05 m/s.

With vertical axis meters the difference between rod-mounted meters may reach about 9 per cent, this being slower for the cable-mounted meter.

When meters are suspended from a cable they may be subjected to oscillations. The effect on meters with ball races is small but on meters with plain bearings the effect is more noticeable and results in a small reduction in accuracy.

2.5.3 Effect of depth of immersion

Meters are generally calibrated at a depth of about 0.6 m. Tests have suggested that at lesser depths the effective pitch of the meter is increased. At a depth of about 0.08 m to the top edge of the meter, the effective pitch may be increased by about 0.25 per cent, at about 0.025 m it is increased by about 0.75 per cent, and at 0.013 m it is increased by about 1 per cent. Current meters should therefore be maintained at a depth below the surface of not less than twice the impeller diameter.

2.5.4 Effect of oblique flow

During calibration on a rod the meter is carefully aligned on the towing carriage so that the relative movement of the water past it is truly axial. If the meter in the field is not aligned with the flow an error will be introduced. With cable suspension the tail unit and free movement of the meter may be relied upon to maintain alignment, but this may not be so with hand-held rod suspended meters, particularly in opaque, slow-moving water.

Since the cup-type meter is unsymmetrical the rate of revolution of the cups may be greater or smaller than the rate when the meter is aligned in the direction of flow, and the effect increases when the angle of obliquity is increased. One type of cup-type meter under-registered by about 2.5 per cent when headed to the left of the axial line when viewed from above, and over-registered by about 2 per cent when headed to the right. When suspended from a boat which rises and falls due to surface waves it may over-register considerably. This is due to the effect of the vertical component of movement on the conical cups, which is additive on both rising and falling motion.

The propeller-type meter always under-registers when held at an oblique angle to the flow in a horizontal plane or when tilted above or below the horizontal. The error in the measurement is very variable and depends on the type of meter and propeller. At angles of about 10 degrees the error may vary from nearly zero to up to about 2 per cent, while at 40 degrees it may vary between about 8 per cent and 60 per cent. When suspended from a boat which rises and falls this type of meter is much less affected than the vertical axis type of instrument.

2.5.5 Effect of drag

It is important to locate a meter accurately at a preselected depth when gauging in flowing water. The accuracy depends on the angle between the suspension cable and the vertical. Although corrections for the angle of inclination of the cable may be made, it is preferable to reduce the effect of the drag of the suspension cable, meter and sinker weight to a minimum.

2.5.6 Effect of turbulence

Meters are usually calibrated in still water but used to measure the speed of turbulent flow. Turbulence superimposes a circular motion on the linear motion of the water past the meter. Little is known of the effect on the calibration characteristics as the effect depends on the size of the eddies and the relative importance of the circular motion compared with the linear motion.

Radial fluctuations cause the meter to under-register the velocity, the effect varying depending on the shape of the blade. The axial fluctuations, however, cause the meter to over-register, the effect increasing with the submerged weight of the blade and the frequency of the fluctuations. Tests show that the effect of turbulence is reduced when the ratio (blade tip radius − hub radius)^2/blade area is small and the blade angle is small also.

The limited information available suggests that the horizontal axis propeller meter may under-register by about 2–3 per cent when running at mid-depth in a rock channel a little over 1 m deep at a speed of about 2 m/s. Vertical axis cup-type meters may be less affected and may be accurate to about 1 per cent.

2.5.7 Accuracy of meters

Some makes and types of meter show a more uniform performance than others, while for those instruments having an interchangeable range of propellers there are differences between them in the precision with which a given range of speeds may be measured. Within these limitations, the accuracy to be expected from a current meter when it is used under conditions similar to those under which it was calibrated depends on whether an individual or group calibration is used.

For propeller-type meters the accuracy at the 95 per cent confidence level is about 1 per cent at 0.1 m/s, improving to about 0.25 per cent at 0.3 m/s and above, but some meters may be less accurate. Near the threshold speed, however, the accuracy may be no better than about 20 per cent.

The accuracy of cup-type meters also varies depending on the type of meter and on the speed. In general, the accuracy is about 1.5 per cent at 0.1 m/s and about 0.3 per cent at speeds in excess of about 1.0 m/s.

The accuracy of propeller-type meters used with a group calibration again varies depending on make and speed. General values are of the order of 7 per cent at 0.08 m/s, 4 per cent at 0.15 m/s and 2.5 per cent at 0.3 m/s and above.

Tests on the accuracy of cup-type meters in the USA show that there is no advantage in calibrating them individually. Similar tests in Britain, however, show that there is a distinct advantage in an individual calibration. Group calibrations are about 3.5 times less accurate than individual calibrations at speeds of about 0.08 m/s, about 6 times at 0.15 m/s, and 6 times at 0.3 m/s and above.

Whe operating conditions are not comparable with those under which the meter was calibrated, additional errors arise. The principal causes of these errors are:

Condition of instrument

Efficient maintanance is the most important factor governing the accuracy of meters, particularly at low speeds. Performance may be markedly affected, especially at low speeds, by damaged propellers or cups, bent spindles, worn or corroded bearings, incorrect oil, misassembly, etc. Careful maintenance in accordance with the maker's instructions is therefore essential.

Method of suspension

Temperature

The calibration information applies at the temperature at which tests were made, generally the ambient temperature of the rating tank. If the temperature in the field is markedly different this may affect particularly the viscosity of the oil with which the bearing of some meters are filled. However, the oil is carfully chosen to minimize this effect, which has been shown to be less than 0.3 per cent at a flow speed of 1 m/s for a temperature difference of 30°C.

Distance from boundary

Research in still water in a rating tank has shown that a propeller-type current meter may approach to within one propeller diameter of the side wall or another meter before significant effect on the indications is noticed. The limiting distance in turbulent flow is not known, but it is recommended that in field use the meter should be kept at least one propeller diameter away from any boundary, particularly at higher speeds.

Depth of immersion

The current meter should be maintained at a depth below the surface of not less than twice the impeller diameter.

Velocity range

The calibration for a given meter applies over a stated range of flow speed only. A large error is likely to result from extrapolating observations beyond the lower limit of calibration, but generally much less so for extrapolation at high speed.

Inclination to flow

If the meter in field use is not aligned with the flow an error will be introduced. With cable suspension the tail unit and free movement of the meter may be relied upon to maintain alignment, but this may not be so with hand-held rod-suspended meters.

Turbulence

Drag affecting depth of immersion

2.6 CHOICE OF METERS

There are many different types of velocity measuring instrument, but only a few are suitable for practical use in rivers. To assist in choosing a suitable instrument the general properties of the more common commercially available instruments are listed.

Horizontal axis propeller meters

(a) Normal speed range 0.15–4 m/s;
(b) propeller diameters usually about 0.11 m and 0.06 m;
(c) some meters are available for measuring direction as well as speed;
(d) may be used from a boat rising and falling with surface waves;
(e) require calibration at intervals but group calibrations are often available for new instruments;
(f) when rod mounted, not seriously affected by oblique flow up to about 15 degrees;
(g) may be used on a wading rod, or from a cable and used from bridges, cableways, static boats or moving boats.

Vertical axis cup meters

(a) Normal speed range 0.15–4 m/s;
(b) may be individually calibrated, but well made and carefully maintained instruments may be used with group calibrations only;
(c) if calibration facilities are not available and provided instrument parts are used, the group calibration may prove adequate;
(d) some instruments which measure direction of flow also are available;
(e) unsuitable for use from a boat rising and falling with surface waves unless suspended in a manner to eliminate the vertical oscillation of the meter;
(f) may be used on rods, or suspended from a cable and used from a bridge or cableway.

Pendulum-type meter

(a) Suitable for use over a wider range of velocities than the previous two types of meter, depending on size and shape of drag body;
(b) best suited to shallower rivers where effect of suspension cable is negligible;
(c) only suitable for use from boats or a bridge to which the measuring device may be fixed.

Floats

(a) May be made from materials locally available;
(b) less accurate than the previous three types of meter;
(c) accuracy depends on type of float, number of floats used, effect of wind, uniformity of channel section over which speed measured, etc.;
(d) normally used when other more accurate instruments are not available or where the velocity is excessive (see Chapter 13);
(e) unless a bridge or boat is available so that floats may be placed in river at a number of points across the channel, the river should be narrow so that they may be thrown into the water from the bank.

Pitot tube

(a) Suited to measurement of high velocities only. If the head difference can be measured to ±2 mm and 5 per cent accuracy is required, the minimum velocity of measurement is about 1 m/s;

(b) accuracy may be improved by using an electronic rather than a water manometer.

(c) if a water manometer is used to measure head difference this must be located by the side of the channel (hence only suitable for use in laboratories) unless provision is made to evacuate some air from above both arms of the manometer when the manometer may be located above the level of the water surface.

Velocity head rod

(a) Generally only used for approximate measurements due to difficulty of measuring head difference accurately;

(b) suitable for high velocities (greater than about 1 m/s) only.

Thrupp's wake method

(a) A simple device intended for approximate measurements only;

(b) convenient for use during field reconnaissance surveys as it can be made on site.

Electromagnetic meters

(a) Accurate over a wider range of velocities than rotating element meters;

(b) measures direction of flow relative to the orientation of the meter;

(c) suitable for rod mounting only;

(b) fairly complex electronic recording instrument required;

(e) respond rapidly to changes in velocity.

2.7 USE OF METERS

2.7.1 Measurement of discharge

The speed of water flowing in an open channel varies in both a vertical and a horizontal direction across a section. In order to establish the average velocity across a section and hence calculate the discharge, knowing the area of cross-section of the channel, it is necessary to locate the horizontal or vertical axis rotating element meter at a number of preselected positions across the channel and elevations above the bed. Various devices, equipment and structures are used to position the meter. These are dealt with in Chapter 1.

2.7.2   Suspension from a static boat

In wide and sufficiently deep rivers, when suspension from a cableway or bridge is not practicable, meters are often suspended from a boat maintained at several fixed distances from the river bank. The position of the boat may be controlled in either of two ways:

(1) The bow may be attached to a cable stretched across the river and tags may be attached to the cable to facilitate locating the boat at the required positions across the river.
(2) The boat may be equipped with an inboard or outboard engine and the driver must then position the boat across the transit line and at the required distance from the bank. This is usually achieved by setting up marker beacons on both banks. The driver maintains the boat's position by holding it such that two beacons on one bank which are on the transit line are in line with each other. The distance from the bank is measured by an observer using a sextant to measure the angle between one of the transit marker beacons and another located at a measured distance upstream.

   For an accurate fix, angles should be between about 30 and 60 degrees. On wide rivers, therefore, it is usually necessary to establish two transit beacons on the transit line and another two beacons at different measured distances upstream of the transit line on both banks. This may require as many as eight beacons, which on very wide rivers may have to be up to 20 m in height for clear visibility.

The first method of fixing a boat's position is the simplest and cheapest, but can be dangerous to the observers if the river carries large floating trees which may damage the boat. It is essential, therefore, to be able to release the boat from the cable quickly in an emergency. Such action then leads to delays in making the observations, because some time must elapse before it is possible to return the boat to the station and reposition it.

   The second method of locating a boat is safer but more expensive. The boat must, however, be adequately powered, depending on the speed of flow of the river, to enable the driver to return to its required position if he has to move downstream to escape damage from floating debris. The length of the boat must also be sufficient for it to operate at the maximum speed of flow which it may encounter. A simple rule for the selection of the length of the boat is

$$v = 1.3\sqrt{l} \qquad\qquad (2.3)$$

where l is the waterline length of boat (m), and v the speed of boat (m/s)— maximum speed of boat should be at least 25 per cent greater than maximum anticipated speed of flow. It is recommended, however, that advice should be sought from an experienced boatbuilder before selecting a boat.

Launches are frequently used in stream gauging but it must be remembered that these have to be fitted with lifting tackle to support the meter and suspension weight over the side of the boat and at a sufficient distance to avoid undue effect when making measurements at shallow depths. A substantial craft is necessary in large deep rivers. Catamarans are frequently used because these have the advantage that the winch for supporting the meter can be located along the centreline of the vessel and the observers can stand to both sides of the meter as it is raised and lowered.

When a boat powered by a motor is used it is often difficult to maintain it exactly on the transit line throughout a measurement of velocity. The practical solution is often to drive the boat forward very slowly, and measurements are begun as it crosses the transit line. It is then allowed to drift slowly astern and again moved forward to cross the transit line, moving forwards a second time. On the second crossing of the transit line, moving in the same direction, the observer stops recording the velocity. Some slight error is introduced but provided the movement of the vessel is not excessive and is slow, the error is negligible.

2.7.3   Suspension from a moving boat

The duration of measurement using the static-boat technique is often lengthy. To overcome this the moving-boat techhnique has been developed. This method is suitable for wide rivers with an almost uniform depth and is described in detail in Section 2.9.

2.8   VARIATION OF VELOCITY ACROSS A CHANNEL

The speed of flow varies across a channel from one bank to the other, and also on a vertical between the bed and the surface.

2.8.1   Horizontal variation of velocity

The variation of speed along a horizontal across a channel depends on the cross-section of the channel and on the geometry of the channel upstream of the measuring section. The relationship between speed of flow and the distance from the bank varies from one river to another, and it is thus necessary when using the velocity area method to determine the discharge to measure speed at a number of points across the channel. In natural channels measurements must usually be made at about 20 points unless the channel is very narrow.

To reduce the variation across a section it is desirable to select a section across which the speed of flow is symmetrical about the axis of the channel. To achieve this the channel upstream should be straight and uniform and at least about 10 channel widths downstream of any bend or non-symmetrical section.

2.8.2   Vertical variation of velocity

Provided that the depth of flow is not less than about 10 times the depth of the equivalent grain roughness of the channel, the variation of the velocity on a vertical between the bed and the water surface may be described by a mathematical equation. The equation most commonly used to relate velocity to elevation above the bed in fully turbulent flow (as occurs in most natural channels) is the logarithmic rough turbulent relationship:

$$v = 5.75 V_* \log \frac{30y}{k} \tag{2.4}$$

where v is the velocity at a point y above the bed; V_* the shear velocity which in wide symmetrical channels is $\sqrt{(gds)}$; g the acceleration due to gravity; d the depth of flow; s the hydraulic gradient; y the height measured above the bed of the channel; and k is the Nikuradse equivalent grain roughness.

Equation (2.4) may be expressed in different forms, e.g.

$$v = v_a + 5.75 V_* \log \frac{y}{a} \tag{2.5}$$

$$v = v_m + 5.75 V_* \log 2.718 \frac{y}{d} \tag{2.6}$$

$$v_m = 5.75 V_* \log 11.05 \frac{d}{k} \tag{2.7}$$

where v_a is the velocity at point a above the bed; v_m the mean velocity over the vertical; and a is the height measured above the bed of the channel.

From equation (2.6), if $v = v_m$, then

$$\frac{v}{d} = 0.37 \tag{2.8}$$

i.e. the velocity at depth 0.63 of the total depth below the water surface is equal to the mean velocity.

It may also be deduced from equation (2.6) that the arithmetic mean of two velocities measured at $y/d = 0.8$ and $y/d = 0.17$ is equal to the mean velocity, i.e. the arithmetic mean of two velocities measured at depths below the surface of 0.2 and 0.83 of the depth of flow is equal to the mean velocity.

It is from vertical velocity distributions similar to those above that the depths at which velocity measurements should be made may be deduced.

A simpler but less accurate relationship between elevation above the bed and velocity is given by

$$\frac{v}{v_a} = \left(\frac{y}{a}\right)^c \tag{2.9}$$

where v, v_a, y and a are as defined above, and c is the coefficient between $\frac{1}{5}$ and $\frac{1}{8}$, usually about $\frac{1}{6}$.

From equation (2.9) the following equations may be deduced:

$$\frac{v_m}{v_a} = \frac{1}{c+1}\left(\frac{d}{a}\right)^c \tag{2.10}$$

$$\frac{v}{v_m} = (c+1)\left(\frac{y}{d}\right)^c \tag{2.11}$$

From equation (2.11) if $c = \frac{1}{6}$ it may be deduced that the measured velocity is equal to the mean velocity when $y/d = 0.4$, i.e. the velocity is measured at depth equal to 0.6 of the total depth below the water surface.

It may also be deduced that the arithmetic mean of velocities measured at $y/d = 0.8$ and 0.18 is equal to the mean velocity, i.e. the average of velocities measured at depths of 0.2 and 0.82 of the total depth below the water surface is equal to the mean velocity.

2.9 THE MOVING-BOAT TECHNIQUE

2.9.1 Principle

In the moving-boat method a propeller current meter is fitted to a gauging launch and the latter made to traverse the stream normal to the flow. The object of the technique is to measure the velocity at numerous points across a river more rapidly than is possible using the static-boat method. The principle is based on measuring the magnitude and direction of the component of the velocity resulting from the river current and the movement of the boat across the river. The speed of the stream is then deduced from these measurements and, knowing the depth of the river, the discharge is calculated.

2.9.2 Application

The method is suitable for large rivers and some estuaries where the depth of flow is nearly uniform across the channel, and there are no large shoals which reduce the depth.

The requirements for a suitable site are similar to those for the static boat method.

2.9.2   Equipment

Set out the transit line across the channel at right angles to the direction of flow, and establish beacons on both banks. The beacons should be sufficiently large and high to be visible from the boat at all points within the river.

The boat should be large enough to carry the equipment, a boat driver, an observer to fix the position using a sextant, and at least one observer to supervise the current meter. The boat should have sufficient power to maintain it on the transit line depending on the speed of the current, and have sufficient waterline length to ensure that the boat does not plane over the surface of the water like a speedboat. The length of the boat may be deduced from equation (2.3).

The boat should be equipped with a recording echo-sounder and for convenience, although this is not necessary, the recorder should also record the speed and direction of the current meter.

The current meter should be rod mounted at a depth of about 1 m below the water surface, and the rod should be carried in bearings and have a pointer to indicate the direction of the meter.

2.9.4   Preliminary measurements

Measure the width of the river by setting out a baseline along one bank perpendicular to the transit line (see Figure 2.1) and measuring the angle TBL by theodolite.

If the river is too wide for the transit marker to be seen from the opposite bank, baselines perpendicular to the transit line should be set out on both

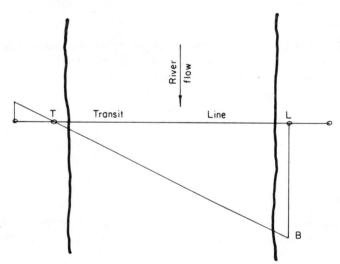

Figure 2.1   The moving-boat method: measurement of river
width by transit line and base line

banks. The boat should be positioned on the transit line, where it can be seen from both banks, and simultaneously the angles between the boat and the baselines should be measured using a theodolite.

The boat should be located at a point on the transit line at a distance from one bank of about 1/20 of the width of the river. Its position should be fixed by observing the angle subtended at the boat by two beacons on one bank. The speed of flow should be measured at a depth of 1 m below the surface and also at depths of 0.2, 0.6 and 0.8 of the depth of water.

This procedure should be repeated at about 19 other points equally spaced across the channel. From these measurements the relation between the velocity at 1 m depth and the mean velocity on each vertical should be established. Similar measurements should be made at two or three other depths of flow, so that at any depth of flow, having measured the velocity at 1 m depth, the mean velocity on that vertical can be deduced.

2.9.5 Technique

The boat is manoeuvred as close to the bank as is safe, depending on the depth of water and the depth of immersion of the current meter and vane. When the current meter and echo-sounder are operating satisfactorily the position of the boat is fixed; one observer using a sextant measures the angle subtended by two of the beacons (Figure 2.2). For a good fix the angle should

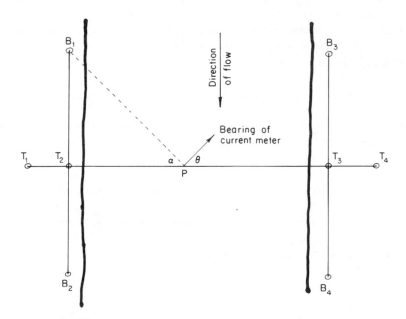

Figure 2.2 The moving-boat method: position fixing

be between 30 and 60 degrees, and the beacons should be positioned to make this possible at any point in the river along the transit line.

The boat is then headed diagonally across the river, and the speed controlled so that it remains on the transit line. At intervals the position of the boat is fixed, while at the same time the speed and bearing of the current meter are recorded.

2.9.6 Computations

The computations to determine the discharge of the river are made in steps, as follows:

(1) Calculate the distances from the transit marker T_2 (Figure 2.2) at which the current meter observations were made: $PT_2 = B_1T_2 \cdot \cot \alpha$

(2) Determine the velocity of the water past the current meter from the recorded number of revolutions of the propeller and duration of measurement.

(3) Calculate the component of each velocity perpendicular to the transit line (Figure 2.2):

$$V_r = V_m \cdot \sin \theta \qquad (2.12)$$

where V_r is the velocity perpendicular to transit line; V_m the measured velocity; and θ is the angle of vane or current meter with transit line.

(4) Determine the depth of river for each point of observation from echo-sounder measurements.

(5) Plot the cross-section of the channel from distances (1) and depths (4).

(6) From preliminary measurements of velocity at different depths determine value of c for use in the equation

$$\frac{V}{\bar{V}} = (c+1)\left(\frac{y}{d}\right)^c \qquad (2.13)$$

where V is the velocity at elevation y above bed; $\bar{V}$ the velocity on a vertical; d the depth of flow; and c is a constant. In absence of field measurements assume $c \doteq \frac{1}{6}$.

(7) Using equation (2.13) calculate mean velocity on each vertical knowing velocity V_r perpendicular to transit line, depth of immersion of current meter and depth of flow.

(8) From mean velocity for each vertical and depth of flow compute discharge per unit width at each point.

(9) Plot discharge per unit width against location across channel. Total area under graph is the total discharge, alternatively see Section 2.9.7.

2.9.7 Example

A typical example of a moving-boat measurement is shown in Table 2.1.

Table 2.1 Sample of computation notes of a moving boat at measurement

River: National River nr River City   Vr Run No. 1–4   Range (L) 22,256   Time: S 7.52 h / F 7.58 h

Measured width: 481.402 m   Computed width: 471.036 m
Width/area adjustment coefficient: 1.022   Number of seconds: 30

1	2	3	4	5	6	7	8	9	10	11	12
Angle θ	L_b (m)	Distance from initial point (m)	Width (m)	Depth (m)	Pulses per second	V_m	Sin θ	$V_m \sin \theta$	Area (m²)	Discharge (m³/s)	Remarks
IP		0									IP to LEW 8.54 m
LEW		8.54	5.95								
20	11.89	20.43	10.99	2.74	250	1.372	0.342	0.470	30.1	14.1	
25	10.09	30.52	14.68	9.45	340	1.857	0.423	0.786	174.6	137.3	
30	19.27	49.79	19.36	11.59	370	2.018	0.500	1.009	223.9	226.0	
29	19.45	69.24	18.37	11.43	340	1.857	0.485	0.899	210.0	188.9	
39	17.29	86.53	17.16	11.28	340	1.857	0.629	1.168	193.2	225.7	
40	17.04	103.57	17.64	10.67	330	1.802	0.643	1.158	187.7	217.5	
35	18.23	121.80	17.64	10.82	330	1.802	0.574	1.034	190.4	196.8	
40	17.04	138.84	15.96	10.06	330	1.802	0.643	1.158	160.7	186.0	
48	14.88	153.72	14.88	9.91	350	1.912	0.743	1.42	147.7	209.8	
48	14.88	168.60	15.44	9.76	340	1.854	0.743	1.378	150.5	207.3	
44	16.01	184.61	15.32	9.60	340	1.857	0.695	1.290	146.8	189.2	
49	14.60	199.21	14.74	9.45	320	1.750	0.755	1.320	139.4	184.0	
48	14.88	214.09	15.96	9.14	330	1.802	0.743	1.338	145.8	195.1	
40	17.04	231.13	16.38	8.69	320	1.750	0.643	1.125	142.1	160.0	
45	15.73	246.86	16.00	8.08	300	1.640	0.707	1.158	129.1	149.5	
43	16.28	263.14	16.54	8.23	330	1.802	0.682	1.229	135.6	166.5	
41	16.80	279.94	15.40	8.23	350	1.912	0.656	1.253	126.4	158.3	

Table 2.1 (*contd.*)

1	2	3	4	5	6	7	8	9	10	11	12
Angle θ	L_b (m)	Distance from initial point (m)	Width (m)	Depth (m)	Pulses per second	V_m	Sin θ	$V_m \sin \theta$	Area (m²)	Discharge (m³/s)	Remarks
51	13.99	293.93	15.64	7.93	330	1.802	0.777	1.399	123.6	172.7	
39	17.29	311.22	16.78	7.62	320	1.750	0.629	1.101	128.2	141.0	
43	16.28	327.50	16.66	7.62	320	1.750	0.682	1.192	126.4	150.6	
40	17.04	344.54	17.52	7.62	330	1.802	0.643	1.158	133.8	154.9	
36	17.99	362.53	15.68	7.47	330	1.802	0.588	1.061	117.0	124.0	
53	13.38	375.91	14.70	7.16	320	1.750	0.799	1.399	105.0	147.0	
44	16.01	391.92	16.40	6.86	320	1.750	0.695	1.216	112.4	136.8	
41	16.80	408.72	17.94	6.86	330	1.802	0.656	1.030	125.5	145.0	
31	19.08	427.80	18.54	6.71	330	1.802	0.515	0.927	124.5	115.2	
36	17.99	445.79	16.89	6.86	320	1.750	0.588	1.030	116.1	119.5	
19	15.79	461.58	16.89	3.66	340	1.857	0.326	0.607	61.8	37.4	
REW	17.99	479.57	8.99	0							
FP	17.99	486.58									
	471.036 m	470.945 m							3905 ×1.022 =3991	4452 ×1.022 (Width/ area adjustment coefficient =4550 ×0.91 (velocity correction coefficient) =4140 m³/s	REW to FP = 7.012 m

Width/area adjustment coefficient: $\dfrac{481.402}{471.036} = 1.022$

IP = Initial point
FP = Final point
LEW = Left edge of waterline
REW = Right edge of waterline
$L_b = V_m \cos \theta t$
where t is the time of travel between observation points

BIBLIOGRAPHY

Alming, K., 1969, June. Calibration of current meters, a comparison, V. L. Report 262, Technical University of Norway.

Beaumont, R. D., 1969, July. Some factors influencing the measurement of water discharge by current meter, South African Council for Scientific and Industrial Research, R. Meg. 298, River Engineering Symposium, Durban.

Biswas, A. K., Beginning of quantitative hydrology, *Proc. Am. Soc. Civ. Eng.*, **94,**(HY5), 1299 (Sept. 1968); **95**(HY2), 770 (March 1969); **96**(HY2), 584 (Feb. 1970).

Biswas, A. K., 1970. *'History of Hydrology'*, North-Holland, Pub. Co., Amsterdam.

British Standards Institution, 1973. Methods of measurement of liquid flow in open channels, Part 8A; Current meters, Current meters incorporating a rotating element, BS 3680: Part 8A, BSI, London.

British Standards Institution, 1973. Methods of measurement of liquid flow in open channels, Part 8B; Current meters, Suspension equipment, BS 3680: Part 8B, BSI, London.

Carter, R. W. and Anderson, I. E., 1963, July. Accuracy of current meter measurements, *Proc. Am. Soc. Civ. Eng.*, **89**(HY4), 105.

Carter, R. W. and Davidian, J., 1968. Techniques of water resources investigations of the USGS General procedure for gauging streams, Book 3, Chapter A6, US Geological Survey, Washington.

Castex, L. and Carvounas, E., 1962, March. Influence de la turbulence sur la mésure du débit en canal à l'écoulement libre au moyen de moulinets, Rapport International Current Meter Group No. 13.

Chaix, B., 1961. Field and laboratory tests to assess the influence of turbulence on the performance of different types of current meter, International Current Meter Group, Report No. 9, National Engineering Laboratory, East Kilbride, Glasgow.

Clayton, C. G. (Ed.), 1972. Modern developments in flow measurement, Proc. Int. Cnf. at Harwell. (September 1971), Atomic Energy Research Establishment, Harwell, and National Engineering Laboratory, East Klbride, Glasgow.

Corbet, D. B. *et al.*, 1956, October. Stream gauging procedure, Water Supply Paper 888, US Geological Survey, Washington.

Folsom, R. G., 1956, October. Review of the Pitot tube, *Trans. Am. Soc. Mech. Eng.*, **73**(7), 1447.

Frazier, A. H., 1974. *Water Current Meters in the Smithsonian Collections of the National Museum of History and Technology*, Smithsonian Institution Press, Washington.

Griffith, R. W., 1941. Discharge by surface floats, *J. Inst. Civ. Eng.*, **15**(4), 284.

Grindley, J., 1970, June. Calibration and behaviour of current meters, INT 80, Hydraulics Research Station, Wallingford.

Grindley, J., 1971. Calibration and behaviour of current meters, effect of oblique flow, INT 87, Hydraulics Research Station, Wallingford.

Grindley, J., 1971, September. Calibration and behaviour of current meters, drag, INT 96, Hydraulics Research Station, Wallingford.

Grindley, J., 1971. Behaviour and calibration of curren meters, effect of suspension, INT 93, Hydraulics Research Station, Wallingford.

Grindley, J., 1971, September. Behaviour and calibration of current meters, accuracy, INT 95, Hydraulics Research Station, Wallingord.

Grindley, J., 1972, January. Calibration and behaviour of current meters, tests in flowing water, INT 99, Hydraulics Research Station, Wallingford.

Harp, J. F., 1974, January. An innovative automatic stream gauging method, *J. Hydrology*, **21,** 1.

Herschy, R. W., 1965. Rating of current meters, Technical Note No. 1, Water Resources Board, Reading.

Herschy, R. W., 1976, May. An evaluation of the Braystoke current meter, Technical Memorandum No. 7, Water Data Unit, Department of the Environment, Reading.

Herschy, R. W., 1976. New methods of river gauging, in *Facets of Hydrology* (Ed. J. C. Rodda), Wiley, New York.

Herschy, R. W., Hindley, D. R., Johnson, D. and Tattersall, H., Effect of pulsation on the accuracy of river flow measurements, Technical Memorandum No. 10, Department of the Environment, Water Data Unit, Reading.

Howe, J. W., 1950. Flow measurement, in *Engineering Hydraulics*, (Ed. H. Rouse), Wiley, New York, and Chapman and Hall, London.

Hydraulics Research Station, 1975, July. Calibration and behaviour of a valeport Brastoke current meter, DE34, HRS, Wallingford.

International Organisation for Standardization, 1973. Liquid flow measurement in open channels, ISO 772: Vocabulary and symbols, ISO, Switzerland.

International Organization for Standardization, 1974. Liquid flow measurement in open channels, ISO 2537: Cup type and propeller type current meters, ISO Switzerland.

International Organization for Standardization, 1976. Liquid flow measurement in open channels, ISO 3455: Calibration of rotating element current meters in straight open tanks, ISO, Switzerland.

International Organisation for Standardization, 1978. Liquid flow measurement in open channels, ISO 4369, The moving boat method, ISO, Switzerland.

Kolupaila, S., Early history of hydrometry in the United States, *Proc. Am. Soc. Civ. Eng.*, **86**(HY1), 1 (Jan. 1960); **86**(HY4), 131 (April 1960); **86**(HY6), 117 (June 1960); **86**(HY7), 33 (July 1960); **86**(HY9), 125 (Nov. 1960); **87**(HY3), 175 (May, 1961).

Leeson, E. R., (1942, 10 November). Effect of non-standard connection on current meter rating, *Water Resources Bulletin*, **162**.

Lorant, M., 1951. The rating of water current meters, *Water and Water Eng.*, **55**, 670.

Ott, A. and Co., 1976. Calibration of current meters, HLe 120/14, Kempton, Bavaria.

Seiffert and Liebs, 1931, 2 December. The question of transferring meter calibrations to water metering, *Wasserkraft und Wasserwirtschaft*, No. 23.

Smoot, G. F., 1970. Flow measurement of some of the world's major rivers by the moving boat method. Proc. International Symposium on Hydrometry, Koblenz, UNESCO/WMO/IAHS, Pub. No. 99, pp. 149–161.

Smoot, G. F. and Carter, R. W., 1968, November. Are individual current meter ratings necessary, *Proc. Am. Soc. Civ. Eng.*, **94**(HY2), (March 1968); and discussion by Schoof, R. R. and Crow, F. R., **94**(HY6).

Smoot, G. F. and Novak, C. E., 1968. Calibration and maintenance of vertical axis type current meters, Techniques of Water Resources Investigations of the USGS, Book 8, Chapter B2, US Geological Survey, Washington.

Smoot, G. F. and Novak, C. E., 1969. Measurement of discharge by the moving boat method, Techniques of Water Resources Investigations of the USGS, Book 3, Chapter A11, US Geological Survey, Washington.

Swiss Federal Water Resources Bureau, 1969. Guide for field work on flow measurement with current meters, Berne, 1967; translated by Water Resources Board, Reading.

Thrupp, E. C., 1906, December. Discussion on paper on water supplies, *Min. Proc. Inst. Civ. Eng.*, **CLXVII**, 271.

Townsend, F. W. and Blust, F. A., 1960, 11 April. A comparison of stream velocity meters, *Proc. Am. Soc. Civ. Eng.*, **86**(HY4).

Whittington, R. B. and E.-Fiki, A. B., 1967, January. The suspended sphere current meter, *Civil Eng. and Public Works Rev., London,* **62**(726), 65.

World Meteorological Organization, 1974. Guide to hydrological Practices, No. 168, WMO, Geneva.

Yarnell, D. L. and Nagler, F. A., 1931. The effect of turbulence on the registration of current meters, *Proc. Am. Soc. Civ. Eng.,* **95,** 766.

CHAPTER 3

Flow Measuring Structures

W. R. WHITE

3.1 INTRODUCTION

[See the following references BSI (1971a, 1971b); Buchanan and Somers (1968); Clayton (1971); Harrison (1969); Helliwell and Bates (1971); IWE (1969); Rodda (1976), and White (1975a).]

3.1.1 Measurement of flow

A flow measuring structure provides a means of recording discharge at a particular location in a closed conduit or an open channel. It is usually installed with the intention of providing a continuous record of flow variations with time and hence it is an alternative to any of the following methods:

(1) The stage–discharge method—see Chapter 1.
(2) The slope-area method—see also Chapter 13.

83

(3) The ultrasonic method—see Chapter 7.
(4) The electromagnetic method—see also Chapter 7.
(5) Certain indirect methods—see Chapter 6.

Occasionally, small flumes and V-notch tanks are used for temporary gauging of flows and under these circumstances they are providing data which could otherwise be obtained by dilution, velocity-area or slope-area techniques.

It is not the purpose of this chapter to compare and contrast the advantages and disadvantages of the various methods available for the measurement of flow but it is useful to indicate the areas where hydraulic structures are particularly attractive. In the laboratory, thin-plate weirs, flumes, orifice plates and Venturi meters cover most of the requirements for flow measurement, the remaining areas being covered by such instruments as hot-wire anemometers, miniature propeller meters, electromagnetic pipe-flow meters and various types of drag meter. In the field, hydraulic structures are much larger and hence involve a much higher capital outlay. For this and other reasons alternative methods play a more significant role in the field than in the laboratory. Hydraulic structures, usually weirs or flumes but occasionally undershot sluices, are most common in the middle and upper reaches of watercourses. In the lower reaches, the width of rivers usually makes the size and hence cost of a structure prohibitive. In any case, the shallow longitudinal slope of the lower reach normally means that the afflux associated with a structure could not be tolerated from a flooding point of view. In the upper reaches of watercourses, high rates of transport of coarse sediment make for difficulties in flow measurement but specialist structures have been developed which can tolerate this type of environment.

In very general terms, conventional hydraulic structures are used in natural watercourses where the longitudinal slope is between 0.003 and 0.000 3, while specialist structures may be used in streams where the longitudinal slope is as high as 0.02. This specification ensures that for typical natural channels the Froude number, Fr, prior to installing a conventional hydraulic structure, is within the range $0.1 < Fr < 0.5$. Specialist structures may be installed in streams where the natural flow conditions are supercritical.

3.1.2   Modular and non-modular flow

The philosophy of using a hydraulic structure to measure flow is founded on the premise that the relationship of discharge to a measure of water level (or pressure) can be forecast, either from basic physical principles or from empirical evidence on performance. The water level is gauged at a prescribed location upstream of the structure, and an equation or graphical relationship is applied to convert this to flow. This is the simplest case where, because the water level downstream of the structure is below some limiting condition (the modular limit), there is a unique relationship between these two quantities. Under these circumstances flow conditions are said to be modular.

If the tailwater conditions do affect the flow, the weir is said to be drowned or operating in the non-modular range. Two independent measurements relating to upstream and downstream water levels are required in order to determine the discharge at the structure under these circumstances.

The dividing line between modular and non-modular flow conditions is defined as the modular limit. It is quoted for each individual structure as the submergence ratio (the ratio of downstream to upstream total heads, both relative to weir crest of flume invert level) when the flow is 1 per cent below the equivalent modular value.

3.1.3 Installation

The complete flow measuring installation consists of an approach channel, the structure itself and the downstream channel. The structure plays the most important role in determining the accuracy of the measured discharge but the condition of the approach and downstream channels also makes a contribution.

The structure should be rigid, watertight and capable of withstanding peak flows without damage and its axis should be in line with the natural direction of flow. The surfaces of the structures should be smooth, particularly in the vicinity of the crest or throat, and the alignment and dimensions of the structure should be set out with care. Parallel and vertical side walls should flank the structure and these should extend upstream at least as far as the head measurement position.

The main requirement for the approach channel is that it should be straight and reasonably regular for a distance upstream of the structure of at least five times its surface width. This ensures that the velocity distribution upstream of the structure is not heavily distorted and, more important, that there is no appreciable variation in water level across the width of the approach channel.

Flow conditions downstream of the structure are important in that they determine tailwater levels which in turn influence the choice of elevation for the structure. It is important, therefore, to survey the downstream channel at the design stage and either measure or calculate the variation of tailwater levels with discharge.

3.1.4 Measurement of heads

The discharge at a gauging structure depends on the geometry of the structure and water levels relative to the structure. The geometry of the structure is usually fixed and hence the accurate determination of its dimensions is relatively straightforward. Water levels relative to the structure (gauged heads) present the engineer with a more difficult problem, particularly at low flows, and as these problems are common to all types of structure it is logical to consider them as a separate item.

The first problem is to determine the best place to measure the water level, the best place being defined as that position which will yield the most accurate and reliable values for computed discharges. The water surface profile upstream of a gauging structure is not horizontal and hence the choice of the location for the water level measurement materially affects computed flows. Immediately upstream of the structure the water surface profile takes the form of a drawdown curve and pressures are not hydrostatic. Further upstream, frictional forces produce a water surface slope towards the structure. In general, the position for measuring the upstream water level, commonly referred to as the gauging section, should avoid the area of drawdown but should not be so far upstream of the structure that frictional losses play a significant role. Common practice is to specify the distance to the gauging section as a multiple of the maximum head to be expected, typical values being two to four times the maximum head. For hydrometric purposes this type of recommendation is probably adequate, but there are one or two legitimate criticisms which may be made of this method. First, the maximum head is meaningless in statistical terms and the designer has to make subjective judgements in assigning to it a numerical value. Secondly, the method presupposes that a location which is satisfactory at the maximum head is satisfactory at all other heads. Since drawdown and friction vary with discharge this is unlikely to be the case. A more logical approach would be to study flow-duration curves at the proposed site for the structure and to assess the range of flows (or heads) where accuracy of measurement is most important. This 'significant' head should be used rather than an esoteric maximum.

Having determined the location for the gauging section it is necessary to consider how and with what type of instrumentation the water level is to be measured. It is usual to measure the water level in a separate gauge well which is connected to the approach channel at the gauging section by a pipe. This pipe may or may not have a perforated coverplate on the outer end. This system reduces the effects of water surface irregularities in the approach channel to the structure, but the dimensions of the well and connecting pipe (and coverplate) must be chosen such that significant differences between river and well water levels do not occur during the critical periods when discharge is changing rapidly with time. An alternative system is to set an open-ended vertical damping tube in the approach channel with instrumentation recording water levels within it. This system is cheaper but causes undesirable disturbances to the flow. Yet another system is to record the water level directly in the approach channel. This method is common in laboratories and is also used in the field either as a check on gauge well levels or as a basic manual method where automatic instrumentation is not considered justifiable.

The computation of discharge depends on a knowledge of water levels relative to the elevation of the structure. In particular, the water levels must be known relative to the crest level of weirs (lowest level for V-notches and flat-V weirs) and the throat invert level of flumes (lowest level for U-shape flumes). This means that the instrumentation for measuring water levels must be accurately zeroed to the level of the structure. The most convenient method is to set metal datum plates on the wing walls of the structure and on the top of the gauge well where applicable. An accurate survey of the structure then relates crest and invert levels to these datum plates. Carefully carried out manual measurements using micrometer, vernier or steel tape gauges then relate the water levels to these datum plates and the readings of the head measuring instrument. Regular checks on zeroes should be carried out, particularly where instruments are moved or replaced. A zero check based on the water level, either when the flow ceases or just begins, is liable to serious errors due to surface tension effects.

3.1.5 Computation of discharge

Discharge equations are usually quoted in terms of total head values, although there are some exceptions to this rule. The reason for this is that both theoretical and experimental studies suggest that total head (or energy) is the more fundamental variable and discharge coefficients are more readily presented within this framework. Data recorded at gauging structures are, however, of gauged heads and total heads must be deduced using one of two methods. The first uses successive approximation techniques and is particularly suited to computer analysis. The method is well known and needs no detailed discussion here. An alternative manual method exists which is particularly useful at the design stage. The approach, known as the coefficient of velocity method, is graphical and considers the structure and flow geometries and their effect on the relationship between gauged and total heads.

As an example, we will consider a straightforward two-dimensional weir, i.e. a weir with a horizontal crest line operating in the modular flow range. We will then extend the argument to the special case of a triangular profile weir operating in the drowned flow range.

Modular flow conditions

The common discharge equation for a two-dimensional weir, in terms of total head, is

$$Q = C_{de}\sqrt{(g)}bH_{1e}^{3/2} \tag{3.1}$$

where $Q(\text{m}^3\,\text{s}^{-1})$ is the discharge; C_{de} the effective coefficient of discharge; $g(\text{m}\,\text{s}^{-2})$ the acceleration due to gravity; $b(\text{m})$ the crest breadth; and $H_{1e}(\text{m})$ is the effective upstream total head.

The corresponding gauged head equation is

$$Q = C_v C_{de} \sqrt{(g)} b h_{1e}^{3/2} \tag{3.2}$$

where C_v is the coefficient of velocity, and h_{1e}(m) is the effective upstream gauged head.

Comparison of equations (3.1) and (3.2) indicates that $C_v = (H_{1e}/h_{1e})^{3/2}$ and that C_v can be related to the weir and flow geometries by the expression

$$\frac{2(C_v^{2/3} - 1)}{C_v^2} = \left\{ C_{de} \frac{h_{1e}}{h_{1e} + P_1} \frac{b}{B} \right\}^2 \tag{3.3}$$

where P_1(m) is the height of weir above upstream bed, and B(m) is the breadth at gauging section.

This functional relationship is plotted in Figure 3.1. Using known values of C_{de}, h_{1e}, P_1, b and B, the ordinate can be evaluated and the value of C_v extracted from the plot. Substitution in equation (3.2) then gives discharge directly.

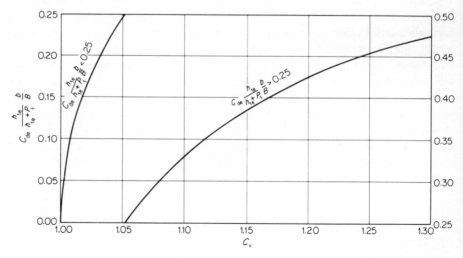

Figure 3.1   Coefficients of velocity for two-dimensional weirs operating in the modular flow range

Non-modular flow conditions

A two-dimensional weir of triangular profile with 1:2 and 1:5 upstream and downstream slopes, respectively, may be used in the non-modular range and the total head and gauged head equations are as follows:

$$Q = f C_{de} \sqrt{(g)} b H_{1e}^{3/2} \tag{3.4}$$

where f is the drowned flow reduction factor and

$$Q = fC_vC_{de}\sqrt{(g)}bh_{1e}^{3/2} \tag{3.5}$$

It has been shown that the drowned flow reduction factor, f, is dependent upon the ratio h_p/h_{1e} and the weir and flow geometries (h_p is the head within the zone of flow separation which forms just downstream of the crest and is recorded as well as the upstream gauged head). The coefficient of velocity, C_v, is dependent upon f and the weir and flow geometries:

$$\frac{2(C_v^{2/3} - 1)}{C_v^2} = \left\{ fC_{de} \frac{h_{1e}}{h_{1e} + P_1} \frac{b}{B} \right\}^2 \tag{3.6}$$

Hence

$$f = \phi^{(i)} \left\{ \frac{h_p}{h_{1e}}, C_{de} \frac{h_{1e}}{h_{1e} + P_1} \frac{b}{B} \right\} \tag{3.7}$$

and

$$C_v = \phi^{(ii)} \left\{ f, C_{de} \frac{h_{1e}}{h_{1e} + P_1} \frac{b}{B} \right\} \tag{3.8}$$

Combining (3.7) and (3.8) gives

$$C_vf = \phi^{(iii)} \left\{ \frac{h_p}{h_{1e}}, C_{de} \frac{h_{1e}}{h_{1e} + P_1} \frac{b}{B} \right\} \tag{3.9}$$

The functional relationship (3.9) is plotted in Figure 3.2. Using known values of h_{1e}, h_p, C_{de}, P_1, b and B, the abscissa and ordinate can be evaluated and the multiple C_vf is extracted from the plot. Substitution of known values in equation (3.5) then gives the discharge.

In the case of a three-dimensional structure, such as a flat-V weir or a U-shape flume, the coefficient of velocity method becomes more complex but is still usable. The method is described by the author elsewhere (White, 1971a).

3.1.6 Calibration methods

There are three basic methods for determining the performance characteristics of flow measuring structures.

Theoretical analysis

Flow conditions at a measuring structure are governed largely by inertia and gravitational forces. Viscous, surface tension and general frictional forces can usually be neglected although they may become significant at low heads, particularly with long structures.

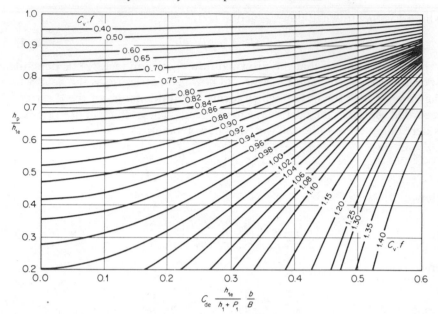

Figure 3.2   Combined coefficients of velocity for Crump weirs operating in the
non-modular flow range

In general, structures of simple geometry and which do not induce flow separation are amenable to theoretical analysis. Calculations based on the conservation of energy and mass, yield discharge equations which can be used with confidence in laboratory and field conditions alike. Long-throated flumes are an example within this category.

Laboratory calibrations

Because inertia and graviational forces are the main influences on the performance of measuring structures, it is possible to model them on a Froudian basis. Laboratory measurements of flows and corresponding heads provide calibration data which can be used on any size of structure, so long as the physical geometry remains identical.

Model calibrations have to be carried out with great care and attention to detail. It is necessary to ensure that the accuracy of the laboratory data is significantly higher than that required by the user. This is because the user will incur additional errors, particularly in the measurement of head, which will add to any errors in the laboratory-derived calibration equation(s). Thus, in the laboratory, it is often necessary to measure flows to an accuracy of 0.2 per cent and heads to an accuracy of 0.01 mm.

Laboratory investigations are best directed towards the development of standard weir designs such that the financial burden of the laboratory work can be spread over a large number of similar structures. Examples of this are to be found in the development of two-dimensional and flat-V triangular profile weirs at the Hydraulics Research Station, Wallingford, UK. Several hundred of these structures are to be found in the UK alone and the weirs have been adopted by both British and International Standards Organizations.

Field calibrations

Structures which are not amenable to theoretical analyses and which have not been the subject of laboratory investigations may be calibrated *in situ*. Discharges and the corresponding water levels are measured on site and correlations between the two are presented in terms of an empirical equation and/or a straightforward tabulation of the results.

The measurement of discharge in the field is not easy and the accuracy of the results may be lower than the accuracy requirement for the structure as a flow measuring device. Thus, *in situ* calibrations which involve standard dilution or velocity-area gauging techniques should be restricted to non-standard structures and have little value in terms of check-calibrating a standard structure which has a theoretical or laboratory rating. These ratings for standard structures are of high accuracy, (typically 2 or 3 per cent of the discharge) and specialist equipment is required to make any sensible field checks.

Experience has shown that the major sources of error in flow measurement with standard structures are in the field measurement of those quantities which occur in the discharge equation(s). Thus, the measurement of the physical dimensions of the structure and the measurement of invert, crest and water levels are of paramount importance.

In the following sections some of the more common types of gauging structures are described. Their usage and discharge characteristics are given in sufficient detail to give designers a working knowledge of their performance but it will occasionally be necessary to refer back to original publications for points of detail.

[See the following references: BSI (1965); Castex (1969); ISO (1975); Kindsvater and Carter (1957); Rehbock (1929); Schoder and Turner (1929); Seitz (1926); Shen (1960) and White (1975b).]

3.2 THIN-PLATE WEIRS

The standard thin-plate weir or notch has a crest profile consisting of a narrow surface at right angles to the upstream face of the plate and a 45-degree champfer at the downstream edge. The upstream edge must be accurately machined to a 90 degree angle. Rounding, rust pitting and damage by debris or sediment can affect the flow and consequently the role of thin-plate devices is restricted to those circumstances where maintenance is good and there is little risk of damage or deterioration. Naturally, if only a rough measure of flow is needed, less stringent standards of manufacture and maintenance can be accepted, but the various equations given later presupposes a stringent regard for the accuracy of the upstream corners of the weir.

Because the performance of a thin-plate device is dependent upon the full development of the contraction below the nappe, any obstructions to the flow as it converges towards the crest can affect the relationship of discharge to upstream level. Thus, projections such as bolt heads, stiffening brackets or other supporting structures cannot be permitted on the upstream face of the plate near the crest. Moreover, the nappe must be at atmospheric pressure on all its surfaces: it has to be well ventilated and cannot be submerged by a tailwater level above crest level without radically altering the flow. If sediment or other deposits accumulate upstream of the weir, the geometry of the approaching flow is affected and this in turn influences the contraction of the nappe and hence the coefficient.

The range of application for thin-plate weirs is restricted, if reasonable accuracy is desired, to the following situations:

(a) clean water, not carrying debris that might damage the crest or sediment that could settle in the approach to the weir;
(b) sites where adequate fall can be made available to ensure free discharge over the whole range to be gauged and where energy dissipation can be achieved satisfactorily.

Although these devices find some use in the field, for example in water supply systems, for the gauging of small compensation water flows, and in commerce, for example in gauging effluents of clean water or containing only dissolved pollutants, they are most frequently used in hydraulic laboratories as a basic method of measuring one of the variables (rate of flow) in a research project or in testing hydraulic machinery. Their attraction is their cheapness, and the ease with which they may be made with simple workshop facilities from readily available sheet material. Thin-plate devices fall into four main categories, depending on the cross-sectional shape presented to the flow:

(1) Full-width weirs occupying, as the name implies, the complete width of a vertical-sided channel. (Positive steps have to be taken to ensure ventilation of the underside of the nappe to achieve atmospheric pressure there.) They present a rectangular cross-section.
(2) Weirs with side contractions. These also present a rectangular cross-section, but do not occupy the full width of the approach channel. Thus the boundary geometry is contracted at each flank, and this in turn makes the nappe contract in a horizontal direction at each side. This type of weir is sometimes called a 'rectangular notch'.
(3) The V-notch. This is a symmetrical triangular version of the thin-plate weir with apex downwards. Its advantage lies in the ability to measure low discharges accurately, and to cover a wide range of flows, albeit small

ones. An included angle of 90 degrees is common, but other angles are permissible for particular situations.

(4) Specially shaped notches. A range of curvilinear tapering shapes has been proposed by different investigators to achieve a linear relationship of head to discharge. Also in this category would come trapezoidal notches, circular holes, etc. These special shapes are mainly of acadamic interest and are seldom used.

3.2.1 Rectangular weirs

Categories (1) and (2) comprise a group known as rectangular weirs. Discharge equations for this group of weirs have been presented by Rehbock (1929), Seitz (1926), Kindsvater and Carter (1957), Castex (1969), Hamilton-Smith (ISO 1438, 1975), and White (1975b). The SIA (Seitz, 1926) and Kindsvater–Carter equations are general equations which apply to full-width, partially and fully contracted weirs. Rehbock, Castex and White refer to full-width weirs and Hamilton-Smith refers to fully contracted weirs. A comparison of these formulae shows that differences of up to 4 per cent in computed discharge can occur, depending on the choice of equation. The following equations are recommended:

Fully and partially contracted weirs (Kindsvater and Carter, 1957)

$$Q = \tfrac{2}{3}\sqrt{(2g)}C_d b_e h_e^{3/2} \tag{3.10}$$

where $Q(\mathrm{m^3\,s^{-1}})$ is the discharge; $g(\mathrm{m\,s^{-2}})$ the accelaration due to gravity; C_d the coefficient of discharge; $b_e(\mathrm{m})$ the effective breadth of notch; and $h_e(\mathrm{m})$ is the effective gauged head.

$$b_e = b + 0.003 \tag{3.11}$$

$$h_e = h + 0.001 \tag{3.12}$$

where $b(\mathrm{m})$ is the breadth of notch, and $h(\mathrm{m})$ is the gauged head.

$$C_d = \alpha + \beta \frac{h}{P} \tag{3.13}$$

where $P(\mathrm{m})$ is the height of crest above upstream bed. Values of α and β depend on the ratio b/B, where $B(\mathrm{m})$ is the upstream breadth as follows:

b/B	0.1	0.2	0.3	0.4	0.5	0.6	0.7	0.8	0.9	1.0
α	0.588	0.589	0.590	0.591	0.592	0.593	0.594	0.596	0.598	0.602
β	−0.002	−0.002	0.002	0.006	0.011	0.018	0.030	0.045	0.064	0.075

Full-width weirs (White, 1975)

$$Q = 0.562 \left\{ 1 + \frac{0.153h}{P} \right\} b \sqrt{(g)} (h + 0.001)^{3/2} \qquad (3.14)$$

(units are metres and seconds).

3.2.2 Triangular (V-notch) weirs

Shen (1960) provides a review of research into the performance of this type of weir, giving data which relate mainly to fully contracted weirs. Kindsvater and Carter (1957) provide additional data for the partially contracted case. The recommended discharge equation is

$$Q = \tfrac{8}{15} \sqrt{(2g)} C_d \tan \frac{\theta}{2} h_e^{5/2} \qquad (3.15)$$

where θ (deg) is the angle included between the sides of the notch

$$h_e = h + k_h \qquad (3.16)$$

For fully contracted weirs, C_d and k_h vary with the notch angle, θ, as follows:

θ (deg)	20	40	60	80	100
C_d	0.592	0.582	0.576	0.576	0.581
k_h(m)	0.002 8	0.001 6	0.001 1	0.001 0	0.000 9

For partially contracted weirs, data are only available for 90-degree notches. k_h may be taken as 0.000 9 (m) but C_d varies in a complex way with h/P, P/B, and θ. Reference should be made to the original publication (Kindvater and Carter, 1957).

3.3 LONG-BASE WEIRS

[See the following references: BSI (1967); Crabbe (1974); Crump (1952); Harrison (1964); Herschy et al. (1977); Singer (1964); White (1966, 1968, 1970a, 1970b, 1971a, 1971b, 1971c); and White and Whitehead (1974)].

For many years there has been a lack of uniformity in the use of the terms 'length' and 'breadth' when applied to the crest of a weir. Thus the breadth of a thin-plate weir was commonly regarded as the horizontal dimension at right angles to the direction of the stream, and the breadth of a broad-crested weir was the horizontal dimension parallel with the direction of the stream. In this text, length is used for horizontal dimensions parallel with the direction of the stream, and breadth for dimensions across the stream. The term broad-crested is avoided and the term long-base is used to signify those structures which have a foundation or base which is relatively long in the direction of flow. This nomenclature accords with the practice adopted by British Standards.

The group of long-base weirs includes what used to be called broad-crested weirs (including both the streamlined and square-edged varieties) as well as weirs with short crests, provided that they have a long base, e.g. triangular profile weirs (including both the two-dimensional and flat-V varieties). These weirs will be described within the context of the revised nomenclature.

3.3.1 Round-nose horizontal-crest weirs

This type of weir has a horizontal surface at the crest which spans between vertical abutments. The upstream corner is rounded across the breadth of weir in such a way that flow separation does not occur up to and including the design maximum head. The radius of the upstream total head, and the length of the horizontal section of the crest should not be less than $1.75H_{max}$. This geometry ensures that flow separation will not occur at the upstream corner and that parallel flow conditions will be developed at some point on the horizontal crest.

At the downstream limit of the horizontal crest there is either a rounded corner, a downward slope or a vertical face. The geometry of the downstream section of the weir does not affect the modular discharge characteristics of the weir but it does affect the modular limit. In general, the more gradual is the expansion of the flow, the higher is the modular limit. In round figures, weirs with vertical downstream faces have a submergence ratio of 70 per cent at the modular limit; a rounded downstream corner can increase this to 75 per cent and a gradual slope downstream of the structure may, under favourable circumstances, produce figures of up to 95 per cent.

The modular flow characteristics of round-nose horizontal-crest weirs can be predicted from first principles because the pressure distribution on the horizontal crest is sufficiently close to hydrostatic for basic critical depth theory to apply. Boundary layer theory enables one to predict the variation of discharge coefficients with the relevant hydraulic flow characteristics, and the geometric and surface roughness properties of the weir itself.

The basic discharge equation for this type of weir may be written as

$$Q = (\tfrac{2}{3})^{3/2}\sqrt{(g)}C_d bH^{3/2} \tag{3.17}$$

The coefficient of discharge is given by the expression

$$C_d = \left\{1 - 2\frac{\delta_*}{L}\frac{L}{b}\right\}\left\{1 - \frac{\delta_*}{L}\frac{L}{h}\right\}^{3/2} \tag{3.18}$$

where δ_*(m) is the boundary layer displacement thickness, and L(m) is the length of the horizontal crest.

The boundary layer displacement thickness varies in a complex way with the length of the crest, the roughness of the crest and the discharge per unit width. However, for most installations with a good surface finish the value of δ_*/L may be taken as 0.003 and equation (3.18) simplifies to

$$C_d = \left\{ 1 - \frac{0.006L}{b} \right\} \left\{ 1 - \frac{0.003L}{h} \right\}^{3/2} \tag{3.19}$$

3.3.2 Rectangular-profile weirs

The rectangular-profile weir has aroused mixed feelings over the past 100 years or so. The protagonists of this type of weir point to its simple geometry which brings the advantage of ease of construction and installation. They suggest that for fieldwork it can be constructed of precast units and the resulting structure is robust and economical. Others are less enchanted with the device and focus attention on its disadvantages, which include (a) a variable coefficient of discharge, and (b) possible damage and rounding of the upstream corner which affects its calibration characteristics.

The crest of the rectangular-profile weir is a horizontal, rectangular plane surface and the upstream and downstream faces are vertical. It is important that the upstream face forms a sharp, right-angle corner where it intersects the plane of the crest. The submergence ratio at the modular limit is generally quoted as 65–70 per cent but there is a dearth of experimental evidence to support these figures.

The basic discharge equations used for rectangular-profile weirs is similar to that used for round-nosed horizontal-crest weirs, viz.

$$Q = (\tfrac{2}{3})^{3/2} F C_d b \sqrt{(g)} H^{3/2} \tag{3.20}$$

As stated earlier the coefficient of discharge, C_d, for a rectangular profile weir is variable. It has been shown that C_d can be taken as a basic (constant) coefficient and that the correction factor, F, is a function of the two ratios h/L and $h/(h+P)$, where L is the length of the crest in the direction of flow and P is the height of the weir relative to the invert of the approach channel. The most recent and reliable data have been provided by Singer (1964) and Crabbe (1974) and their functional relationships are shown in Figure 3.3.

Singer suggested a basic coefficient, C_d, of 0.848 and Crabbe's results showed a value of 0.855, approximately 1 per cent higher. The factor, F, can be taken as unity within the limited range defined by the limits

$$0.08 < h/L < 0.3 \tag{3.21}$$

and

$$0.18 < h/(h+P) < 0.36 \tag{3.22}$$

When h/L exceeds 0.33 and/or $h/(h+P)$ exceeds 0.36, the factor F rises and this is called the variable coefficient range. The variations of the correction

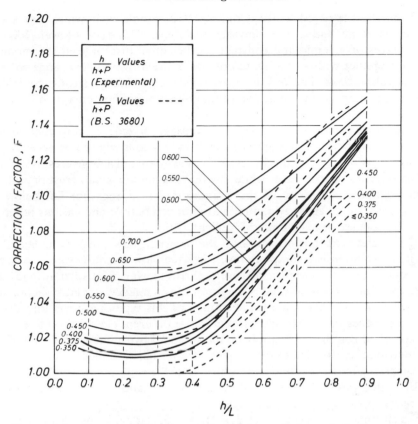

Figure 3.3 Correction factors on the basic coefficient of discharge for rectangular-profile weirs

factor, F, obtained by Singer and Crabbe as a multiple of the basic coefficient of 0.848, are compared in Figure 3.3. For deep approach conditions, $h/(h + P) < 0.36$, Crabbe indicates correction factors which exceed those used by Singer. At low values of h/L the difference is 1 per cent, but at $h/L = 0.85$ the discrepancy is about 3.5 per cent. This trend is reversed for shallow approach conditions, $h/(h + P) > 0.60$, because of the convergence of Crabbe's curves with increasing h/L. It is suggested that Crabbe's data should be used for determining correction factors, F, in the variable coefficient range and that a basic coefficient, C_d, of 0.848 is most appropriate.

3.3.3 Triangular-profile weirs

Triangular-profile weirs have become very popular throughout the world and are used in both the laboratory and the field. Many longitudinal profiles have been investigated and there is not room here to describe them all. Comments will thus be restricted to the Crump (1952) profile weir which has a $1:2$ upstream slope and a $1:5$ downstream alope. This type of weir has the advantages of a stable and constant coefficient of discharge in the modular range, together with an option to operate in the drowned flow range without *in situ* calibrations. The $1:5$ downstream slope produces a strong, controlled hydraulic jump in the modular range which simplifies energy dissipation problems.

Two dimensional weirs have the advantage of ease of construction and simplicity in operation but are not suited to locations where it is necessary to measure a wide range of discharge. High flows demand a large crest breadth if afflux is not to be excessive, while low flows demand a small crest breadth in order that the sensitivity of the gauging structure does not fall below an acceptable figure. These two criteria are not compatible and various attempts have been made to overcome this difficulty.

If it is assumed that operation in the drowned flow range is undesirable, then one solution is a compound weir in which low discharges are measured by containing flow within a relatively narrow lower crest section, and high afflux at peak flows is avoided by incorporating a much wider crest section at a higher elevation. The main disadvantage of this type of weir is that divide piers are necessary between crest sections at different elevations if there is to be complete confidence in applying the standard two-dimensional calibration to individual crests. These divide piers are unsightly, collect debris and add considerably to the cost of the structure. If divide walls are not included in the design, then three-dimensional flows are created and laboratory calibrations of each proposed structure are necessary.

An alternative method of avoiding large afflux at peak flows is to operate a simple single-crested weir in the drowned flow range. This method involves the use of two gauges rather than a single gauge recording upstream water levels, the second measurement being either the downstream water level or, preferably, the pressure within the separation pocket which is formed just downstream of the weir crest. This latter position provides a more sensitive means of measuring flow and has the enormous advantage that it is not affected by the geometry of the downstream section of the weir installation.

These two possible ways of measuring a wide range of flows without incurring difficulties at either end of the range have, on several occasions, been combined and compound weirs have been operated in the drowned flow range. However, there are several objections to this type of weir: the need for divide piers, as stated above, loss of accuracy when the head over the upper crest section is small, and the tedious (and not very precise) method of evaluating drowned flows over a compound structure.

It was consideration of these difficulties which led to the development of the flat-V weir. It could, with careful selection of crest breadth and cross slope, provide sensitivity at the low flows without the need for divide piers and also minimize afflux during flood conditions. Operation in the drowned flow range, if feasible, would further increase the versatility of the structure.

Two-dimensional weirs

The discharge equation for two-dimensional weirs operating in the modular flow range takes the form

$$Q = C_{de}\sqrt{(g)}bH_{1e}^{3/2} \qquad (2.23)$$

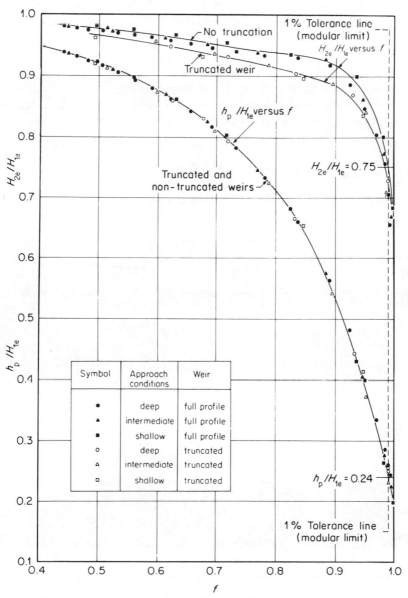

Figure 3.4 Drowned flow reduction factors for Crump weirs

where

$$H_{1e} = H_1 - k_h = h_1 + \gamma \frac{v^2}{2g} - k_h \qquad (3.24)$$

and $v(\mathrm{m\ s^{-1}})$ is the average velocity at gauging section, and γ is the Coriolis energy coefficient. The coefficient of discharge, C_{de}, takes a constant value of 0.633 and the head correction factor, k_h, can be taken as 0.000 3(m).

In the drowned flow range a drowned flow reduction factor, f, is introduced. Thus,

$$Q = C_{de} f \sqrt{(g)} b H_{1e}^{3/2} \qquad (3.25)$$

The drowned flow reduction factor is determined from a measurement of the downstream water level or (preferably) from the pressure head, h_p, measured within the separation pocket which forms just downstream of the crest. Thus,

$$f = \phi^{(i)}\{h_p/H_{1e}\} \quad \text{or} \quad \phi^{(ii)}\{H_{2e}/H_{1e}\} \qquad (3.26)$$

These relationships are shown graphically in Figure 3.4. The modular limit for this type of weir occurs at a submergence ratio of 0.75. Figure 3.5 shows the two-dimensional weir on the River Pang at Pangbourne.

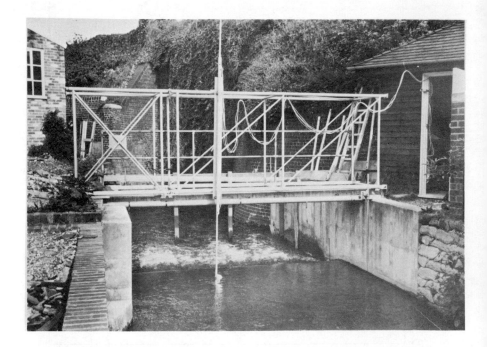

Figure 3.5 Two-dimensional triangular-profile weir on the River Pang at Pangbourne

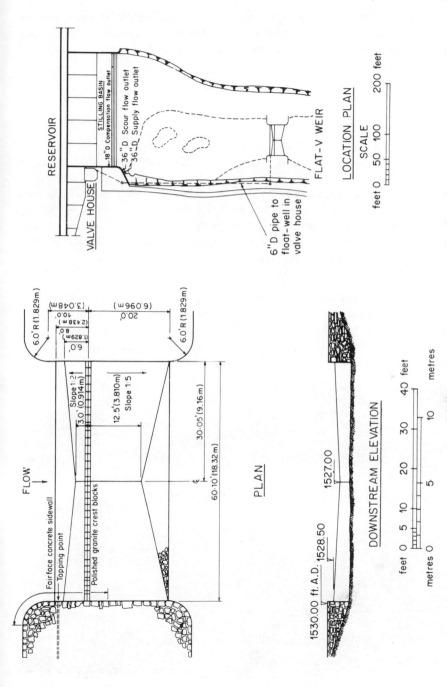

Figure 3.6 Flat-V weir on the River Tees at Cow Green

Flat-V weirs

Two modular flow discharge equations are obtained for flat-V weirs; one for conditions when flow is confined within the V and a second, at higher discharges, where the presence of the vertical side walls has to be taken into account. These are as follows:

$$Q = 0.8C_{de}\sqrt{(g)}nH_{1e}^{5/2} \qquad \text{when } H_{1e} \leq H' \tag{3.27}$$

and

$$Q = 0.8C_{de}\sqrt{(g)}nH_{1e}^{5/2}\{1 - (1 - H'/H_{1e})^{5/2}\} \qquad \text{when } H_{1e} > H' \tag{3.28}$$

where $C_{de} = 0.63$ for weirs with a $1:2/1:5$ longitudinal profile; $H'(\text{m})$ is the difference between lowest and highest crest elevation; and n is the crest cross-slope (1 vert.: n horiz.).

Flat-V weirs are occasionally used in the drowned flow range with a crest tapping. Details of the method are given by the author elsewhere (White, 1971a). The modular limit for Crump profile flat-V weirs occurs at submergence ratios between 0.67 and 0.78. Figures 3.6 and 3.7 show the flat-V weir on the River Tees at Cow Green.

Compound weirs

A typical compound weir is shown in Figures 3.8 and 3.9. Individual weir sections have different crest elevations and they are separated by divide piers. This ensures that two-dimensional flow is preserved at each section and, as a consequence, calibrations can be derived from the basic equations for two-dimensional weirs.

Figure 3.7   Flat-V weir operating at the V-full stage

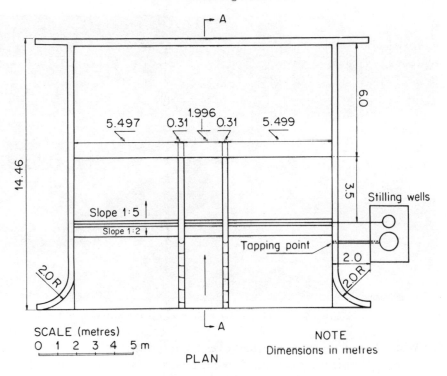

SCALE (metres)

0 1 2 3 4 5 m

PLAN

NOTE

Dimensions in metres

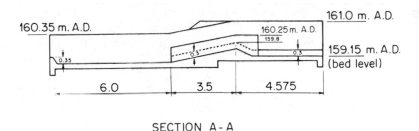

SECTION A-A

Figure 3.8 Compound weir on the St John's Beck at Thirlmere

General practice is to record upstream water levels at one position only. Hence, in calculating total river flow it is necessary to make assumptions about the relationships between flow conditions at the various individual crest sections. Three possibilities have been investigated and these may be summarized as follows:

(1) Total head level is assumed constant over the full width of the weir. It is obtained by adding, onto the observed static head, a velocity head based on the mean velocity of approach to the structure as a whole.

Figure 3.9   Specialist check calibration at Thirlmere

(2) Water level is assumed constant over the full width of the weir. Thus, the observed static head is applied to each crest section (taking into account differences in crest levels) and individual total heads are obtained by adding velocity heads appropriate to each crest section.
(3) Total head level is assumed constant over the full width of the weir. It is obtained by adding, on to the observed static head, the velocity head appropriate to the individual crest section at which the water level is observed.

Field experiments covering a wide range of conditions have indicated that the third method is the most accurate (White and Whitehead, 1974).

Several compound weirs have been built with crest tappings in the lowest crest section. This facilitates operation in the drowned flow range but, because of the complex geometry of compound weirs, the accuracy of flow measurement in the drowned flow range is reduced considerably (Herschy *et al.*, 1977).

3.4   LONG-THROATED FLUMES

[See the following references: Ackers (1961); Ackers and Harrison (1963); BSI (1974; Chow (1959); Harrison (1965); and Harrison and Owen (1967)].

The discharge in an open channel may be measured by means of a flume, consisting essentially of contractions in the sides and/or bottom of the channel forming a throat. When the reduction in cross-sectional area exceeds a certain value, critical depth occurs in the constriction and the stage–discharge relationship becomes independent of conditions in the channel downstream of the structure. The device is then said to be 'free-flowing' and constitutes a critical depth-measuring flume.

If the throat is prismatic over a significant length in the direction of flow, critical depth will always occur at a similar section and the calibration of the flume becomes amenable to theoretical derivation. Such a flume is said to be long-throated, and three basic types are to be found in general use: rectangular-throated, trapezoidal-throated and U-throated.

The type of flume which is used depends on several factors, such as the range of discharge to be measured, the accuracy requirements, the head available, and the estimated sediment load of the stream. The rectangular-throated flume is probably the simplest geometry to construct and its calibration is absolutely straightforward. The trapezoidal-throated flume is more appropriate to large installations, particularly where a wide range of discharge is to be measured with consistent accuracy. This shape of throat is particularly suitable where it is necessary to work to a given stage–discharge relationship. The U-throated flume is useful for installation in a U-shaped channel or where discharge is from a circular section conduit. It has found particular application in sewers and at sewage works. The versatility of all three types may be enhanced by raising the invert of the throat above upstream bed level.

Critical depth theory, augmented by experimental data, may be used to deduce the basic equations for free discharge through long-throated flumes. The simple theory relates to the frictionless flow of an ideal fluid, and an additional coefficient has to be introduced in practice, either based on experiment or deduced by considering boundary layer development within the throat of the flume. Assuming that the velocity distribution is uniform within the throat and that there is no significant curvature of the streamlines, considerations of minimum energy yield the general expression

$$Q = (gA_c^3/w_c)^{1/2} \qquad (3.29)$$

where A_c(m) is the cross-sectional area of flow at the critical section, and w_c(m) is the water surface width at the critical section.

Equation (3.29) gives discharge in terms of depth at the critical section, but in practice it is necessary to measure heads upstream of the throat of the structure. Thus, an essential step is to convert the formula into one which relates upstream gauged head to discharge. The effective total head, H_e, measured relative to the invert of the flume, relates to the conditions at the critical section as follows:

$$H_e = d_c + \frac{A_c}{2w_c} \qquad (3.30)$$

where d_c(m) is the depth of flow at the critical section.

Using equations (3.29) amd (3.30), a theoretical calibration for a gauging structure over the full range of discharge can be derived by considering flow conditions in the throat of the flume and deducing corresponding heads and discharges. The principle of the method is to select a series of values of d_c, the critical depth in the throat, and calculate corresponding values of Q and H_e. Total heads are then converted to gauged heads using successive approximation techniques (BSI 1974).

An explicit relationship may be derived for rectangular-throated flumes from equations (3.29) and (3.30) as follows:

$$Q = (\tfrac{2}{3})^{3/2} b_e \sqrt{(g)} H_e^{3/2} \tag{3.31}$$

where b_e(m) is the effective width of the throat.

Shape coefficients have to be introduced for trapezoidal and U-throated flumes and the explicit method becomes more cumbersome. Graphical solutions are, however, available.

The modular limit of long-throated flumes depends to a large extent on the severity of the expansion downstream of the throat of the flume. A rapid expansion results in the modular limit occurring at a submergence ratio of around 75 per cent, whereas a very smooth and gradual expansion can increase this figure to as high as 95 per cent.

3.5 SHORT-THROATED FLUMES

Short-throated flumes (Skogerboe and Hyatt, 1967; Skogerboe *et al.*, 1967) have contracted sections which are not long enough to develop parallel flow conditions where critical depth occurs. In practice this means that theoretically derived calibrations are difficult if not impossible to obtain and that emirically derived coefficient values vary significantly with the flume and flow geometries. These flumes are, however, somewhat cheaper than their long-throated equivalents and have become popular in some parts of the world as flow-measuring devices. The most common types of short-throated flumes, developed and used mainly in the USA, are cut-throat and Parshall flumes.

3.5.1 Cut-throat flumes

Cut-throat flumes have a plain horizontal invert. The inlet consists of a converging section (usually 1 laterally to 3 longitudinally) and there is a divergent outlet section (usually 1 laterally to 6 longitudinally). There is an abrupt change of direction of the flume walls between the inlet and outlet sections and the narrowest position is usually referred to as the neck.

Cut-throat flumes may be used in both the modular and non-modular flow regimes. The distinguishing difference between the two is the occurrence of critical depth in the vicinity of the neck of the flume, although the exact location cannot be stated. In the modular flow regime, water levels are measured near the upstream end of the inlet section. In the non-modular flow regime, an additional measurement has to be made near the outlet of the divergent section. It must be stressed, however, that individual calibrations are related to specific structures with specific head-measuring positions. Lack of space precludes the provision of comprehensive lists of performance data and readers are referred to more specialist publications (Skogerboe and Hyatt, 1967; Skogerboe *et al.*, 1967).

3.5.2 Parshall flumes

A Parshall flume consists of an entry section, a short throat and an exit section. The entry section converges in plan (usually at around 1 in 5) and has a horizontal invert. The throat has parallel walls but the invert falls at a slope of 1 on 2.5. The exit diverges in plan (usually at around 1 in 5) and the invert rises at a slope of 1 on 3 to meet the downstream bed level.

As with the cut-throat flume, the Parshall flume may be used in both the modular and non-modular flow ranges and the modes of operation are similar. A second head measurement is again required in the non-modular range of flows and it is usually taken towards the downstream end of the throat. Parshall flumes are, however, predominantly used in the modular flow range. Readers are again referred to more specialized publications for the details of the design and performance of this type of flume (Skogerboe *et al.*, 1967).

REFERENCES

Ackers, P., 1961, July. Comprehensive formulae for critical depth flumes, *Water and Water Engineering*, **65**(785), 296–306.
Ackers, P. and Harrison, A. J. M., 1963. Critical depth flumes for flow measurement in open channels, Hydraulics Research Station, Wallingford, UK, Hydraulics Research Paper No. 5, HMSO.
British Standards Institution, 1965. Measurement of liquid flow in open channels: thin-plate weirs and Venturi flumes, BS 3680: Part 4A.
British Standards Institution, 1969. Measurement of liquid flow in open channels: long-base weirs, BS 3680: Part 4B.
British Standards Institution, 1971a. Measurement of liquid flow in open channels: the measurement of liquid level (stage), BS 3680: Part 7.
British Standards Institution, 1971b. Measurement of liquid flow in open channels: water level instruments, BS 3680: Part 9A.
British Standards Institution, 1974. Measurement of liquid flow in channels: flumes, BS 3680: Part 4C.
Buchanan, T. J. and Somers, W. P., 1968. Techniques of water resources investigations of the United States Geological Survey: Stage measurement at gauging stations, Chapter A7, USGS, Washington.
Castex, L., 1969. Quelques nouventes sur les deversoirs pour la mesure de debits, *La Houille Blanche*, No. 5, pp. 541–548.
Chow, V. T., 1959. *Open Channel Hydraulics*, McGraw-Hill, London.

Clayton, C. G. (Ed.), 1971, September. Modern developments in flow measurement, Proceedings of International Conference, Harwell.

Crabbe, A. D., 1974, October. Some hydraulic features of the square-edged broad-crested weir, *Water Services*, **78**(944), 354–358.

Crump, E. S., 1952, March. A new method of gauging stream flow with little afflux by means of a submerged weir of triangular profile, *Proceedings ICE*, Paper No. 5848, pp. 223–242.

Harrison, A. J. M., 1964, November. Some comments on the square-edged broad-crested weir, *Water and Water Engineering*, **68**(825), 445–448.

Harrison, A. J. M., 1965, August. Some problems concerning flow measurement in steep streams, *J. Inst. Water Engineers*, **19**(6), 469–477.

Harrison, A. J. M. and Owen, M. W., 1967, February. A new type of structure for flow measurement in steep streams, *Proceedings ICE*, **36**, 273–296.

Harrison, A. J. M., 1969, April. Factors governing the choice of a hydraulic structure for flow measurements, International Commission on Irrigation and Drainage, 7th Congress, Mexico.

Helliwell, P. R. and Bates, A. V., 1971. An investigation into the costs of streamflow data collection, Dept. of Civil Engineering, University of Southampton.

Herschy, R. W., White, W. R. and Whitehead, E., 1977. Crump weir design, DOE Water Data Unit, Technical Memorandum No. 8.

Institution of Water Engineers, 1969, September. River flow measurement, Proceedings of Symposium, Loughborough University of Technology.

International Standards Organisation, 1975. Liquid flow measurement in open channels: thin-plate weirs, ISO: 1438, Geneva.

Kindsvater, C. E. and Carter, R. W., 1957, December. Discharge characteristics of rectangular thin-plate weirs, *Proceedings ASCE*, **83**(HY6), Paper 1453.

Rehbock, T., 1929. Wassermessung mit Scharf Kantigen Ueberfallwehren, *Z. Ver. dtsch Ing.*, **73**, 817–823.

Rodda, J. C. (Ed.), 1976. *Facets of Hydrology*, Wiley, London.

Schoder, E. W. and Turner, K. B., 1929. Precise weir measurements, *Transactions ASCE*, **93**, 999–1110.

Seitz, E., 1926. Normen fur wassermessungen bei Durchfuhring von Abnahmerer-suchen an Wasserkraftmaschinen, *Schweizerische Bauzeit.*, **88**(1), 17–18.

Shen, J., 1960, July. Discharge characteristics of triangular notch thin-plate weirs, US Dept. of Interior, Geological Survey, Draft for ISO.

Singer, J., 1964, June. Square-edged broad-crested weir as a flow measuring device, *Water and Water Engineering*, **68**(820), 229–235.

Skogerboe, G. V. and Hyatt, M. L., 1967, December. Rectangular cutthroat flow measuring flumes, *Proceedings ASCE*, **93**(IR4), Paper No. 5628.

Skogerboe, G. V., Hyatt, M. L. and Eggleston, K. O., 1967, February. Design and calibration of submerged open channel flow measurement structures, Utah State University, Logan, Utah, Report WG 31–32.

White, W. R., 1966. The flat-V weir, Hydraulics Research Station, Wallingford, UK, Report No. INT 56.

White, W. R., 1968, January. The flat-vee weir, *Water and Water Engineering*, **72**(863), 13–19.

White, W. R., 1970a. The Crump profile flat-V weir, Hydraulics Research Station, Wallingford, UK, Report No. EX 473.

White, W. R., 1970b. The triangular profile Crump weir—a re-examination of discharge characteristics, Hydraulics Research Station, Wallingford, UK, Report No. EX 477.

White, W. R., 1971a, March. The performance of two-dimensional and flat-V triangular profile weirs, *Proceedings ICE*, Supplement Paper No. 7350s.

White, W. R., 1971b, March. Flat-vee weirs in alluvial channels, *Proceedings ASCE*, **97**(HY3), 395–408.

White, W. R., 1971c, September. A field comparison of the calibrations of a flat-vee weir and a set of electromagnetic flowmeters, *Water and Water Entgineering*, **75**(907), 340–344.

White, W. R., 1975a, September. Field calibration of flow measuring structures, *Proceedings ICE*, **59,** Part 2, Paper 7821, pp. 429–447.

White, W. R., 1975b, December. Thin-plate weirs, Hydraulics Research Station, Wallingford, UK, Report No. INT 152.

White, W. R. and Whitehead, E., 1974. Field calibration of a compound Crump weir on St. John's Beck, Cumbria, Hydraulics Research Station, Wallingford, UK, Report No. INT 135.

CHAPTER 4

Dilution Methods

K. E. WHITE

4.1 INTRODUCTION

Most of the gauging methods described in other chapters can be used to measure the discharge at a chosen cross-section irrespective, within reason, of the behaviour of water or the characteristics of the channel far upstream. Dilution gauging requires that a whole reach, possibly many kilometres long, satisfies a number of criteria and for this reason some preliminary studies may be necessary to determine whether gauging by dilution methods is possible. Ultimately, where information about the local surface geology and ground-water infiltration or bed leakage is sparse, the only way to establish the feasibility of dilution gauging may be to attempt it. Dilution gauging is likely to be successful at the first attempt in well-defined channels such as sewers, and in pipes flowing full of water, but every stage of gauging in natural channels requires considerable care and all results must be examined critically to assess their accuracy, which can be very poor even when samples are capable of being analysed with very good precision and repeatability.

The outstanding advantage of dilution gauging is that it measures the flow in an absolute way because the discharge is calculated from measurements of volume and time only; tracer concentrations need be determined only in dimensionless relative readings. The methods are often used for the purposes of calibration and they may be regarded as costly but powerful techniques for occasional use. Dilution methods may provide the only effective means of estimating the flow in shallow rock-strewn rivers or when rivers are in extreme conditions of flood or drought.

Dilution gauging techniques are outlined in publications of the International Organization for Standardization (1973, 1974) and national standards, for example those of the British Standards Institution (1967, 1974). The aim of this chapter is to complement the information contained in the standards which, of necessity, has to be given in a concise form. In particular, the basic principles are examined in detail to illustrate the range of conditions under which the method is applicable—conditions which may be difficult to find or to prove in some cases owing to the vagaries of nature. Because of its fundamental importance the selection of reach to achieve satisfactory mixing of the tracer is discussed in some detail. Substances for use as tracers are listed in the standards, and this information is amplified by a general discussion of the relative merits of various tracers, but methods for their assay are not included. Fieldwork preparations, site operations, and the subsequent laboratory analyses and computations are outlined in so far as it is necessary to indicate the aspects which can lead to loss of gauging accuracy. Some equations are used in the text, but it is hoped that these will not deter the non-mathematical reader since sufficient explanation has been included to explain their implications. As far as possible, terminology is consistent with the ISO 'Vocabulary and symbols' (1978). Otherwise meanings should be inferred from the context in which the terms are used.

Dilution gauging has gradually developed as a method in widespread use over the past 50 years or so. A very detailed account of the methods, using common salt, was given by Groat (1915). He and his colleagues identified many of the pitfalls in the method as a result of very painstaking experimental work more than half a century ago. The paper is recommended to anyone concerned with river flow gauging and particularly to those gauging teams using common salt as the tracer—it is a fascinating account over 300 pages long, and one of his remarks is worth quoting as it is just as true today, particularly with regard to large natural channels, as it was in 1915:

> Turbine testing and discharge gauging have been more of an art than a science. Results have depended more for accuracy on the particular person who directed the work than on any method heretofore used. It has been a common occurrence that there has been a complete failure to interpret the data correctly, with the result that much of the work in discharge measurements has been unsatisfactory. When results have been shown to be accurate, we may rest assured that the person conducting the tests has arrived at a conclusion only by the most painstaking effort and perseverance.

4.2 PRINCIPLES

There are two basic methods of dilution gauging: the constant-rate injection method and the integration (gulp) method. The principles of dilution gauging can be well illustrated by considering the constant-rate injection method in the first instance, but it is not intended to keep the subsequent discussion of the two methods separate as the hydrological aspects of mixing in open channels are the same and are equally important for both. The dispersion of a tracer, injected into a channel, is discussed below in order to define the conditions under which the basic equations for calculating the flow may be applied.

In the constant-rate injection method a tracer solution, of concentration c_1, is injected continuously, at a volumetric rate q, for a period (on which guidelines are given later) such that an equilibrium concentration, c_2, is established for a finite time at a sampling station downstream. Then the mass rate at which tracer enters the test reach is $(qc_1 + Qc_0)$, where Q is the discharge and c_0 is the background tracer concentration in the river water. On the assumption that satisfactory mixing of the tracer with the entire flow across the section has taken place by the time the water has reached the sampling station (discussed later), the rate at which the tracer leaves the test reach is $(Q+q)c_2$. Equating these two rates gives the discharge

$$Q = \left(\frac{c_1 - c_2}{c_2 - c_0}\right)q \qquad (4.1)$$

An approximate form of this equation is often quoted in the literature but should not be used unless it can be shown that q compared with Q, and c_2 compared with c_1, are so small that the desired accuracy will not be affected. The reason for some caution here is that c_2 represents the absolute plateau level, including the background contribution c_0, and a correction for this background must not be made in the numerator. Also, the 'background' referred to in the above equation is solely the concentration present in the water before the addition of tracer, and this contribution must be distinguished from the so-called 'background' components of analytical measurements which arise within the instruments or are caused by other external influences.

As well as thorough mixing, the method requires that all of the tracer passes through the sampling station—normally the cross-section at which the measurement of flow is required. However, when the flow divides, because of islands or abstractions, into two streams, either may be sampled, but the calculated flow then refers to the cross-section just upstream of the division.

In the integration method of dilution gauging, a quantity of tracer of volume V_1 and concentration c_1 is added to the river, often by a simple, steady emptying of a flask of tracer solution, and at the sampling station the passage of the entire tracer cloud is monitored to determine the relationship between concentration and time. The discharge is calculated from the equation

$$M_1 = c_1 V_1 = Q \int_{t_1}^{t_2} (c_2 - c_0)\, dt \qquad (4.2)$$

where M_1 is the quantity of tracer, t_1 is a time before the leading edge of the tracer cloud arrives at the sampling point, and t_2 is a time after all the tracer has passed this point—here the word 'point' rather than 'cross-section' is used deliberately. In this equation, the derivation of which will be given later, it is clear that only the values of the concentration above the total sum of background effects are required to be integrated. The conditions of satisfactory mixing are just as stringent for this method as for the constant-rate method and they depend on the turbulent dispersion of tracer, which will now be examined for both cases simultaneously since a constant-rate injection may be regarded as a series of closely spaced 'gulp' injections. It may be useful to note that the period of injection in the integration method is not important and an abortive constant-rate gauging may often be rescued by applying integration principles. (The integration method has been termed, even in standards, the 'sudden injection method', but this term is being replaced as it is misleading. There is no necessity to ensure a 'sudden' or 'instantaneous' release of tracer.)

When tracer is injected into flowing water each element is subject to convective transport by time-averaged velocities, $\bar{U}_x$ (longitudinally), $\bar{U}_y$ (vertically), and $\bar{U}_z$ (laterally), and the rate of change of concentration at any point with time due to convection will be given by

$$-\frac{\partial c}{\partial t} = \bar{U}_x \frac{\partial c}{\partial x} + \bar{U}_y \frac{\partial c}{\partial y} + \bar{U}_z \frac{\partial c}{\partial z} \qquad (4.3)$$

Smaller scale turbulent-mixing processes will also be changing the concentration within each elementary volume, depending on the mixing coefficients D_x, D_y, and D_z as follows:

$$\frac{\partial c}{\partial t} = D_x \frac{\partial^2 c}{\partial x^2} + D_y \frac{\partial^2 c}{\partial y^2} + D_z \frac{\partial^2 c}{\partial z^2} \qquad (4.4)$$

The dispersion pattern resulting from these two processes, for any given initial conditions, is difficult to illustrate because very complex three-dimensional forms evolve, even for a single-point short-term release, as a result of variations in velocity over the cross-section. Although it is important to appreciate that these fundamental processes are taking place it is possible to simplify the theoretical considerations because it is stipulated for dilution gauging that the tracer must have dispersed and mixed with (or 'sampled') every element of flowing water across the channel, i.e. that vertical and lateral dispersion processes must have been substantially completeted. The above equations may therefore be replaced by a dispersion equation in one dimension (along the channel) and this has an exact solution, giving the form of the tracer concentration profile downstream of the 'mixing distance'. The equation may be written in the following form by simplifying and combining equations (4.3) and (4.4):

$$\frac{\partial \bar{c}}{\partial t} = \bar{D}\frac{\partial^2 \bar{c}}{\partial x^2} - \bar{U}_x \frac{\partial \bar{c}}{\partial x} \tag{4.5}$$

where $\bar{c}$ and $\bar{U}_x$ are now the mean tracer concentration and velocity, respectively, in the cross-section and $\bar{D}$ is the bulk dispersion coefficient. The solution of this equation, for the theoretical initial conditions of a sheet of tracer existing across the injection section, at distance $x = 0$ and time $t = 0$, is

$$\bar{c} = \frac{\Delta M_1}{2\sqrt{(\pi \bar{D}t)}} \exp\left[-\frac{(x - \bar{U}_x t)^2}{4\bar{D}t} \right] \tag{4.6}$$

where ΔM_1 is the mass of tracer injected per unit area of the cross-section. When mixing has been completed vertically and horizontally, therefore, equation (4.6) predicts that the tracer profile should be of symmetrical Gaussian form. The experimental profile should be of similar form to demonstrate the criterion that mixing has been sufficiently effective for the dilution gauging equations to be applied. In the case of the constant-rate injection method the Gaussian profile will characterize the shape of the leading edge of the concentration–time curve between background and plateau levels and may be generated by differentiating the tracer curve (strictly this should be carried out with respect to distance at a given instant of time).

The tracer balance equation for the integration method may now be derived by simply considering an elementary interval of time dt at the sampling station. As $\bar{c}$ is the mean concentration across the section, the amount of tracer passing through the section in interval dt is $\bar{c}Q$ dt and therefore we may write, for constant Q,

$$M_1 = Q\int_0^\infty \bar{c}\, \mathrm{d}t \tag{4.7}$$

Equations (4.7) and (4.2) are basically the same for the reasons outlined below. The tracer 'flow-through' curve is often derived for a point in the cross-section, possibly just below the surface mid-stream. The two concentrations $\bar{c}$ and $(c_2 - c_0)$ at any individual sampling point, equation (4.2), will not be the same, nor will they have the same relative distribution with respect to time, as illustrated in Figure 4.1; however, this does not matter provided that both distributions integrate to the same total and are reasonably symmetrical in form to prove the mixing criteria, another check on which will be that the distributions for all points in the cross-section will have the same duration, but not necessarily at coincident times of day.

The duration of passage of tracer should be similar for all sampling points in the section, under conditions of steady flow, because the non-symmetrical arrow head taken up by the tracer front will be reflected in similar form at a

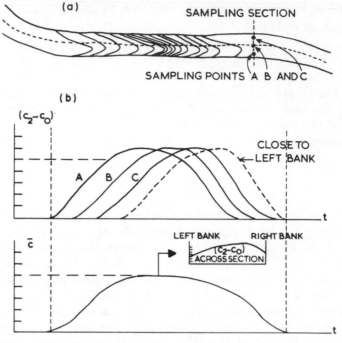

Figure 4.1 (a) Plan view of tracer cloud and sampling points; (b) tracer 'flow-through' curves illustrating satisfactory conditions for flow measurement. The area under all five distributions shown should be the same

later time for tracer-free water following the tracer cloud. This will be true when the 'tail' reaches the sampling section, even if the tail arrow head has a different form at its upstream position at the time the tracer front reaches the sampling section. Figure 4.1 shows only the two-dimensional case for clarity but the 'bow' of the tracer front may be sub-surface and again, in the vertical plane, the clear-water front following the tracer should superimpose exactly on the tracer front at some later time, which is the same for all points in the section.

In natural channels the time taken for satisfactory mixing to be achieved may be considerable, and the subsequent tracer curves may have to be recorded over a period of many hours. It is therefore necessary to consider the criterion of good mixing and the consequences of sampling the tracer tail many hours after sampling the tracer front. (The problem of estimating the required mixing distance from channel characteristics will be dealt with in Section 4.4.)

Equation (4.6) gives the distribution of tracer along the channel at instant t from the time the idealized 'injection' was made, and may be used to define the length of channel $(x_2 - x_1)$ over which the tracer has dispersed, as shown in

Figure 4.2 (x_2 and x_1 define the reach within which most of the tracer is effectively contained—equation (4.6) predicts some tracer at all values of x). The time taken for the volume of water bounded by x_1 and x_2 to pass the sampling station will be given by T, where

$$T = \frac{(x_2 - x_1)}{\bar{U}_x}$$

(4.8)

and the flow Q must remain steady for a period of time longer than T.

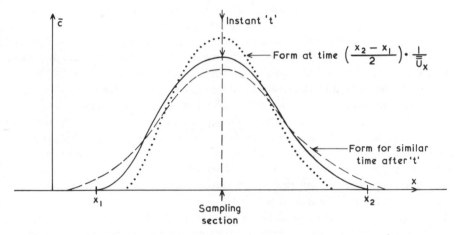

Figure 4.2   Tracer distribution along the channel prior to, at, and after instant 't'

Equation (4.2) corresponds exactly with experimental practice, and integration of the curve of $c_2 - c_0$ plotted against t gives the flow. However, the criterion of mixing is not that the plot of $\bar{c}$ against t is symmetrical, but that the plot of $\bar{c}$ against x is symmetrical (equation (4.6)). It can be noted from Figure 4.2 that the leading edge of the rising limb of the tracer curve is determined at time $(x_2 - x_1)/2\bar{U}_x$ before, and the end of the 'tail' is found a similar period after, the instant 't', when the forms of the distribution are as shown (exaggerated) by the dotted lines in Figure 4.2.

Clearly the profile of c_2 against t recorded experimentally at the sampling section will be slightly skew, even when perfect mixing has been achieved, particularly when $\bar{D}$ has a high value (wide shallow rivers), because the tracer curve will be a composite of the rising limb from the narrower distribution of Figure 4.2 and the falling limb of the curve which shows greater dispersion. A critical assessment of the profiles recorded at several points in the sampling cross-section may therefore be required to establish that mixing is satisfactory. It is of value therefore to be able to estimate or calculate $\bar{D}$ from experimental data. Also, experimental verification that Q is constant is necessary since a skew tracer distribution due to incomplete mixing could be recorded as a more symmetrical one if the mean velocity is increasing with time. The value of $\bar{D}$ may be calculated from two concentration–time profiles (variance σ^2) taken in a similar way at two cross-sections A and B (Fischer, 1968):

$$\bar{D} = \frac{\bar{U}_x^2(\sigma_B^2 - \sigma_A^2)}{2(t_B - t_A)} \tag{4.9}$$

where t_B and t_A represent the mean times for the passage of the tracer cloud. These should correspond approximately to the times at which the tracer concentration peaks are recorded at each station. At cross-sections where satisfactory mixing has not yet been established the tracer curve will be asymmetrical with distance and will appear even more so when recorded as a function of time. In small turbulent streams a symmetrical concentration–time profile may be obtained within a very convenient short reach, but if $\bar{D}$ is very low the tracer flow-through period will be very short and this may cause some difficulty if multiple samples are to be taken to define the tracer curve.

Attempts to predict $\bar{D}$ from the channel characteristics have only met with partial success. Use of the equation $\bar{D} = 10 R_h U^*$, where R_h is the hydraulic radius and U^* is the friction velocity, will provide a first estimate of $\bar{D}$ for small turbulent streams. The various formulae that have been proposed to give a better estimate of the much higher values of $\bar{D}$, which are found experimentally as the ratio of width to depth increases, have been listed by Bansal (1971) who discussed the variation of $\bar{D}$ with flow geometry. Liu (1977) has tested the application of more recent empirical formulae and has proposed a more complex equation which appears to be applicable to rivers of any size and allows $\bar{D}$ to be predicted to an accuracy at least of the right order of magnitude in most cases, including wide shallow rivers.

Although the tracer concentration profile does not have to be integrated in the constant-rate injection method, the dispersion characteristics of the channel define the period of injection necessary to establish plateau conditions at the sampling station. A constant-rate injection may be considered as a series of gulp injections, as shown in Figure 4.3. The injections (1 to n) will give a series of overlapping tracer profiles which will lead to an integrated resultant tracer form by adding ordinates. Without presenting the mathematics it is clear that, in the limiting case where the elementary injections follow continuously, the minimum period of injection to obtain plateau conditions will be T, i.e. $(x_2 - x_1 / \bar{U}_x)$, because in the period T after the arrival of tracer the concentration of tracer will increase from the addition of tracer elements, but subsequently each gain from an elementary profile will be matched by a corresponding loss from an earlier profile. A similar argument will apply, at the end of the plateau, before the start of distribution 'n'. The value of T may not be known when planning the gauging exercise and it is therefore important to note, from Figure 4.3, that to establish plateau conditions for a time period T_p the duration of the injection must be at least $(T + T_p)$. The

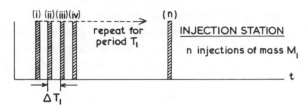

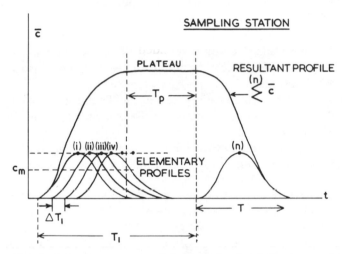

Figure 4.3   Constant-rate injection simulated by repeated gulp injections to show relationship between plateau duration T_p and tracer flow-through period T

duration of injection will therefore depend on $\bar{D}$ which, for some wide natural channels, will mean very long periods—possibly too long to justify the additional quantity of tracer needed for the application of the constant-rate method in preference to the integration method. For a given order of accuracy this can be estimated from Figure 4.3, since if a quantity M_1 is suitable for the integration method, then for a plateau at a concentration corresponding to the peak of the gulp method it would be necessary to use a quantity of $M_1(1 + 2T_p/T)$ approximately.

4.3   WATER TRACERS

The various substances chosen to act as water tracers are selected for the possession of properties which provide ease of detection at low concentrations. It is possible to envisage other forms of tracer, such as heat, which may be introduced without adding any substance to the flow. Also some non-miscible tracers might be possible, such as bacterial or phage tracers (Wimpenny *et al.*, 1972), if the component cells behave as individual tracer elements and still exist in statistically significant concentrations at the sampling station. The main types of tracer in use will be discussed below but the choice, in practice, may well be decided by the detection apparatus available to the gauging team. For example, many modern water-quality laboratories have an atomic absorption spectrophotometer, which may well be very satisfactory for determination of the tracer lithium, whereas access to a fluorimeter or nucleonic counting equipment might prove more difficult and therefore make the application of other tracers less convenient or more costly. The ideal properties of the tracer are as follows:

(a) it should not be adsorbed on suspended solids, sediments, bed and bank materials, sample containers, or piping and vessels in the instrumentation, or lost by evaporation;
(b) it should not react chemically with any of the surfaces given above, or with substances in solution under the pH and other chemical conditions existing along the channel;
(c) it should be stable under environmental factors existing for the envisaged time-of-travel, and in particular it should not show photochemical decay in sunlight;
(d) it should not have any harmful effects on human health or adverse effects on flora and fauna in the channel, particularly fish;
(e) it should be readily detectable above the background level at concentrations which are compatible with the accuracy desired and the quantity of tracer it is convenient to inject.

The chemical tracers to consider first are simple anions, because most natural mineral surfaces are negatively charged thereby discouraging losses by adsorption. The traditional tracer is chloride in the form of common salt, but bromide ions may be considered equally suitable from their proven conservative properties. The use of fluoride may be limited to non-calcareous streams because of the low solubility of its calcium salt, whereas iodide has been shown to be unsatisfactory in some applications. In tests reported by Neal and Truesdale (1976) the fraction of iodide, or iodate, at a level of 0.04 mg/l, sorbed by sediments, was found to range from negligible quantities to almost total uptake depending on the suspended-sediment loading and the composition of the sediment. Very high uptakes were found to occur on peat and natural peat/clay sediments in their associated fresh waters. These tests confirmed indications of a similar nature obtained much earlier by Eden *et al.* (1952) when reservoir water containing added iodine-131 was passed through a laboratory-scale slow sand filter. The activity of the effluent rose slowly over a period of several days to about half of that in the water before filtration. When filtration of water without tracer was resumed, iodine-131 was evident for many days. As sand filters trap a layer of organic debris, the implication of these results is that iodide should not be selected for work of high accuracy unless the levels of organic sediments or suspended matter can be shown experimentally to have a negligible effect on tracer recovery.

The concentration of the halide ions down to about 0.02 mg/l can be determined by ion-specific electrodes under ideal conditions. Samples may have to be treated to reduce the effects of interfering ions, and considerable time and patience on the part of the operator may be required to obtain stable and repeatable readings. As ion-specific electrode systems develop they will probably replace concentration determinations based on electrical conductivity changes, but at present the latter method with common salt as the tracer provides reliability, repeatability and simplicity, the disadvantages being that approximately 10–100 kg of salt per cubic meter per second of discharged, depending on the techniques used and accuracy desired, will be required.

Salt (NaCl) has a solubility of 357–360 g/l at temperatures from 0°C to 40°C, respectively, and for concentrations up to about 50 mg/l the relationship between conductivity and concentration is almost linear (approximately 1.86 μS/cm per mg/l at 18°C). The minimum concentration of NaCl which may be measured with an error of ±1 per cent is given in ISO 555/I (International Organization for Standardization, 1973) as 1 mg/l when the conductivity of the natural water is 100 μS/cm, and as 10 mg/l when it is 1000 μS/cm. Clearly, for accurate work, steady background concentrations are required and these may not occur in rivers receiving sewage effluents or in winter months with intermittent runoff from salted roads. Also, the background levels of chloride in rain may be very variable and may range from a few to several tens of milligrams per litre. It is important to remember that tracer materials in bulk, and common salt in particular, may be chemically impure. Salt may also contain additives to improve its physical properties or for health reasons.

The other chemical tracer which has been used, mainly in the UK, for dilution gauging in recent years is lithium, mostly as lithium chloride. Cations should, in general, be suspect with regard to cation exchange on many common minerals or bacterial slimes, but lithium behaves in a conservative way, probably because it forms the largest of the simple hydrated alkali cations. Lithium chloride has a solubility of 637 g/l at 0°C and the element lithium can be detected in concentrations down to about 10^{-4} mg/l in the laboratory with specialized flame photometers. The range of the simple portable filter flame photometer will extend to below 1 mg/l, depending on the amount of interference from other alkali ions in the samples.

At the concentrations used for dilution gauging, lithium does not present any hazard to man. It is used in medical practice for its tranquillizing action. However, it could possibly have adverse effects on fish. Tests with yearling rainbow trout in hard water (Department of the Environment, 1971) indicated that the 35-d median lethal concentration was 1.4 mg/l. When dilution gauging is undertaken, tracer concentrations at such levels only exist for transient periods, but some caution in the use of lithium is necessary.

In several tracer studies conducted by the Water Research Centre in the UK, where lithium has been used as part of a tracer 'cocktail' in retention studies, it has been shown that its passage may, on occasions, be slightly delayed relative to other tracers such as tritiated water, bromide or chloride, and the explanation may be the occurrence of very short-term adsorption and desorption processes. An example is shown in Figure 4.4 (Department of the Environment, 1976) for a polluted river. In view of the discussion on mixing criteria in the previous section, it is important to note that adsorption and exchange reactions will cause a skew tracer recovery curve, as illustrated in Figure 4.4, even if the mixing characteristics of the channel are otherwise ideal. It is clear that, as with iodide, the use of lithium for accurate gauging work should be considered only after jar tests on sediments taken from the channel have shown that it is unlikely to be adsorbed. Measurements by the Water Research Centre of the concentration of lithium in UK rivers show that most of those into which effluents flow will have background levels ranging up to about 0.05 mg Li/l, whereas springs and unpolluted waters will usually contain less than 5×10^{-4} mg Li/l.

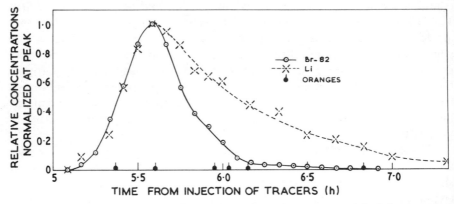

Figure 4.4   Example of tracer distributions 12 km downstream of the injection of a tracer cocktail (six oranges also released to assess their effectiveness as time-of-travel indicators)

Another chemical tracer that has been in widespread use, particularly in Switzerland and France, is chromium in the form of sodium dichromate which has a solubility allowing up to about 600 g/l to be used in practice. Unpolluted natural waters mostly contain negligible concentrations of chromium ions and colorimetric analysis permits detection down to about 0.2 mg/l, with further sensitivity provided by solvent extraction techniques. However, chromium is a very undesirable pollutant, as indicated by the proposed limit for concentrations in drinking water of 0.05 mg/l. It may also have an adverse effect on some fish. Tests conducted at the Water Research Centre indicate median survival times for rainbow trout ranging from 10 up to about 50 days at a concentration of 10 mg/l and about 1 day at 100 mg/l. Dichromate may not be a conservative tracer in polluted waters because its reduction to the trivalent state would cause losses by adsorption or precipitation. Other chemicals that have been used for dilution gauging include sodium nitrite and manganese sulphate, although the use of these, as of bromides and fluorides, is not thought to be widespread.

Long ago, Joly (1922) pointed out the tactical merits of radioactive tracers for dilution gauging:

> I propose to utilize, in determining river discharge, the extraordinary accuracy with which radioactive measurements can be effected. A very simple form of electroscope suffices to determine a quantity of radium to the billionth part of a gram. The apparatus costs a very few shillings. If, now, in place of introducing salt by the hudredweight into the river, we flow into the river a few litres of a solution containing a trace of radium, and taking samples downstream examine them by the electroscope, the discharge of the river may be determined.

It took almost half a century for the use of radiotracers to be regarded as part of the normal range of techniques available to the engineer or hydrologist. No one today would consider using radium, as there are dozens of short-lived radiotracers which have since been developed and tested.

Radioactive tracers have considerable advantages when high discharges are considered. The injection solution may range in concentration up to tens of curies per litre and most isotopes are accurately detectable down to tens of picocuries per litre, so that great dilutions are possible as well as flexibility with regard to injection techniques. When chamicals are used, large systems may require such large volumes of saturated solution that tactical difficulties with injection may occur and layering of the dense solutions in the channel may restrict the degree of mixing. Two radioisotopes have proved their value for accurate gauging at any rate of flow; these are bromine-82, which may be obtained as irradiated potassium bromide tablets, and tritium in the form of tritiated water (HTO). Surface waters at the present time contain background levels of tritium of several tens of picocuries per litre (about one order of magnitude above natural levels) caused by the contamination of the hydrosphere from nuclear bomb tests in the 1960s.

Background levels of bromine-82 are usually zero, but there may be some variations in the counter background because of fluctuations in the cosmic and terrestrial radiation and the presence in the water of natural or bomb-fallout isotopes and possibly waste isotopes discharged from hospitals. Because both the decay process and the detection process are digital in nature, radiotracer concentrations can be determined with exact linearity at all concentrations. Radiotracers emitting gamma-rays have the advantage of being detectable by remote means. This allows 'dirty' samples to be analysed directly without pretreatment. Additionally, factors such as pressure, temperature, and the presence of other substances cannot affect the precision with which tracer concentrations are determined. Even chemical reactions may not matter so long as the tracer stays in solution.

Bromine-82 is short-lived, with a half-life of only 35.4 h. It is readily detectable *in situ* and has been used in a wide range of hydrological studies with great success. It disappears at a predictable rate by radioactive decay and cannot cause prolonged radioactive pollution or contamination. However tritium, although it can be considered the ideal water tracer, since it is in the form of a water molecule, is long-lived with a half-life of 12.3 years. It must therefore be considered as a pollutant and used only for particularly important studies where short-lived isotopes cannot be used or where the time-scale of mixing justifies its use. Since it cannot be detected *in situ* unless very specialized apparatus is available, the sampling programme must be adequate to ensure that the benefits of its use are secured. The concentrations of both isotopes, at the sampling station, can be kept less than an order of magnitude below drinking water tolerance levels set for these isotopes by the International Commission on Radiological Protection (1959).

It should be noted that when potassium bromide is irradiated to produce bromine-82, other isotopes are formed, principally bromine-80m (half-life 4.6 h), bromine-80 (half-life 18 min), and potassium-42 (half-life 12.4 h). In order to reduce the influence of these radioisotopes to a practical minimum, the potassium bromide tablets are removed from the reactor at least three days before the time of injection. Checks of isotope purity are made by gamma-ray spectrometry and by decay measurements.

Other isotopes to consider are sodium-24 (half-life 15 h) and iodine-131 (half-life 8.0 d) but, as mentioned earlier, preliminary tests of both of these would have to be made to assess the possibility of adsorption losses. The use of radioisotope generators for automatic dilution gauging could be attractive for some situations. A long-lived parent isotope is trapped on an ion-exchange column and the short-lived daughter may be eluted to form the tracer injection, whenever required. Prototype equipment has been used for pipe-flow measurements by the velocity-area principle using the very short-lived daughter, barium-137 (half-life 2.6 min), of caesium-137 which has a half-life of 30 years. For frequent short-term measurements, or mixing tests, over a period of one to two weeks, excellent *in situ* detection is possible using iodine-132 (half-life 2.3 h) generated from tellurium-132 which has a half-life of 78 h.

The detection sensitivity attainable with radioisotopes can be utilized, without the administrative delays or public apprehensions associated with radiotracers, by means of post-gauging neutron-activation analysis, in which the tracer levels are derived by neutron irradiation of the water samples. The prime contender for the tracer would probably be bromide injected at much higher concentrations, in chemical terms, than when used directly in radioisotope form. High concentrations allow the shortest possible time of sample irradiation and thus reduce the level of activation products of other elements in the sample which may interfere with the detection of the isotope required. Costs would probably confine the technique to the analysis of a composite or bulk sample obtained during tracer passage and, of course, a second analysis of the injection solution (comparative dilutions may be unnecessary with this method).

Fluorescent dyes have been used as mixing indicators and as tracers for gauging for almost as long as common salt. In particular the green dye fluorescein (uranin) has been the traditional indicator. There has been more concern in recent years about pollution of rivers by organic materials and the use of dyes will be subject to greater scrutiny as time passes, particularly if fears that have been expressed about the carcinogenic properties of some of them are shown to have foundation. There are, of course, similar fears about radioisotopes, but a tracer such as bromine-82 disappears quickly in the environment by its own decay, whereas dyes and chemicals may have a very long effective half-life in the environment.

Fluorescent dyes will not be discussed at length here because, in most cases, they are not satisfactory for dilution gauging, although Smart and Laidlaw (1977) have shown that one or two may be convenient to use for qualitative time-of-travel, retention, or dispersion studies. Their main disadvantages are photochemical decay and a tendency to adsorb (at very low concentrations) on suspended solids, or surfaces including sample containers. Skew tracer-recovery curves with partial tracer loss will occur in most natural systems. However, should this type of tracer be most convenient to a user who is not seeking high accuracy then it would still be prudent to make specific plans (second sampling station) to assess tracer loss. The two most reliable tracers are probably Rhodamine-WT and Sulpho-Rhodamine B Extra. These can be injected at concentrations of tens of grams per litre and detected at tens of nanograms per litre, depending on the extent to which background substances fluoresce in water. Although fluorescent dyes can be detected in samples that are colourless to the human eye, they may cause taste problems and are capable of colouring fish flesh should fish become contaminated near the injection point. Modern instrumentation employing lasers makes it possible to detect dyes down to picogram per litre levels, but the potential advantages of such sensitivity may be outweighed by concern over the conservative properties of the tracer.

To summarize, tracers are used on a 'horses for courses' basis and the winning 'horse' would be chosen by the strength of public opinion, the magnitude of the discharge, the sensitivity of instrumentation available, the cost, the remoteness of the site, and the accuracy required. When the highest accuracy is sought, and the field conditions are such that the accuracy will be limited only by the tracer characteristics, there are very few tracers, the best of which are chloride, bromide (bromine-82), and tritium (or stable isotopes such as deuterium or oxygen-18 as a tracer incorporated in the water molecule), which have proved to be reliable. In some countries, particularly India, the USSR and the UK, while chemical or fluorecsent tracers may be in use routinely for local checks, radioactive tracers are used for calibration work where accuracy for long-term application of the results is essential.

4.4 SELECTION OF MEASURING REACH

The length of the reach required for satisfactory mixing, according to the criteria discussed in the section on principles, can be many kilometers. The criteria are based on the assumption that the dispersion coefficient does not change along the reach between the cross-sections used for the injection of tracer and for sampling. The measuring reach should therefore be of similar form and nature over the whole of its length. For example, if a tracer has become well mixed vertically and laterally at a certain point along a channel which then increases considerably in width, say on the approach to an estuary or weir, and the tracer fronts advance preferentially along a particular axis, then lateral dispersion processes would again become effective in diluting the tracer, and sampling for flow measurement would not be appropriate even though mid-stream samples might indicate a tracer profile approximating a Gaussian distribution. The helical flow induced by a bend may be advantageous in increasing the rate of lateral mixing, thus decreasing the length required for the measuring reach. Meandering reaches may be satisfactory, provided the channel and flow pattern are such that there are no dead zones.

The criteria for mixing also make it clear that there should be no ingress of water along the measuring reach because such an occurrence may be thought of as the introduction of a negative tracer, which therefore must also be mixed completely upstream of the sampling station. However, a mixing reach can be selected independently of the position of the injection station when it is not convenient to make the tracer injection at the head of the reach selected. For example, a reach between two tributaries may be selected as the measuring reach, whereas it may be convenient to inject the tracer into either the main stream, or the tributary, upstream of the confluence. Under these conditions, the tracer-injection geometry and period should be regarded as those existing at the confluence. Loss of water or tracer (e.g. by interaction with sediments) in the reach must be avoided or the basic tracer balance equation will no longer hold. In practice, loss of tracer by seepage through the bed may be most difficult to detect and the effect will be reflected in an erroneous high flow result with no apparent reason why its accuracy should be doubted. It is therefore necessary on occasions to select a subsidiary measuring reach, downstream of the first, where the flow and suface geology are similar. If the flows determined at both stations are the same then the tracer has behaved conservatively with respect to losses by adsorption, precipitation, etc., and there has been no significant ingress of water between the stations and in all probability there has been no ingress of water in the primary reach. The only way to check whether egress of water is taking place is to measure the flow by a non-tracer method. Egress of water is equivalent to diversion of some of the flow through a bifurcation and the flow measured, therefore, is representative of the discharge upstream of the zone of egress and not the volume transported past the sampling cross-section. Repeat gaugings using the sampling section but longer upstream reaches may help to elucidate the problem of leakage.

A formula for predicting the length of measuring reach, from the channel characteristics, was presented by Fischer (1967). He derived an equation for estimating the dispersion coefficient $\bar{D}$ in a channel of large width-to-depth ratio and deduced the time-scale of the convective period, i.e. that period in which disperion is not properly described by the one-dimensional equation given earlier as equation (4.5):

$$\text{convective period} = \frac{0.3l^2}{R_h U^*} \tag{4.10}$$

where l is the distance between the thread of maximum velocity, and the furthest distant point in the cross-section, R_h, is the hydraulic radius (area divided by wetted perimeter), and U^* is the shear velocity $\sqrt{(2gR_hS)}$, where S is the slope of the energy line but may be taken as the slope of the water, and g is the acceleration due to gravity. He analysed experimental data and showed that the mixing criteria for a point or line source were established after times greater than six theoretical convective-mixing periods which, by using the mean velocity $\bar{U}_x$, provides a guide to the mixing distance, L: $L > 6 \times \text{convective period} \times \bar{U}_x$, i.e.

$$L > \frac{1.8l^2 \bar{U}_x}{R_h U^*} \tag{4.11}$$

If this distance is not available, the reach selected should be at least one-half L, as given by equation (4.11), since Fischer found that the growth rate of the tracer–distribution curve variance became approximately linear after three convective periods.

McQuivey and Keefer (1976a) gave values of L for several rives at various flows. Their data are shown in the first seven columns of Table 4.1, which illustrates the considerable length of reach required for wide rivers. Note the six-fold difference in L (seventh column) for the two rivers of width 47.2 m, presumably due to the differing l values (not presented).

In another paper (McQuivey and Keefer, 1976b) a specific test in the Mississippi using a large quantity of dye (1815 kg Rhodamine-WT, 20 per cent solution) is described in which flow-through curves were obtained at nine

Table 4.1 Examples of prediction of length of channel (*L*) for complete mixing using data from McQuivey and Keefer (1976a); equation (4.11) compared with equation (4.12)

River	Width (m)	Discharge (m³/s)	R_h (m)	U^* (m/s)	$\bar{U}_x$ (m/s)	(4.11)	Mixing distance '*L*' (km) using equation: (4.12) $C'=15$	(4.12) $C'=50$
Comite	12.2	0.98	0.27	0.043	0.311	8.1	1.8	15.0
	15.8	2.38	0.37	0.049	0.369	12.9	2.2	18.4
Monocacy	35.1	2.38	0.30	0.043	0.213	12.9	13.7	112
	36.6	5.18	0.37	0.049	0.320	16.1	12.1	99
	47.2	18.2	0.79	0.070	0.442	24.2	9.4	77
Clinch	47.2	9.1	0.85	0.067	0.229	4.0	8.7	72
	59.4	84	2.13	0.104	0.671	4.8	5.5	45
Wind	71.6	56	1.07	0.116	0.853	19.3	160	131
	80.7	224	1.83	0.168	1.524	29.0	11.8	97
Sabine	103	118	1.58	0.046	0.564	76.6	22.4	184
	128	385	3.51	0.061	0.643	128.8	25.6	128
Missouri	185	378	1.19	0.067	0.930	104.7	52.1	427
	197	938	3.05	0.079	1.530	120.8	42.4	348
Mississippi	778	10 192	15.24	0.055	0.838	193.2	129	1 058
	867	22 400	16.76	0.058	1.478	289.8	149	1 226

locations over a distance of 288 km. The flows were 22 430 m³/s at the injection point and 22 600 m³/s at the lower end of the reach, i.e. similar to the last flow given in column 3 of Table 4.1. It was estimated that vertical mixing had been accomplished in about 16 km and lateral mixing in about 80 km. Even if estimates based on equation (4.11) tend to be high, they are valuable in giving a maximum value of the length of reach to aim for. In practice, the maximum length available is often decided by other factors such as the length of reach available between tributaries.

There are a number of empirical formulae for estimating the mixing distance *L*. The one given in the ISO standards, for a point injection in a straight reach for mixing to be within 1 per cent of complete homogeneity, is Rimmar's formula:

$$L = \frac{0.13b^2C'(0.7C' + 2\sqrt{g})}{gd} \tag{4.12}$$

where *b* and *d* are the average width and depth of the channel, respectively, and *C'* is the Chezy coefficient of roughness which varies from about 15 to 50 for rough to smooth bed conditions. This equation is very useful because, even if little is known about the roughness of the bed, lower and upper limits can be placed on *L* for guidance. The values of *L* for *C'* values of 15 and 50 have been calculated for the rivers given in Table 4.1 using R_h for the average depth, and the results are shown in the last two columns of the table. It is interesting to note that the values may increase or decrease as a function of river discharge and width. Day (1975) reported experimental results indicating similar findings for mountain streams of discharge up to $15 \, m^3/s$ and width up to 22 m. There is evidence that the mixing length of larger systems increases in a more predictable manner as the width and discharge increase to a stage where dilution gauging may become impracticable (probably in the range 10^4–$10^5 \, m^3/s$). If there is little chance of finding a reach even approaching the estimated length required, then a multi-point injection technique may in some cases be a solution, as outlined in Section 4.5.

To summarize, it is necessary to select a length of channel of similar geometrical form from end to end having no ingress or egress of water, pools, or bays and inlets of large magnitude relative to the channel dimensions. Rock sills causing waterfalls without pools upstream or downstream should not matter and, in general, bends should be advantageous in decreasing the mixing distance. Reaches with bifurcations should not be chosen but islands often cannot be avoided, a trivial example being those due to bridge piers. If only rudimentary data on width, depth, and type of channel are available then there is little choice other than to use Rimmar's formula. However, if it is possible to obtain information on the energy slope or water surface slope and the factor l, possibly from preliminary survey work, then equation (4.11) should also be used for further guidance. A valuable compilation of photographs of river channels with their roughness coefficients has been produced by Barnes (1967).

It is clear from the last two columns of Table 4.1 that too high a value for Chezy C' should not be chosen because the selected value is effectively squared in the computation of L in Rimmar's formula. Rimmar (1960—NEL translation) presented values for the numerical constant applicable to a range of completeness of mixing and some of these values are given in Table 4.2. The appropriate value should be selected to replace the value 0.13 when guidance on the mixing length for more or less accurate work is required.

Table 4.2

Completeness of mixing	Numerical coefficient to use in Rimmar's formula
10	0.074
5	0.091
1	0.130—value usually quoted— see equation (4.12)
0.1	0.187
0.01	0.244

4.5   TRACER INJECTION TECHNIQUES

When the integration method is used the injection procedure should in most cases be very straightforward, but the sampling programme must be thorough and exact. When the constant-rate injection method is used the injection procedure must be exact and may be complex, whereas the sampling programme should be straightforward. More expertise will therefore be required at the sampling station for the integration method and at the injection station for the constant-rate injection method, which may also require the preparation of specialized constant-rate injection equipment at the laboratory.

It was pointed out in the section on principles that when the integration method is used the duration of the injection is unimportant, but to lessen the chance of density effects and to reduce the maximum concentrations of the

Figure 4.5   Injection of Br-82 from mobile platform into the River Thames for dilution gauging by the extended gulp technique

tracer at the tracer release point an extended injection is preferable. There is no requirement to achieve an exact rectangular injection profile—a steady emptying of the bottle or container will suffice. It is more important to ensure that all of the tracer enters the flowing water and that the quantity injected can be accurately deduced from initial and final measurements. Any tracer spilled or blown away by the wind will lead to the calculated discharge being erroneously high.

It is worth while taking some precautions against loss of tracer and the best way to do this is to carry out a complete 'dummy run', possibly using a weak dye solution, in order to identify any problems and potential spill-risk points that may be present. The tracer should be released as near the water surface as possible and when the injection is made from a high bridge a plastic pipe should be used to convey the tracer down into or near the water surface, as shown in Figure 4.5. These precautions are particularly relevant when tritium is used. Tritiated water should be released below the surface if possible to avoid any evaporation and splashing which could represent direct losses of tracer.

The types of apparatus that have been developed for the purpose of injecting tracer at constant rates are illustrated and described in national and international standards, e.g. International Organization for Standardization (1973), and details will not be given here. The basic requirement is an orifice, to govern the rate of injection, connected to a constant-head supply tank. Otherwise a metering pump may be used. Two ingenious and effective containers which maintain a constant head are the Mariotte vessel and the floating siphon.

For larger quantities of tracer, a reservoir tank may be used to supply a constant-level tank employing a regulating weir to maintain a steady hydrostatic head on the orifice. An orifice is less sensitive to variations in temperature than a capillary. Whatever device is used, some regular short-term periodicity in delivery can be tolerated, since the effects will be smoothed in the channel during the preliminary mixing stage.

Small metering pumps, either of the peristaltic or reciprocating-piston type, are usually preferred for radioisotope injections because of the small total volumes involved and the necessity to minimize the residual volumes of tracer. It is beneficial to keep radioisotope volumes small to miminize the cost and weight of lead shielding. The injection rates may be as little as a few cubic centimetres per minute and it is convenient to use a hosepipe with a flush of water to transport the isotope to the channel. The plastic pipe supplied from the tank towards the front of the truck shown in Figure 4.6 serves this purpose.

It is important to design the injection equipment in such a way that the rate of delivery of tracer into the river can be determined on site. Volumes, viscosities, and therefore injection rates, will vary with temperature and when electric metering pumps are employed the stability with time of the power

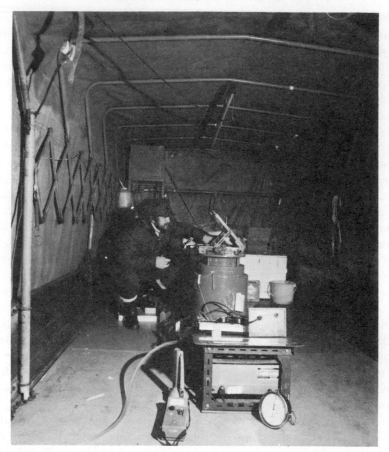

Figure 4.6   Injection equipment inside truck showing plastic injection
hose

supply must be checked. The injection rate must be measured at the beginning and end of the period of injection and considerable skill may be
required to carry out these operations accurately, particularly for radioisotopes where a compromise must be made between taking adequate samples
for timing as well as volume (or mass) accuracy and keeping the sample as
small as possible to reduce the radiation dose received by the operator. White
et al. (1975) have described a technique whereby the sample is collected
within the lead shielding, via a T-piece just downstream of the peristaltic
metering pump, for a measured period of time. The sample is then transferred
immediately into a previously prepared flask of water of several litres capacity
which forms the first dilution stage of the serial dilutions necessary to
compare the injection solution with that of the samples taken downstream.
The dilution volume provides some self-shielding and the problem of handling, weighing and diluting a small aliquot of 'hot' solution is avoided.

In most cases a single tracer-release outlet near mid-stream is chosen, but mid-stream should not be regarded as essential, particularly when a bridge support, which might cause local back and eddy currents, occupies this position. Injection below the centre of the arch of the bridge beneath which most of the flow occurs may then be appropriate. It may be advantageous to divide the tracer dosing line into two where the flow beneath a bridge is effectively formed into two channels. The criteria for good mixing were based on theoretical initial conditions of tracer 'injection' as a sheet across the upstream section. It must be remembered that the technique, often suggested, of making use of a multi-point injection to reduce the mixing distance applies to a reduction in distance compared with a single point injection, i.e. compared with the estimate of L given by equation (4.12) but not compared with the requirements of basic theory. However, if it is possible to determine the relative discharge of the channel as a function of width and make a multi-point injection distribution to match, then the mixing distance may be reduced considerably in those cases where vertical mixing takes place long before complete lateral mixing, and this is usually so for wide shallow rivers. This approach simulates dividing the river longitudinally into a number of component channels and adding tracer to achieve similar tracer conditions in each channel at some point downstream. In the limit, if a matrix of tracer injection nozzles were distributed with width and depth such that the injection rate at each point matched the elementary discharge crossing each component elementary area of the cross-section, then the mixing distance would effectively be zero for the constant-rate method. (A flow gauging structure based on heating elements, using heat as the tracer, could be developed along these lines.)

An estimate of the potential benefits of multi-point injection can be demonstrated by considering the case of a wide river of uniform depth such that the discharge per unit width does not vary across the section. Consider a tracer injection from N points; the channel may then be assumed to consist of N equal channels, side by side. Clearly the characteristic length l used in equation (4.11) will be N times smaller and so the mixing distance should be N^2 times shorter. Therefore even a twin-injection technique might reduce the mixing distance by a factor of 4 and, for similar reasons, if a bank injection of tracer is necessary, then the mixing distance should be about four times greater than that for a mid-stream injection.

Some experimental data on the degree of mixing downstream of five-point injections was given by Cole (1969) and by Sanderson in the discussion on Cole's paper. It was found that, at distances where the percentage mixing P was greater than about 95, the advantage of multi-point injection was only marginal except where data on the relative discharge as a function of width were available and the injection rate was matched to this. In the latter case, the results suggest that five injection points gave an order of magnitude decrease in mixing distance compared with a single-point injection and this result would begin to approach the predicted 25-fold reduction. The perceeentage mixing was defined in terms of the absolute mean deviations of concentrations c_a, $c_b \ldots c_n$ for samples taken across the section; at equally-spaced intervals from the mean concentration $\bar{c}'$:

$$P = 100\left(\frac{1 - |c_a - \bar{c}'| + |c_b - \bar{c}'| + \ldots + |c_n - \bar{c}'|}{n\bar{c}'}\right) \qquad (4.13)$$

Also, in the discussion of the same paper, Ward illustrated the difference between mixing lengths predicted by Fischer's formula (equations 4.11) and Rimmar's formula (equation 4.12) as a function of the friction factor of the stream bed.

A complex arrangement, constructed for the injection of common salt, was described in detail by Groat (1915) who used a frame of fixed dosing pipes with holes drilled at regular intervals so that injections could be made in rows at six depths simultaneously to provide a complete matrix of injection points. Various configurations and sizes of holes in the distribution pipes were investigated to obtain good distribution of tracer between the pipes, and one general design conclusion was that better distribution and mixing of tracer was obtained with fewer high-pressure jets than more numerous jets from holes subject to less hydrostatic head. To overcome the problems posed by dilution gauging within a weir structure the Water Research Centre has used multi-point injection.

Another method used to decrease the mixing distance is artificial mixing in the stream downstream of the injection point. Submerged pumps to jet water laterally may be beneficial, or boats with outboard motors may be used to improve lateral mixing—only trial and error, both with regard to the degree of mixing obtained and the optimum position at which to use it downstream of the injection point, will show the effectiveness of such techniques. In small streams, mixing by compressed air may be feasible.

4.6 SAMPLING TECHNIQUES

The procedure for sampling will need to be as complex as that warranted by the importance of the gauging and the accuracy demanded. The essential aspect to consider for each programme of sampling is how well the samples collected at individual points in the cross-sections represent the charac-teristics of tracer passage for every point in the section. Clearly, the more points that can be sampled, the more thorough will be the verification that the degree of mixing (defined in the previous section) is compatible with the accuracy required. A single sampling point can suffice but should, in general, be regarded as inadequate. Sampling is costly in terms of labour in the field and laboratory and a satisfactory compromise for many rivers is to select three (sometimes four) sampling points, e.g. mid-stream, left bank and right bank. To make the bank positions reasonably critical, in order to assess lateral mixing, suitable positions (to avoid debris, weeds and unrepresentative shallows) might be about 10 per cent of the width out from the bank. A fourth sampling point should be selected when part of the channel is of sufficient depth to justify checking the degree of vertical mixing. For example, relatively cold groundwater leaking into the channel could be detected by an analysis of the tracer contents of samples collected near the bed. On deep rivers suitable sampling points might be about 1 m above the bed, to avoid sediments, and 1 m below the surface. Such positions are also suitable for positioning *in situ* radioisotope detectors for efficient and stable detection geometry. Otherwise the depths at which samples are taken may be chosen as shown in Figure 4.7.

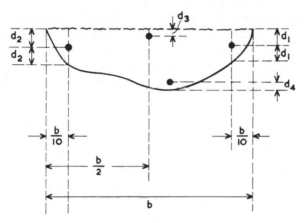

Figure 4.7 Sampling section defining minimum number of sampling positions for dilution gauging in natural channels (d_3 and d_4 are one-tenth max. depth or 1 m on very deep rivers)

The sampling effort needed to apply the constant-rate method might only be small, compared with the effort for the gulp method, when *in situ* detection is possible. It is then simply a case of observing the detector read-out and taking a series of samples at each sampling position when equilibrium conditions are indicated—possibly five samples at intervals of about 5 min for a plateau lasting one half-hour. The tracer recording may be examined to assess the mixing criteria. However, if *in situ* detection is not possible, then, for accurate gauging, just as many samples may be necessary as for the gulp method in order to verify that a plateau has been reached and that the curve of tracer concentration from background to plateau to background is symmetrical.

When the tracer curves for either method are determined by taking multiple samples it is important that each sample is taken from the same depth and position across the section, otherwise different distributions are being sampled (Figure 4.1). A fixed sampling tube and abstraction pump may be used for this purpose but it is essential that each sample is collected in a very short time compared with the rate at which the concentration of tracer is changing. When automatic samplers are used, the samples will be taken at equal intervals of time, but equal spacing in time is not essential; all that is required is a series of samples, with actual water abstraction times accurately recorded, in order to define the concentration–time curve. In most cases it will be necessary to have at least 20 samples of measurable concentration to define the tracer curve adequately. Automatic samplers should be checked for systematic errors due to dead water trapped at the tube ends, cross-contamination possibilities and timing errors.

Multiple samples lead to a point-by-point definition of the tracer curve, as illustrated in Figure 4.8(a). The tracer concentrations between sample points must be interpolated by a best-fit procedure. A more complex technique, to overcome this difficulty, is to collect samples at a constant rate for consecutive periods. Each sample then represents the average concentration for the

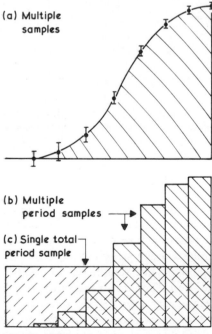

(a) Multiple samples

(b) Multiple period samples

(c) Single total period sample

Figure 4.8 Resultant tracer curve areas to integrate for three sampling techniques

period it was collected and this procedure will lead to a plot of histogram type, as shown in Figure 4.8(b), where it is very easy to calculate the area under the curve. Clearly, as indicated, this idea could be extended so that only one bulk sample is collected to determine the area under the curve, $\int c \, dt$.

A single bulk sample is a worthwhile consideration as a back-up measure, even when multiple samples are taken, as it gives an independent result. It is risky to rely on such a technique alone because no knowledge of the tracer flow-through profile is obtained, but it is very economical and may be valuable for a person who is gauging on his own. The problem is to decide when to start and end the period of sampling when no *in situ* detection is possible. When the quantity of tracer used has been adequate (or more than adequate) it does not matter if the sample is started a long time before the arrival of tracer and ended some time after the whole of the tracer has passed, so long as the background concentration and discharge are steady with time, because the concentration of the integrated peak plus background sample should be clearly discernible above the background concentration.

A number of more complex sampling techniques to apply at a single sampling point has been suggested, both in the standards and in the literature, such as starting or ending a series of total bulk samples at different times to establish the flow by successive approximations if all the tracer had not passed by the time sampling ended. It is somewhat illogical to increase effort for such complex sampling schemes because the additional effort might be far more profitably employed by sampling at a second cross-section in order to provide a check on the behaviour of the tracer.

Alternatively, effort can always be profitably devoted to sampling more points in the sampling cross-section in order to verify mixing. In cases where it is not possible to obtain samples across the section, then several sets of samples should be taken, at points as far apart as possible relative to the mixing distance, along the bank. If the same discharge is determined at the chosen points it may be assumed that the degree of mixing was satisfactory at the first station and that the probability that the tracer is confined along one bank is low, because otherwise lateral dispersion would cause a systematic variation in the results with distance downstream.

In small, narrow and very turbulent streams, where there is no doubt about mixing efficiency and where conditions are almost ideal for mixing within a short reach, then single injection and sampling points may be perfectly adequate and a straightforward simplified step-by-step procedure for the gauging operation can be used for routine operations, as shown in the example given by White *et al.* (1975).

4.7 PREPARATIONS AND FIELD ACTIVITIES

In most cases it is prudent to spend at least one day in the local area to obtain data on flows, flow frequencies, etc., to get copies of plans and maps, to find out about any past results or experience (e.g. times-of-travel of pollutants) gathered on the river, to assess the nature of the channel and whether ingress or egress of water is likely, to take some rudimentary measurements and photographs to aid planning, to select accessible cross-sections for pilot tests, and to learn of interested parties to be kept informed about the exercise and particularly the use of tracers.

The next stage of preparations, except for small turbulent streams, will be pilot surveys to inspect the river channel in detail in order to select the measuring reach. A boat or helicopter may be used for this purpose, depending on the size of the river, but an inspection from road bridges will not suffice as it will not reveal information on inlets, branches, temporary abstractions, and effluents. Maps, however large the scale, do not necessarily provide relevant information and are often out of date. The approximate velocity distribution in a typical cross-section of the measuring reach should be determined and the maximum velocity 'thread' should be located to determine l (equation 4.11). The hydraulic radius must be estimated from measurements to establish the average width, depth and form of the channel. These activities will enable the shear velocity U^* and the discharge to be estimated under the prevailing conditions, which should be recorded photographically if stage staff gauges are not available. It may be possible to carry out some preliminary mixing tests and to check the nature of effluents to assess their effect on tracers. Time-of-travel estimates can sometimes be obtained using oranges (Figure 4.4), and aerial photography of the behaviour of turbid tributaries or effluents can reveal mixing characteristics along the river.

The information gathered allows detailed planning and administrative preparations to proceed. The nature and purpose of the exercise should be confirmed and interested parties notified. Permission must be obtained to use the tracer and any mixing indicator, such as dye, in the quantities envisaged (authorization is required by national legislation in the case of some tracers). There may be matters of way-leave and insurance to deal with and the supply of tracer must be organized.

The fourth stage to consider is that of the preparations at the laboratory to make up tracer solutions, to check tracer purity, to calibrate injection or sampling equipment, to obtain suitable sample containers and check them for any possible effects on, or interaction with, the tracer at the low concentrations envisaged, to check detection equipment, and to prepare spares—the quantity of which will depend on how remote the field site is from the laboratory.

The next stage is the actual gauging operation and this will start with checks on the degree of similarity of the flow conditions to those existing when the pilot survey was carried out. It is important to establish that the discharge is steady, e.g. by employing portable recording level gauges, and to liaise with river management teams to ensure that the gauging will not be upset by the operations of locks, sluices, etc. If the flow is not steady, a decision must be taken on whether to proceed in order to obtain data for future attempts, or to postpone the exercise. It may be necessary to gather flow data for tributaries and weather forecasts to aid this decision. Gilman (1975) showed that serious loss of accuracy will not be incurred by gauging on the recession limb of a hydrograph, but increasing or fluctuating flows may cause considerable errors. It will, of course, still be necessary to prove that mixing conditions are satisfactory and to determine the variation of flow with time in order to assess the tolerance limits for the results using guidelines given in his paper.

The decision on whether to attempt to gauge the flow under prevailing conditions will depend also on the objective; if some discharge information is needed where previously none existed, then it might be worth proceeding even under unsatisfactory conditions, but if the objective is to establish whether a gauging structure is giving correct indications to say ±2 per cent, then it would be wise to postpone gauging until the flow is steady.

4.8 ANALYSIS AND ASSESSMENT OF GAUGING ACCURACY

The analysis of samples and fieldwork data must be carried out to a degree determined by the success or otherwise of the field operations. The follow-up work and studies at the laboratory, or base, can be very time-consuming and therefore costly, and a decision must be taken on whether the samples and information gathered will give an acceptable gauging result or be used to provide mixing and time-of-travel data to aid the planning of repeat attempts. On completion of the field operations the following procedure is therefore suggested:

(1) Store all samples taken at the sampling station in a secure place away from possible contamination by concentrated stocks of tracer and check that every sample has a secure label showing place, position, time, etc.
(2) Prepare the analytical equipment and check its operation with laboratory standards.
(3) Carry out an approximate analysis on selected samples to establish the times-of-travel of tracer and the maximum concentrations encountered.
(4) Check the background samples to assess their stability during the period of field measurements.
(5) Determine the relative tracer concentrations in the sets of samples to a precision compatible with the findings of (3) and (4) above. (Several samples should also be analysed to check on absolute levels of tracer and the presence of any foreign substance which might interfere with the relative readings.)
(6) Plot graphs of tracer concentration against time, using the same axes for results applicable to different sampling positions, if possible. Critically

examine the curves with other data, such as those indicating flow steadiness during the gauging, and assess the efficiency of mixing. Decide whether a valid discharge result can be derived.

(7) Use the large-volume background samples to make up serial dilutions of the injection solution to about the same concentration as that of the samples. (An alternative procedure that may be used for short-lived radioisotopes is explained below.)

(8) Report the calculated discharge with the estimated uncertainties as required by national standards.

It is more important to discuss in detail examples of how errors may be caused, than to describe the analysis of samples for various tracers. In any case there are too many tracers to cover and a manual could be written on the detailed precautions for each type of tracer. 'Standard Methods' (American Public Health Association *et al.*, 1975) gives procedures for the determination of the concentration of chemical tracers; in addition, for common salt consult Groat (1915) or Rimmar (1960), and for iodide (where suitable) consult Truesdale and Smith (1975). There are relatively few manuals for fluorescent tracers, the most detailed probably being collected into a book chapter (US Geological Survey, 1965). Smart and Laidlaw (1977) should also prove useful. The literature on radioactive tracers is vast, but an understanding of fundamental concepts for accurate low-level counting may be obtained from the comprehensive laboratory manual prepared by Chase and Rabinowitz (1967) or the text of Watt and Ramsden (1964). An extremely well-prepared and concise general account of radioactivity measurement is given in an ICRU report (International Committee on Radiation Units and Measurements, 1972).

The first cause of errors is in the determination of volumes and the effect of temperature on both the expansion of solutions and the volumetric capacity of measuring flasks. Even when using the highest grade of glassware, the errors between different operators can be high. It is good practice therefore, whenever possible, and particularly when very accurate gauging is required, to measure all quantities of solution by weight. Modern top-loading electronic balances capable of weighing 1–10 kg quantities to an accuracy of better than 0.01 per cent are available. They are expensive but have a long useful life and form the ideal complement to the accurate balances normally available in chemical laboratories for weights up to about 100 g. Use of these balances for serial dilutions lessens the chance of spillage and saves time, as only nominal volumes need to be prepared before accurate determination by weighing them. It must be remembered that if all the samples are analysed at a room temperature which is different from the river temperature (e.g. 20°C and near zero, respectively), then the calculated discharge applies at room temperature and would need to be corrected for the temperature difference unless the result is quoted in gravimetric terms.

Temperature variation may also be a cause of error in the determination of tracer concentrations. An example is the temperature coefficient of fluorescence of the Rhodamine dyes, which is about 3 per cent per degree centigrade. It is good practice therefore to let all samples, dilutions, standards and the instrumentation reach a common temperature prior to measurement. It would be possible to list very many temperature effects on detectors, instrumentation components, etc., as well as physical, optical and chemical effects on the solutions, and any gauging team must regard temperature variation as a potential enemy, to be kept under surveillance at all times.

Aspects of the measurement of time need considerable attention, not only with regard to cost. Accurate time-of-day records are essential when records from different sources have to be compared, and accurate determination of time intervals may be of fundamental importance to the gauging, particularly for the constant-rate injection method where the rate of injection is determined at the injection site by taking samples over a short period of time. For the integration method, times of day must be accurate to the desired tolerance. If the timepiece is reading consistently fast or slow, this may not introduce any error in the determination of the area under the tracer curve, so the same timepiece should be handed over on change of sampling personnel. Very seldom do the watches of two persons record exactly the same time even if synchronized at the start of the day. Serious timekeeping errors become obvious when the concentrations are plotted and a step occurs in an otherwise smooth curve. Avoidance of such problems is better than correction on the examination of results. For a tracer curve 'p' hours long, timekeeping to $\pm 3.6p$ seconds should keep errors from this cause to less than 0.1 per cent.

In order to calibrate the rates of injection of tracer, problems not unlike those of timing sporting events must be considered. Human reaction times should be assessed and at least two persons may be necessary to determine time periods accurately. One person can operate the controls and one timepiece and the other person can operate two timepieces. If automatic flow-diverter devices are used to collect the sample, then there is the possibility of electronic timing; otherwise the times at which the edge of the sampling orifice crosses the flow must be assessed. Clearly the longer the period of time for sample collection, the lower the proportional error, but a compromise is necessary, particularly with radioactive tracers, because too large a sample may be inconvenient to handle. In most field situations two periods of at least 100 s will be necessary, both at the beginning and end of a constant-rate injection, to provide data from which to assess the injection rate and its tolerance accurately. Electronic timers sometimes record repeatable units of wrong duration, and devices relying on the frequency of the electricity supply for timing should be checked for errors.

The next source of error to consider is concerned with the mixing process in making up serial dilutions. Groat (1915) gave a guideline when he carried out some tests to track down apparent sources of error. He found that an ordinary stoppered litre flask with a long neck, about 1.5 cm in diameter, required from 25 to 30 double inversions, with time for the air to rise to the surface in each position, in order to secure a uniform mixture of 10 ml of nearly saturated salt solution with about 990 ml of distilled water. At the Water Research Centre it has been found that the addition of diluent to the flask last, or the use of magnetic or air-bubble stirring techniques, can help the process. Again, the critical procedure is to carry out the operations in duplicate or triplicate, i.e. repeats of the serial dilutions to assess both systematic and random errors. If there is little urgency for results, magnetic stirrers can be used to advantage to mix each dilution more thoroughly overnight. Also, if time is available when short-lived radioactive tracers are used, it is possible to utilize the physical decay of the isotope instead of dilution (for bromine-82 a 100-fold decrease in concentration takes 235 h) and the compromise can only be assessed by careful consideration of the errors involved in both processes. The purity of the isotope should be checked and the decay of one sample checked over the decay period used.

Groat (1915) presented tables, graphs and numerical examples to correct for the volumetric shrinkage that occurs when two salt solutions of differing densities are mixed. Although this effect may be trivial in particular cases, it should be included (but has not been in this account) in rigorous presentations of the basic dilution gauging equations and when calculating the values of comparative dilutions.

If considerable quantities of chemicals are used, the total amount of tracer to be used for one injection should be taken from the same supply batch whenever possible. Systematic errors may arise because of differences in the chemical composition of the salt used. As an example, the readings of flame photometers used to determine lithium will depend on the concentration of other substances present. When the concentrations of tracer in the samples have been found they should be plotted, as shown in Figure 4.8(a), with the analytical error bars. At this stage, one or two results may look doubtful and the samples can then be re-read. Obvious 'flyers' which may be caused by cross-contamination are also eliminated from the results at this stage. The remaining results, to be used to compute the flow, are averaged for the constant-rate method or used to calculate the area under the curve in the integration method. The area may be found using a technique compatible with the accuracy desired of the gauging, i.e. by counting squares, weighing areas cut from graph paper, by use of a planimeter, or by numerical methods with the aid of a computer.

Computer techniques are extremely valuable when short-lived radioactive tracers are used because decay corrections may be carried out as part of the computation. Another advantage is that where the results show that some tracer was still passing when sampling stopped, the area under the falling limb of the curve can be extrapolated using the last few results. Some pocket calculators are capable of performing this task, and when such predictions are necessary it is convenient to end the integration when the predicted concentration reaches three standard deviation above the mean background level. Florkowski *et al.* (1969), using tritiated water as the tracer, have shown experimentally in a number of rivers that the falling limb of tracer curves is very close to exponential form and that the extrapolation of the curve will cover an area a_t, where

$$a_t = \frac{c_t t_{1/2}}{0.693} \qquad (4.14)$$

c_t is the concentration of the last sample, and $t_{1/2}$ the half-time of the concentration decay read from the linear portion of a semilogarithmic plot of the falling limb of the tracer curve.

When dilution gauging results are compared with discharge results produced by other methods such as weirs and flumes, it must be remembered that dilution methods determine the flow of water only and this may not be the same as the flow of fluid along the channel. The fluid comprises the diluent water and suspended solids. In some parts of the world, rivers can often contain very high loads of sediment. Beverage and Culbertson (1964) have given examples where the flow consists of over 40 per cent suspended solids; they also suggest that the degree of mixing may be considerably reduced as the suspended-solids loading is increased. These considerations are of little concern if the quantity of water, as a resource, is being evaluated but clearly the density and composition of the fluid will be important when comparing data.

It was pointed out that a fundamental requirement for dilution gauging is that the flow is steady. If the discharge should change appreciably during the measurement, systematic positive or negative errors will occur because of the change in storage volume and retention time of elements of water in the measuring reach. The magnitude of the errors has been calculated by Gilman (1975) using residence–time distribution models. The error was found to be least when the discharge varies smoothly and slowly, as on the recession limb of the hydrograph, and when the mean transit time between the injection and sampling cross-sections is least. This latter finding implies advantage in minimizing the mixing distance at those sites where unsteady flows are probable. The data gathered during the fieldwork from gauging stations and stage readings along the reach should be examined to assess the degree of steadiness of flow before, at, and following the time of gauging. Gilman's work may then, if necessary, be used to estimate the errors. Similar errors will be caused if the storage volume in the reach changes for any other reasons, such as those discussed in Chapter 6.

It is important that the background sample should be representative of the background concentration of tracer during the gauging. A change in discharge may be accompanied by a change in the background concentration, or in coloration, which may be interpreted by the method of analysis as a change in the tracer concentration (colorimetric analysis is especially prone to this type of interference). Any apparent change in the colour or sediment load of the stream with time should be noted, and repeat background samples taken.

4.9 COSTS

Dilution gauging is a labour-intensive activity and the costs of tracer and equipment will, in most cases, be a relatively minor part of the total cost involved for the stages outlined in Section 4.7. During the early planning stages the objectives should be fully discussed, particularly the accuracy to be aimed for. The effect of accuracy on operations, and therefore cost, should be fairly constant for accuracies poorer than about 5 per cent but may markedly increase below this, depending on how complex the channel is. A gauging team cannot, of course, assess the accuracy actually achieved until the end of operations, but if there is a specific target, then additional fieldwork must be carried out to investigate potential tracer loss and the degree of mixing must be investigated, by extra sampling, in sufficient detail to match the limits of accuracy decided on. At the highest accuracies, additional effort becomes necessary to check the linearity, precision and calibration of items, such as balances, used during the gauging work.

Costs of the entire process involved in dilution gauging have not been discussed in the literature because of the complex issues involved. For very small channels and flows the fieldwork may represent a considerable fraction of the total cost, whereas for large systems the preparatory and analytical stages may dominate the costs. Dilution gauging accuracies of about 0.5 per cent may be taken as the lower limit attainable and relative costs in the range 0.5–5 per cent may be reflected by a proposed relationship of the form:

$$\text{dilution gauging cost for natural channels} = y(Q)\{1 + k_L(5 - \Delta)^2\} \quad (4.15)$$

where Δ is the percentage accuracy sought in the range 0.5–5 per cent; $y(Q)$ is the total gauging cost (any chosen units of currency) for routine operational procedures giving, at discharge Q, an accuracy of 5 per cent or worse; and k_L is a constant depending primarily on the length and complexity of the mixing reach.

When conditions for gauging are very good, a gauging team may achieve an accuracy of better than 5 per cent using routine procedures. On the other hand, the accuracy might be only 20 per cent, but this could prove most useful and need not reflect in any way on the competence of the gauging team. On occasions, for example, when flood or drought flows occur or when the discharge is unsteady, a result which enables the flow to be quantified even to an accuracy of 20–40 per cent can be valuable. One obvious additional task in working to a selected value of Δ will be the sampling at a second cross-section for reasons specified earlier.

The constant k_L may be taken as about 0.1 for uncomplicated small systems, higher values being appropriate for reaches through complex industrial areas or regions of uncertain surface geology, particularly where such factors are compounded with the necessity for a mixing distance of many kilometres.

The symbol for the cost of normal operations, $y(Q)$, is in the form shown because the cost will clearly be a function of discharge, since times-of-travel and tracer passage times will be greater and almost every stage of the procedures outlined will take longer as the discharge increases. One component of $y(Q)$ will be tracer cost and Figure 4.9 indicates in a general way the variation of tracer costs with discharge. Actual examples of costs will not be given as they become outdated so rapidly. It is not possible to give an accurate idea of where the cross-over between radioactive and other tracer costs may be, but the cost and convenience in handling chemicals may become very much less attractive in the range 50–500 m^3/s. For guidance on the order of magnitude for short times-of-travel, about 100 mCi of radiotracer, 10 g Rhodamine-WT, 20 per cent solution, 100 g lithium and kilogram quantities of common salt per cubic metre per second of flow will be required.

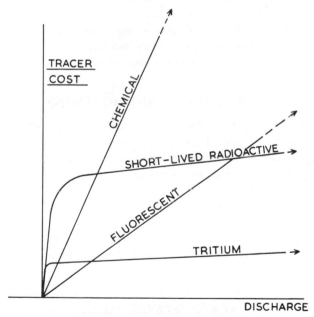

Figure 4.9 The cost of various types of tracer as a function of discharge

Equipment costs, averaged over a suitable number of years, are often small compared with labour costs. The items of capital equipment which may be considered essential, for highly accurate gauging, are two top-loading balances (say £3000 total at 1978 prices), tracer detection equipment (say £10 000), equipment with portable power supplies for injection and sampling (say £2000), and instruments for water level and velocity determinations (say £2000). Most of these items will be useful for other purposes as well as dilution gauging. It must be emphasized that the items listed above as a guideline apply to cases where the accuracy will not be limited by the equipment. It is possible to conduct gauging with the most elementary equipment at small discharges. A range of ancillary apparatus is also necessary, including sample containers in quantity, timing devices, ropes, tubing, chain, extension leads, life-jackets, and other safety equipment, protective clothing, etc. A boat with trailer, engine, etc., may be required. Also, in order to prevent any chance of cross-contamination, separate vehicles should be used to transport solutions, samples and equipment used at the injection and sampling stations. The costs of some of these supporting items for fieldwork is sometimes surprising; for example, a single length of rope or plastic tubing can each cost about £100.

The maintenance costs attributable to dilution gauging are mostly small because much of the equipment is used for other purposes. When *in situ* detection equipment is used, there can be considerable maintenance costs associated with problems of underwater connectors and general insulation breakdown problems, particularly in freezing or wet-weather conditions. The major aim must be to keep equipment dry and clean in covers to prevent sand or dust, and airborne salt particles in coastal areas, getting into the apparatus.

4.10   CONCLUDING REMARKS

It has been shown that there are many combinations of techniques that may be used for dilution gauging. There are two basic injection techniques, several sampling techniques and a large number of possible tracers of three main types—chemical, fluorescent, and radioactive. Gauging teams tend to have their own preferred methods which they know suit their range of conditions best, and it is not possible to recommend any particular combination of techniques. However, on balance, more information is gathered using the constant-rate injection method and, where this method is possible, the chance of serious loss of accuracy should be less than when the integration method is used.

ACKNOWLEDGEMENTS

This chapter is published with the permission of the Director of the Water Research Centre UK. The author acknowledges the contributions made by many past and present colleages at the centre, and by fellow members of BSI and ISO committees in discussions on the principles and practices of flow gauging.

REFERENCES

American Public Health Association, American Water Works Association, and Water Pollution Control Federation, 1975. *Standard Methods for the Examination of Water and Wastewater*, 14th edn, APHA, Washington.

Bansal, M. K., 1971. Dispersion in natural streams, *J. hydr. Div. Am. Soc. civ. Engrs*, **97**(HY11), 1867–1886.

Barnes, H. H., 1967. Roughness characteristics of natural channels, US Geol. Surv. Wat. Supp. Pap. No. 1849, US Govt Printing Office.

Beverage, J. P. and Culbertson, J. K., 1964. Hyperconcentrations of suspended sediment, *J. hydr. Div. Am. Soc. civ. Engrs*, **90**(HY6), 117–128.

British Standards Institution, 1967. Methods of measurement of liquid flow in open channels: Part 2, Dilution methods; 2C, Radioisotopes techniques, BS 3680; Part 2c, London (revised version at advanced stage of preparation—as ISO 555/111).

British Standards Institution, 1974. Methods of measurement of liquid flow in open channels: Part 2, Dilution methods; 2A, Constant-rate injection, BS 3680; Part 2A, London.

Chase, G. D. and Rabinowitz, J. L., 1967. *Principles of Radioisotope Methodology*, 3rd edn, Burgess, Minneapolis.

Cole, J. A., 1969. Dilution gauging by inorganic tracers; notably the plateau method using dichromates, Symposium on River Flow Measurement held at Loughborough University of Technology, (10–11 September), Institution of Water Engineers, London.

Day, T. J., 1975. Longitudinal dispersion in natural channels, *Wat. Resour. Res.*, **11**, 909–918.

Department of the Environment, 1971. *Water Pollution Research 1970*, p. 61, HMSO, London.

Department of the Environment, 1976. *Water Pollution Research 1973*, pp. 94–98, HMSO, London.

Eden, G. E., Downing, A. L. and Wheatland, A. B., 1952. Observations on the removal of radioisotopes during purifictation of domestic water supplies: 1. Radio-iodine, *J. Instn Wat. Engrs*, **6**, 511–532.

Fischer, H. B., 1967. The mechanics of dispersion in natural streams, *J. hydr. Div. Am. Soc. civ. Engrs*, **93**(HY6), 187–216.

Fischer, H. B., 1968. Dispersion predictions in natural streams, *J. sanit. Engng Div. Am. Soc. civ. Engrs*, **94**(SA5), 927–943.

Florkowsi, T., Davis, T. G., Wallander, B. and Prabhakar, D. R. L., 1969. The measurement of high discharges in turbulent rivers using tritium tracer, *J. Hydrol.*, **8**, 249–264.

Gilman, K., 1975. Application of a residence-time model to dilution gauging with particular reference to the problem of changing discharge, *Bull. IAHS*, **20**, 523–537.

Groat, B. F., 1915. Chemi-hydrometry and its application to the precise testing of hydro-electric generators, *Proc. Am. Soc. civ. Engrs*, **41**, 2103–2427.

International Atomic Energy Agency, 1968. *Guidebook on nuclear techniques in hydrology*, Tech. Rep. Ser. No. 91, IAEA, Vienna.

International Commission on Radiological Protection, 1959. Publication 2, Recommendations of the international commission on radiological protection, Report of Committee II on permissible dose for international radiation, Pergamon Press, Oxford.

International Committee on Radiation Units and Measurements, 1972. Measurement of low-level radioactivity, ICRU Rep. No. 22, Washington.

International Organization for Standardization, 1973. Liquid flow measurement in open channels; Dilution methods for measurement of steady flow; Constant-rate injection method, ISO: 555/I, Switzerland.

International Organization for Standardization, 1974. Liquid flow measurement in open channels; Dilution methods for measurement of steady flow, Part II, Integration (sudden injection) method, ISO: 555/II, Switzerland (revised version under preparation).

International Organization for Standardization, 1978. Liquid flow measurement in open channels, Vocabulary and symbols, ISO: 772, Switzerland.

Joly, J., 1922. On a new method of gauging the discharge of rivers, *Proc. R. Soc., Dublin*, **16,** 489–491.

Liu, H., 1977. Predicting dispersion coefficient of streams, *J. envir. Engng Div. Am. Soc. civ. Engrs*, **103**(EE1), 59–69.

McQuivey, R. S. and Keefer, T. N. 1976a. Convective model of longitudinal dispersion, *J. hydr. Div. Am. Soc. civ. Engrs*, **102**(HY10), 1409–1424.

McQuivey, R. S. and Keefer, T. N., 1976b. Dispersion—Mississippi river below Baton Rouge, La., *J. hydr. Div. Am. Soc. civ. Engrs*, **102**(HY10), 1425–1437.

Neal, C. and Truesdale, V. W., 1976. The sorption of iodate and iodide by riverine sediments: Its implications to dilution gauging and hydrochemistry of iodine, *J. Hydrol*, **31,** 281–291.

Rimmar, G. M., 1960. Use of electrical conductivity for measuring discharges by the dilution method, National Engineering Laboratory Translation No. 749 (from *Trudy GGI*, **36**(90), 18–48).

Smart, P. L. and Laidlaw, I. M. S., 1977. An evaluation of some fluorescent dyes for water tracing, *Wat. Resour. Res.*, **13,** 15–33.

Truesdale, V. W. and Smith, P. J., 1975. The automatic determination of iodide or iodate in solution by catalytic spectrophotometry, with particular reference to river water, *Analyst, Lond.*, **100,** 111–123.

US Geological Survey, 1965. Measurement of discharge by dye-dilution techniques, chap. 14, *Surface Water Techniques*, US Geol. Surv., Washington DC.

Watt, D. E. and Ramsden, D., 1964. *High Sensitivity Counting Techniques*, Pergamon Press, Oxford.

White, K. E., Belcher, A. S. B. and Lee, P. J., 1975. Absolute flow measurements to obtain depth against discharge for sewers using bromine-82 and lithium as tracers, Conference on Fluid Flow Measurements in the Mid-1970s, National Engng Lab., Glasgow.

Wimpenny, J. W. T., Cotton, N. and Statham, M., 1972. Microbes as tracers of water movement, *Wat. Res.*, **6,** 731–739.

CHAPTER 5

Indirect Methods

H. H. BARNES, Jr. and J. DAVIDIAN

5.1   INTRODUCTION

5.1.1   Overall objectives of surface-water data collection

The broad objective of any governmental agency concerned with the surface-water resources of its jurisdiction is a general knowledge of the occurrence and the availability of surface water. This general knowledge must be based on individual gaging station records, either of complete or of partial-record stations, and further supplemented by any additional data needed to define unusual events such as floods or droughts, wherever they may occur.

Essential to any comprehensive analysis of water resources is the establishment of a well-defined stage–discharge rating curve throughout the entire spectrum of stages (discharges) experienced or expected at each gaging station. Not only is it necessary to convert recorded stages to discharges at each site, but there is an increasing need to determine stages corresponding to a range of discharges of selected magnitude, such as the 10-, 25-, or 100-year flood.

In addition to considerations such as expected useful life and cost analyses, the design of many structures such as bridges, culverts, or dams requires information on the magnitude as well as frequency of certain discharges. Information on the stages corresponding to these discharges is also needed for channel conditions both with and without the constricting structure in place. In the case of road embankments, the stage information is needed for a consideration of cheaper designs of lower elevation, which would permit occasional inundation with limited damage. Builders and operators of reservoirs and power stations require reliable stage–discharge information, also.

There is an increasing demand for information on high stages corresponding to large discharges. Flood maps, showing areas of inundation, are one important product of such information. Many local communities base land-use, building, or tax regulations pertaining to flood plains or other areas affected by high water on such inundation maps. Flood-insurance rates are also necessarily based on the stage, and resultant inundation, caused by a discharge of a certain magnitude such as the 100-year flood.

Specifically, then, an important part of the job of evaluating surface-water resources consists of developing the upper ends of stage–discharge rating curves at gaging stations and the determination of peak discharges for historical floods at ungaged sites.

5.1.2 Deficiencies of flood records

The job of measuring peak stages and of defining the upper portions of stage–discharge rating curves has been somewhat neglected until recent years.

Peak discharges occurring during severe localized floods have not always been measured, particularly those not occurring at gaging stations. This lack has been due to (1) inadequate manpower to deal with extraordinary conditions during the flood period, and (2) the limited manpower, resources, and techniques for post-flood methodology of measuring discharge.

Stage–discharge relations at the upper end of rating curves at gaging stations are not as well defined as the median portions of the ratings. A significant effort is expended in this analysis of intermediate-stage records. At the same time ratings are frequently extended logarithmically far beyond the highest measurement, and high-stage discharges have been published with little consideration given to the possible error. Yet many of the most important uses of surface-water records are concerned only with the high stages and discharges.

5.2 REASONS FOR MAKING INDIRECT MEASUREMENTS

Measurements of discharge by direct methods, such as current meters, are still the best and cheapest way of measuring high discharges. There are, however, some very good reasons why indirect measurements are either desirable or necessary:

(1) Many intense local storms occur and are past before streamgagers learn of them.
(2) Even with warning of flood conditions, there is usually a lack of sufficient personnel needed to cover the necessary peaks both at regular gaging sections and at miscellaneous points during the time that the water remains high.
(3) During extreme floods measuring sites for current-meter measurements may be either lacking or inaccessible.
(4) On small drainage areas, where peaks are likely to be flashy, it may be difficult or even impossible to obtain a reliable current-meter measurement, even though a streamgager were to be present with equipment set up at the time. The growing attention to, and emphasis on, gaging small catchments makes it necessary to resort to other methods for measuring such peaks.

(5) Gaging stations' are frequently established at some time after an important flood, and it is desired to recover this previously occurring peak discharge. An indirect method of measuring would be the only means of doing so.

(6) It is desirable to recover an extreme flood at a discontinued station for the added regional coverage this information would provide.

The necessity for indirect methods of measurement becomes apparent following any widespread flood of unusual magnitude. It becomes nearly impossible to carry on the normal streamgaging program in the flooded area. Travel and communication during and immediately following the flood period come nearly to a standstill; even if it is possible to reach gaging sites, the structures from which measurements would normally be made might be destroyed, or otherwise inaccessible. Consequently, few direct measurements of peak, or near-peak, flows are made. For floods of such extreme and widespread proportions, however, it is imperative that discharges are adequately defined.

5.3 ALTERNATIVES TO CURRENT-METER MEASUREMENTS

Where it has been impossible or infeasible to measure peak discharges directly, there are various alternatives available to compute a peak flow following a flood event.

The next best thing to the 'direct' measurement (the current-meter measurement), is the 'indirect' measurement. This latter is indirect only in the sense that it is made after the flood has passed. It utilizes known principles of hydraulic engineering and actual data collected in the field, which represent conditions at the time of the peak. The various types of such indirect measurements will be discussed in detail later.

The most direct of these indirect methods, such as the velocity-area, the slope-area, contracted-opening, flow-through-culverts, and flow-over-dams measurements, are the most valuable in that they use field data at the very site and apply established hydraulic principles to that data to obtain a discharge.

Other methods of obtaining peak discharges are progressively less direct and rely on progressively less actual field data. Among these are the conveyance-slope method, based on peak gage height, channel cross-section properties, and current-meter measurements at lower stages. The step-backwater method utilizes channel cross-section properties to establish or extend stage–discharge rating curves. Flood-routing is another method of computing a peak discharge at one point in a river by transferring a known peak at another point in that river to that site.

Other methods, almost wholly indirect, involve correlations with peak discharges at other locations or other streams. The unit-hydrograph method uses rainfall records and past discharge hydrographs. In recent years, various types of so-called rainfall-runoff models are being used to extend rating curves at both undeveloped and urban basins.

5.4 VELOCITY-AREA METHOD

The velocity-area method of determining flood discharge is used at the regular high-water measuring cross-section. It is necessary to determine the peak stage at this section from an analysis of high-water marks along both banks of the stream in the vicinity. The prime requisite is that the stream bed be known to be fairly stable at high stages. This can be done by comparing cross-sections from current-meter measurements at various stages.

The procedure is to establish a stage–area curve up to the flood stage, and to develop a stage–velocity curve up to the highest discharge measurement and then extend it, also, to the flood stage. These curves may be developed as bulk values for the entire cross-section, by using total cross-sectional area, and mean cross-sectional velocity. They may also be developed for individual subsections of the cross-section if sufficient and reliable current-meter data are available. The product of the values of area and velocity selected from their respective curves at the flood stage, or any other stages, gives the corresponding value of total cross-section discharge, or of individual subsection discharges which can be added together.

In establishing the area and velocity curves, all available determinations of area and velocity from discharge measurements made at the high-water measuring section are plotted to the same scale of gage heights and each curve is developed up to, or slightly beyond, the point where it is defined by measurements. The areas for high stages are computed from the cross-section profile of the measuring section. The extension of the velocity curve above the point of definition by measurements is based on studies of flow characteristics, as described by Corbett *et al.* (1945). These characteristics are the slope of the stream bed or water surface; the presence of contractions, bends, dams, and vegetal growth; the possibility of backwater from downstream tributaries; the extent of overflow areas; the depth of water on such areas.

The successful application of this method requires a thorough analysis and careful appraisal of fundamental flow factors and a knowledge of the channel conditions at the river-measurement station, especially in regard to the manner in which the width and depth of the channel vary with changes in stage; also, the factors affecting changes of velocity with changes in stage, such as rapids or falls, which may tend to increase velocities, or contractions of channel downstream from the gage, which may tend to decrease velocities. The cross-section of the channel at flood stage, including all overflow and bypass channels, should be determined by instrumental surveys. Pertinent conditions, such as backwater from downstream tributaries and changing influences of contracted sections of the channel below, must be visualized and their effects appraised as accurately as possible from available information.

The conditions most favourable for the accurate extension of a rating curve consist of well-defined rapids or riffles below the station at all stages and a uniform increase of channel cross-section as the stage increases, without abrupt changes in area or addition of overflow channels—in other words, a general uniformity of those channel conditions which control the stage and discharge relations at the gage.

Radical changes in downstream conditions that control the velocity or stages, or abrupt increases in the area of channel cross-section such as result from overflow banks, may interfere seriously with the reliable application of these methods of analysis.

The reliability of discharge computations using this method decreases as the extension of the velocity curve is carried far beyond the range defined by current-meter measurements. Methods for extending a stage–discharge curve are also presented in Sections 5.11 and 5.12.

5.5 GENERAL FEATURES OF INDIRECT MEASUREMENTS

5.5.1 Common attributes

Indirect measurements of flood discharge are made following the actual occurrence of the peak discharge. This is possible only because floods generally leave high-water marks which enable the recovery of the water-surface profiles at the time of the peaks. Once the profiles are determined from a survey of high-water marks, the problem is one in the hydraulics of flow.

The general hydraulic principle exists that in any reach of channel there is a relation between discharge, the water-surface profile, and the hydraulic characteristics of the channel. This holds for all types of channel, although the relationships are better defined for the simpler types. A simple type, for example, would be a uniform slope-area reach; a complex type might be a drop-inlet structure. The resulting discharge is more reliable for those conditions which are better defined by theory and experience.

The general procedure in any type of indirect measurement is (1) to make a survey of high-water marks to define the slope or fall between two points, (2) to take cross-sections and otherwise detail the channel geometry, and (3) to define the roughness of the reach.

The water-surface profile is an important element in an indirect measurement. Not only must its elevation be known, but even more important, its changes within the reach of channel under consideration. These changes in profile are primarily the result of (1) energy losses due to bed roughness, runup, drawdown, etc., and (2) acceleration. For a relatively uniform reach of stream channel, the change in water-surface profile results largely from bed roughness. At sudden contractions, such as bridges, culverts, and dams, the surface reacts to the influence of acceleration; that is, the change in profile reflects primarily a change in energy from potential to kinetic.

The channel geometry is adequately measured in the case of slope-area measurements by cross-sections; for other measurements a complete survey of all physical details, such as bridges, abutments, piers, culverts, and dams, must also be included.

In addition to the large-scale geometric features of the channel, the retarding features of the channel grouped under 'roughness' must be evaluated. Selection of the roughness coefficient (n) is a critical step in a slope-area measurement, so much so because personal judgement cannot be entirely eliminated. Moreover, for a natural channel the coefficient is descriptive of losses other than bed roughness. Bank irregularities, channel meanders and curvature, overhanging trees, and many other retarding influences that defy quantitative description, are present. Tables of n values, as found in many hydraulic texts, provide little assistance to the inexperienced, unless very large errors are permissible.

A useful and reliable reference index of n values has been compiled by the US Geological Survey (see Barnes, 1967). For streams representing a wide variety of conditions, and where peak discharges were known, slope-area measurements were obtained for the purpose of computing n values. Comprehensive photographic coverage in colour was obtained for each definition reach. The resultant file of n-verifications enables the less experienced engineer to select an n value for a channel under consideration by a near-realistic and visual comparison of that channel with similar channels having defined coefficients. Values of n ranging from 0.028 to 0.075 are included in this reference file.

An additional source of useful information on values of n is Chow (1959). This reference includes black-and-white photographs, tables of ranges of typical roughness coefficients for a wide variety of materials and channel beds, and a very comprehensive discussion of various factors that influence the value of n. These factors include surface roughness; vegetation; channel irregularity; channel alignment, bends, and meanders; size and shape of channel stage and discharge; and the effects of silting, scouring, suspended material, and bed load.

A study by Limerinos (1970) shows that values of n in stream channels may be related to some characteristic size of the stream-bed particles, and to the distribution of particle sizes. These two elements involving particle size are combined into a single element weighting characteristic particle sizes. The investigation was confined to channels with coarse bed material to avoid the complication of bed-form roughness that is associated with alluvial channels composed of fine bed material.

5.5.2 Indirect measurements classified

Indirect methods of river discharge measurement are grouped into five major categories for ease of application and reference. These are: (1) slope-area, (2) contracted-opening, (3) flow-through-culverts, (4) flow-over-dams and embankments, and (5) critical-depth method. The five classifications, although somewhat arbitrary, have been found convenient in setting up field and office procedures. Occasionally an indirect measurement will involve a combination of methods or another method of solution outside of these general classes.

The slope-area method is the most frequently used, especially on the large rivers, primarily because a reach of channel suitable for a slope-area reach can usually be found. Contracted-opening, culvert, dam, or critical-depth sites are used whenever conditions are favourable. The change in water-surface profile through a slope-area reach results primarily from channel roughness, and hence the ability to select proper roughness coefficients is a measure of the accuracy of the computed discharge. The other methods involve abrupt contractions or changes in stage—hence, the changes in water-surface profile reflect mainly changes in energy form and the value of the roughness coefficient becomes less important.

5.6 SLOPE-AREA METHOD

5.6.1 Theoretical basis

In the slope-area method, discharge is computed on the basis of a uniform-flow equation involving channel characteristics, water-surface profiles, and a roughness or retardation coefficient. The change in water-surface profile for a uniform reach of channel represents losses caused by bed roughness. In this section the theory and applicaton of the slope-area method for the measurement of peak flows is presented. In Chapter 9 an examination of the method is made for continuous measurement in a river having severe aquatic growth, and in Chapter 13 the use of the method to measure past flood peaks in arid regions is discussed.

In application of the slope-area method, any one of the well-known variations of the Chezy equation might well be used. Perhaps the most commonly used of these is the Manning formula. This formula was originally adopted because of its simplicity of application. Many years of experience in its use have now been accumulated. That it has become a medium through which reliable results can be obtained justifies its use.

The Manning eauation, written in terms of discharge, is

$$Q = \frac{1}{n}AR_h^{2/3}S^{1/2} \qquad \text{(in metric units)} \qquad (5.1a)$$

$$Q = \frac{1.486}{n}AR_h^{2/3}s^{1/2} \qquad \text{(in imperial units)} \qquad (5.1b)$$

where Q is the discharge; A the cross-sectional area; R_h the hydraulic radius; S the friction slope; and n is the roughness coefficient.

Manning's formula, as originally developed, was intended only for uniform flow where the water-surface profile is parallel to the stream bed and the area and hydraulic radius remain constant throughout the reach. In spite of these limitations, and lacking a better solution, the formula is used for the non-uniform reaches that are invariably encountered in natural channels. The only justification for such use is that of necessity; however, several factors in the formula are modified in an attempt to correct for non-uniformity.

A description of the method used in computing river discharge by the slope-area method is discussed in conjunction with Figure 5.1, a definition sketch of a two-section reach which is gradually contracting in the direction of the flow. Cross-sections 1 and 2 are selected as being representative of the channel between 1 and 2.

The energy equation for a reach of non-uniform channel between section 1 and section 2 shown in Figure 5.1 is

$$(h + h_v)_1 = (h + h_v)_2 + (h_f)_{1-2} \quad k\,(\Delta h_v)_{1-2} \qquad (5.2)$$

where h is the elevation of the water surface at the respective sections above a common datum; h_v the velocity head at the respective section $= \alpha V^2/2g$; h_f the energy loss due to boundary friction in the reach; Δh_v the upstream velocity head minus the downstream velocity head, used as a criterion for

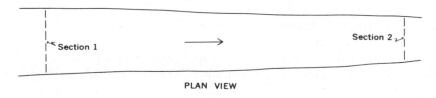

PLAN VIEW

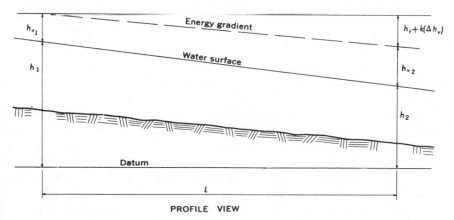

PROFILE VIEW

Figure 5.1 Definition of a slope-area reach

expansion or contraction of reach; $k(\Delta h_v)$ the energy loss due to acceleration or deceleration in a contracting or expanding reach; and k is the energy loss coefficient.

The term $(1.486/n)AR_h^{2/3}$, or $(1/n)AR_h^{2/3}$ (metric units), contains the several factors descriptive of channel characteristics and is defined as 'conveyance' (K). For the reach 1–2, because it is not truly uniform, conveyance is expressed as the geometric mean of the conveyances of the two sections:

$$K = \sqrt{(K_1 K_2)} \tag{5.3}$$

For the contracting reach, the drop in water-surface profile (Δh) is not entirely a result of friction loss (h_f), but also reflects the acceleration between sections 1 and 2. The friction slope (S) to be used in the Manning equation is thus defined as

$$S = \frac{h_f}{L} = \frac{\Delta h + \Delta h_v - k(\Delta h_v)}{L} \tag{5.4}$$

where h is the difference in water-surface elevation at the two sections, and L is the length of the reach.

The discharge formula, therefore, may be condensed to

$$Q = \sqrt{(K_1 K_2 S)} = K\sqrt{(S)} \tag{5.5}$$

where K represents the average conveyance of the reach, and S the energy gradient, or friction slope.

In this manner Manning's formula is applied to gradually varied flow in natural channels. Admittedly, assumptions have been made that greatly oversimplify the complex functions inherent in non-uniform flow. This treatment, however, provides a logical method to which experience can be related, and when a given problem is within the range of accumulated experience, it has been found that reliable results can generally be obtained.

The practice of using reaches that are contracting gradually and to a moderate degree in the direction of flow is the recommended procedure. Contracting reaches have been found to yield a consistent scale of values in the roughness-investigation programme (Barnes, 1967). On the other hand, expanding reaches have not, probably because of unknown energy losses and velocity distributions associated with diverging flow patterns. Uniform reaches are not intentionally avoided, but in reality can rarely be found. By searching for slightly contracting reaches, the engineer can at least avoid expanding reaches. Contracting reaches also are favoured for other reasons. For example, any errors involved in estimating n affect the computed discharge to a lesser degree than in expanding reaches, owing to the effect of the velocity-head correction involved in computing the energy gradient (S).

The computation of the slope by equation (5.4) involves the determination of the water-surface elevation and velocity heads at each section and an evaluation of the loss due to contraction or expansion. Water-surface elevations are taken from the profile defined by high-water marks as described earlier. The velocity head (h_v) at each section is computed as

$$h_v = \frac{\alpha V^2}{2g} \qquad (5.6)$$

where V is the mean velocity in the section, and α is the velocity-head coefficient. The value of α is assumed to be 1.0 if the section is not subdivided. The value of α in subdivided channels is computed as

$$\alpha = \frac{\sum K_i^3 / a_i^2}{K_T^3 / A_T^2} \qquad (5.7)$$

where the subscript i refers to the conveyance or area of the individual subsections, and T to the area or conveyance of the entire cross-section.

The energy loss due to contraction or expansion of the channel in the reach is assumed to be equal to the difference in velocity heads at the two sections (Δh_v) times a coefficient k. If Δh_v is zero or negative, the reach is considered to be contracting, with full energy recovery. Thus, the coefficient k is given a value of zero. If Δh_v is positive, the reach is considered to be expanding, with a partial recovery of energy, where k is ordinarily taken to have a value of 0.5 for 50 per cent energy recovery. Both this procedure and the value of the coefficient k are questionable, however, in the case of expanding reaches and thus major expansions are avoided, if possible, in selecting sites for slope-area measurements.

The value of Δh_v is computed as the upstream velocity head minus the downstream velocity head, and thus the friction slope to be used in the Manning equation is computed algebraically as

$$S = \frac{\Delta h + \Delta h_v (1 - k)}{L} \qquad \text{(when } \Delta h_v \text{ is positive)} \qquad (5.8a)$$

and

$$S = \frac{\Delta h + \Delta h_v}{L} \qquad \text{(when } \Delta h_v \text{ is negative)} \qquad (5.8b)$$

5.6.2 Selection of site

The most important element of a slope-area measurement is the choice of a suitable reach. Proximity to the gaging station may occasionally be the deciding factor between a good reach located so far as to have substantial tributary inflow and channel storage problems, and a less desirable reach having the advantage of being closer to the gage.

The next consideration for a reliable slope-area computation is the quality of the high-water marks. An otherwise ideal reach having excellent hydraulic qualities is worthless without good high-water marks. Some good reaches are sometimes rendered useless when heavy rains follow the peak and destroy marks, or when extensive channel work is done by bulldozers during removal of debris or of sand and gravel depositions.

A reach should be as uniform as possible in shape and conveyance throughout its length; it should be slighly contracting rather than expanding; it should be as straight as possible, including the approach and getaway channel; it should be free of any possible effects of channel configuration or of man-made structures, both upstream and downstream, that could affect the distribution of flow within the study reach. The entire reach should be either wholly within the subcritical-flow regime or wholly within the supercritical-flow regime, with the water-surface profile never crossing the critical-depth line in a hydraulic drop or a hydraulic jump. Because of the wavy nature of the water surface in flows that are near critical, or are in the supercritical range, the high-water profile may be difficult to find or to interpret; it is preferable, therefore, to choose a reach in which the flow is subcritical.

If a reach with bends is unavoidable, the presence of the bends will influence the total reach length. It is best, then, to select a reach such that it includes one or more channel bends, with end sections in the straight portions of the channel.

The flow should preferably be confined to a simple trapezoidal channel because *n*-verification studies have been done in such channels. Compound channels can be used, however, if they are properly subdivided. The roughness values in the reach should be within the range of those that have been verified (Barnes, 1967).

The selected reach should have a fall greater than the range of error due to alternate interpretations of the high-water profile, or to uncertainties regarding the computation of velocity head.

Although the accuracy of a slope-area measurement increases as the reach length increases, there are other considerations for the total length: the geometry of the channel, the denseness of foliage, and the practical difficulties of surveying long reaches of river channel.

At least one of the following criteria should be met in the selection of the slope-area reach:

(a) The length of the reach should be equal to or greater than 75 times the mean depth in the channel.
(b) The fall in the reach should be equal to or greater than the velocity head.
(c) The fall in the reach should be equal to or greater than 0.15 m.

5.6.3 High-water profiles

The high-water marks should be chosen on banks along lines parallel to the flow so that they represent the water surface and not the energy grade line. They should be avoided where the probability of local effects such as pile-up or drawdown exist. Marks on the ground or those left by quiet water are best. They should be taken at the highest level reached (not at lower levels left by the receding water).

The best types of mark include drift, silt lines, wash lines, seed lines on trees or poles, silt or stain lines on and in buildings, etc. Where the debris consists of mounds of material, the top of the mound should be used if the material is consolidated, and the shoreward toe of the mound if the material is loose.

Marks should be located through the reach, for a considerable distance above and below the reach, and particularly in the vicinity of the ends of cross-sections.

Regardless of where the actual reach is located with respect to the gaging station, high-water marks should also be found in the immediate vicinity of the gage installation.

5.6.4 Selection of cross-sections

Cross-sections are placed at intervals throughout a total length of reach to represent samples of the geometry and conveyance. A minimum of three cross-sections through the total reach is recommended. The more cross-sections there are, the better the channel conditions are represented. The purpose is to subdivide the total reach into several sub-reaches that are, between any two cross-sections, as practicably uniform as possible insofar as shape and conveyance are concerned. They should be located to enable accurate evaluation of energy losses. With reference to Figure 5.1, which represents a typical sub-reach, the cross-sections should be located at such intervals that the energy gradient, the water-surface profile, and the stream-bed slope are all as nearly parallel to each other, and as close to being straight lines, as possible. If any channel feature causes one of these three profiles to curve, break, or run unparallel to the others locally, this is a clear indication that the particular sub-reach should be further subdivided with another cross-section (the exception being, of course, that a slight contractions is desirable, and preferable to a slight expansion).

With the foregoing in mind, the engineer should first plot the high-water and the thalweg profiles and make his on-site selection of cross-section locations, giving consideration to several general criteria discussed below.

The cross-sections should be as nearly as possible at right angles to the flow. On large streams it may be necessary to break the cross-section at one or more points, so that each subsection will be oriented roughly perpendicular to the flow in that part of the cross-section.

Cross-sections should be located at major breaks both in the high-water and in the thalweg profiles. They should also be placed at points of minimum and maximum cross-sectional width and/or area.

If the high-water profiles suggest a series of regular waves, cross-sections should be located at comparable parts of the waves—both at the crest, or both at the trough. Similarly, if the stream meanders, the cross-sections should be located at comparable parts of the bends.

It is assumed that discharge is distributed across the section in accordance with the distribution of conveyance. This means that the cross-section is fully effective and that there are no parts of the cross-section that are extremely sluggish, have dead water, or even have negative velocities. For this reason, the ends of the reach should not be near man-made or natural constrictions which would cause expansions, eddying, or backwater effects that would affect the distribution of flow. A sudden deepening of the channel may also represent non-effective area.

Cross-sections should be placed more frequently in reaches where the conveyance changes greatly as a result of changes in width, depth, or roughness. Because friction losses within sub-reaches are computed with a geometric-mean conveyance, equation (5.3), the relation between the conveyances at ends of a sub-reach should satisfy equation (5.9):

$$0.7 < (K_1/K_2) < 1.4 \qquad\qquad (5.9)$$

Sharp changes in shape, even though total area and roughness are very similar, may call for additional cross-sections because of the change in hydraulic radius. More cross-sections could also be required in a reach where the end sections are identical in total area, conveyance, and shape, but nevertheless have a very large difference in the way the conveyance is distributed across each section. If one of the cross-sections requires more subdivision than the other, its velocity-head coefficient, equation (5.7), could be several times larger than that of the other, thus influencing the respective velocity-head terms, and the computed value of discharge.

If a reach containing a tributary is unavoidable, the cross-sections should be placed such that the tributary enters the middle of a sub-reach.

The criteria above suggest that the effects of almost all the undesirable features of non-uniform, natural stream channels can be lessened by taking more cross-sections. Whereas this is true in that the channel geometry would be better described, consideration must also be given to the time, cost, and effort necessary to survey additional cross-sections, and the expected gain in accuracy of the computed discharge. A balance must be set between the number deemed desirable to have, and the number deemed practical to provide. These criteria, then, serve to call attention to the considerations behind the need for them, and to help the engineer exercise his judgement.

5.6.5   Additional field information

Cross-sections are useful only when the high-water elevations are well defined at the ends of the section. After the locations of the sections are chosen, enough high-water marks should be found to assure this.

A field sketch, and descriptions of the stream bed, banks, flood plains, and the types of roughness, and values of n should all be part of the field data. Many photographs, showing approach and getaway conditions, individual sub-reaches, and each cross-section are also necessary.

Values of n (Barnes, 1967) were determined for unit channels of trapezoidal shape. Assignation of n values should, therefore, be for similar cross-sectional shapes. Rather than assign different roughness values to parts of a trapezoidal-shape cross-section—one value of n for the centre portion and streambed, and another value of n for the banks—the engineer should visually 'weigh' the various types of roughness and determine a composite n value for the unit channel.

Sometimes the shape of the channel is such that a subdivision of the cross-section is required, as, for example, an overbank, flood-plain panhandle in conjunction with the main-channel subsection. Occasionally, however, it is difficult to decide whether the overbank portion does indeed require subdivision. There are two useful criteria applicable to such a situation; either one can be used:

(1) The overbank distance, L, is measured, along the cross-section from the edge of the water to the edge of the main channel. The depth, y, is measured at the edge of the main channel where overbank flow begins. If the ratio L/y is equal to or exceeds 5, the overbank portion should be subdivided.
(2) The maximum depth in the main channel, D_{max}, is compared with the average overbank depth, d_b. If $D_{max} \geq 2d_b$, then the overbank portion is subdivided.

Subdivision for changes in roughness, alone, is permissible on a wide flood plain, as, for example, between a cultivated field and a stand of heavy brush and trees. As long as the individual subsections result in 'rectangular' shapes in which the depth is fairly constant (the hydraulic radius is approximately equal to the depth), such subdivisions should be done.

The importance of proper subdivision cannot be overemphasized. Not only would inappropriate values of hydraulic radius and conveyance be computed, but there can be a significant effect on the velocity-head coefficient, equation (5.7), as well. Points of possible subdivision, as well as n values and photographs, should all be part of the field survey.

5.6.6 Office procedures

After high-water profiles and cross-section data are plotted, all cross-section properties (conveyance, area, velocity-head coefficient, hydraulic radius), and the fall in water-surface elevation between all cross-sections, are determined. The length of each sub-reach is determined from a plan plot of the reach. If the channel is curving and is within the main channel, the length is measured on the curve along the centreline of the channel. At bends, and for cross-sections having overbank flows, if either a straight line or the curved line would not be appropriate, the centroid of conveyance can be computed for each cross-section and for each subsection. Water that is entirely within the main channel would have the main channel length; water flowing along the flood plains would travel a more direct route between sections. By weighting the different lengths in proportion to the approximate amount of water flowing in each portion, or in proportion to the conveyances of the subsections, the engineer can compute a representative length for the reach.

The solution for discharge is based on either the Manning formula, equation (5.5), or the energy equation, equation (5.2).

This solution is implicit because the unknown discharge is on both sides of the equation; a trial-and-error procedure is required. The discharge term can be isolated for a direct solution as shown in Table 5.1.

Table 5.1 Discharge equations for use in slope-area measurements

Number of cross-sections	Equation
Two sections	$Q = K_2 \sqrt{\dfrac{\Delta h}{\dfrac{K_2}{K_1}L + \dfrac{K_2^2}{2gA_2^2}\left[-\alpha_1\left(\dfrac{A_2}{A_1}\right)^2(1-k) + \alpha_2(1-k)\right]}}$
Three sections	$Q = K_3 \sqrt{\dfrac{\Delta h}{\dfrac{K_3}{K_2}\left(\dfrac{K_3}{K_1}L_{1.2} + L_{2.3}\right) + \dfrac{K_3^2}{2gA_3^2}\left[-\alpha_1\left(\dfrac{A_3}{A_1}\right)^2(1-k_{1.2}) + \alpha_2\left(\dfrac{A_3}{A_2}\right)^2(k_{2.3}-k_{1.2}) + \alpha_3(1-k_{2.3})\right]}}$
Multiple (n) sections	$Q = K_n\sqrt{\left(\dfrac{\Delta h}{A+B}\right)}$, where

$$A = K_n^2\frac{L_{1.2}}{K_1 K_2} + K_n^2\frac{L_{2.3}}{K_2 K_3} + \cdots + \frac{K_n^2 L_{(n-2)\cdot(n-1)}}{K_{(n-2)}K_{(n-1)}} + \frac{K_n^2 L_{(n-1)\cdot n}}{K_{(n-1)}K_n}$$

$$B = \frac{K_n^2}{A_n^2 2g}\left[-\alpha_1\left(\frac{A_n}{A_1}\right)^2(1-k_{1.2}) + \alpha_2\left(\frac{A_n}{A_2}\right)^2(k_{2.3}-k_{1.2})\right.$$

$$+ \alpha_3\left(\frac{A_n}{A_3}\right)^2(k_{3.4}-k_{2.3}) + \cdots + \alpha_{(n-1)}\left(\frac{A_n}{A_{(n-1)}}\right)^2$$

$$\left. \times(k_{(n-1)\cdot n} - k_{(n-2)\cdot(n-1)}) + \alpha_n(1-k_{(n-1)\cdot n})\right]$$

After discharge has been computed, that value should be used to compute the subsection discharges for subdivided cross-sections, the corresponding velocities and Froude numbers, and the mean section velocities and Froude numbers for all sections. Although the slope-area method is valid for both tranquil and rapid flows, the state of flow must not change within the overall reach from one to the other because of uncertain energy losses.

A thorough discussion of the slope-area method of indirect measurement of discharge is available in Benson (1968) and Dalrymple and Benson (1967).

5.7 CONTRACTED-OPENING METHOD

A highway or railroad crossing of a river channel is generally so constructed as to impose an abrupt width constriction upon flood flow. At such constrictions the contracted-opening method of measuring peak discharge can be applied.

5.7.1 Theoretical basis

At an abrupt width constriction in a reach of channel, as shown in the definition sketch of Figure 5.2, the change in water-surface profile between an approach section (1) and contracted section (3) results largely from the acceleration within the reach. Because the reach is so short, friction loss is of little importance; thus, the effect of possible errors in selecting roughness coefficients (n) is greatly minimized.

The drop in water surface between sections 1 and 3 is related primarily to the corresponding change in velocity. The combined energy and continuity equations between sections 1 and 3 result in the discharge formula:

$$Q = CA_3 \sqrt{\left[2g\left(\Delta h + \alpha_1 \frac{V_1^2}{2g} - h_f \right) \right]} \qquad (5.10)$$

in which Q is the discharge; C the coefficient of discharge; A_3 the gross area of section 3 (this is the minimum section parallel to the constriction between the abutments and is not necessarily at the downstream side of the bridge); Δh the difference in elevation of the water surface between sections 1 and 3; $\alpha_1(V_1^2/2g)$ the weighted average velocity head at section 1, where V_1 is the average velocity, Q/A_1, and α_1 is a coefficient which takes into account the variation in velocity in that section; and h_f is the head loss due to friction between sections 1 and 3.

In addition to information on high-water elevation, channel geometry, and bridge dimensions, the most important element needed for the solution of equation (5.10) is the coefficient of discharge, C. A very comprehensive investigation of the mechanics of flow through single-opening width constrictions in open channels has been reported by Kindsvater and Carter (1955). The resultant methodology for computing peak discharge through

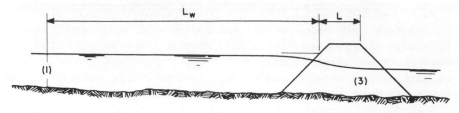

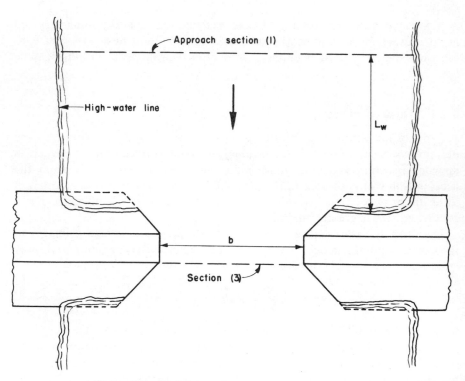

Figure 5.2   Definition of an open-channel constriction

bridge waterways has been thoroughly described by Matthai (1967) and Benson (1968). The following is therefore a general discussion of the procedure recommended for evaluating each term in the discharge equation (5.10).

5.7.2   Field information

The approach section, section 1, is a cross-section of the natural, unconstricted channel upstream from the beginning of drawdown. As shown in

Figure 5.2, section 1 is located one bridge-opening width, b, upstream from the contraction. It includes the entire width of the valley perpendicular to the line of flow. An approach section should always be taken, even when ponded approach sections are indicated.

Sites where the main channel meanders severely in the approach to the contraction should be avoided. The energy losses cannot be correctly evaluated, for example, at a site where the meandering main channel intersects the approach section more than once.

The contracted section, section 3, defines the minimum area on a line parallel to the contraction. It is generally between the bridge abutments.

When a scour hole occurs under a bridge, a section between the bridge abutments may not be the minimum section. Under extreme conditions of scour, the upstream lip of the scour hole may be the physical condition, rather than bridge geometry, that determines the headwater elevation. Such sites should be avoided because the coefficient of discharge for such a condition has not been evaluated, and it is furthermore difficult to determine the amount of scour at the time of the peak.

The area, A_3, to be used in the discharge equation is always the gross area of the section and is the area below the level of the free water surface as determined by the methods described in a subsequent section. No deductions are made for the areas occupied by piles, piers, or submerged parts of the bridge if they lie in the plane of the contracted section. However, the conveyance, K_3, is computed with the area of piles, piers, or submerged parts deducted from the gross area of section 3. Also, the wetted perimeter used to compute the hydraulic radius, R, will include the lengths of the sides of the piles, piers, or bridge surfaces in contact with the water. The mean velocity, V_3, is computed using gross area, A_3.

The field data should include complete details of the bridge geometry so that both plan and elevation drawings can be made. Items to be measured (a steel tape should be used rather than scaled distances from a plan drawing or dimensions from construction plans) include: wingwall angles and lengths, length of abutments, positions and slopes of the embankments and abutments, roadway elevation, top width of embankment, details of piers, piles, and spur dikes, and elevations of the bottom of girders or beams spanning the contraction. Photographs of all of the above are essential for review purposes.

In order for coefficients derived from laboratory studies to be applicable, the water-surface levels at sections 1 and 3 should be determined the same way.

The water-surface elevation at section 1 is determined from a high-water-mark profile near the water's edge along each bank of the stream from above section 1 to the upstream embankment. If no high-water marks can be found at the approach section and there is a large degree of contraction, the approach-section elevation can be assumed to be equal to the elevation at each edge of water along the upstream embankment.

The water-surface elevation for section 3 is obtained on the downstream side of the embankment adjacent to the abutment, regardless of the location of section 3. Water-surface profiles along the downstream face of the embankment are useful in establishing this elevation.

The water-surface elevations of sections 1 and 3 are computed by averaging the elevations on each bank. The fall in water surface, as defined by high-water marks between sections 1 and 3, should not be less than 0.15 m. The fall should also be at least four times the friction loss between sections 1 and 3. It is therefore advisable to avoid long bridges situated in heavily wooded flood plains.

The expressions for velocity head and velocity-head coefficient, α, were defined in equations (5.6) and (5.7).

The friction loss in equation (5.10) refers to that between sections 1 and 3. The distance between the two sections is divided into two reaches, the approach reach from section 1 to the upstream side of the bridge opening, and the bridge-opening reach. The conveyance at the upstream side of the bridge opening is assumed to be the same as at section 3. The total head loss due to friction can be obtained from the equation

$$h_f = L_w(Q^2/K_1K_3) + L(Q/K_3)^2 \qquad (5.11)$$

where L_w is the length of the approach reach; L is the length of the bridge opening; and K_1, K_3 are the total conveyances of sections 1 and 3.

When the approach reach has heavy brush and the reach under the bridge is relatively clear, the weighted conveyance computed from the conveyances of sections 1 and 3 will not be correct. A better approximation of the friction loss may be obtained if $L_w(Q^2/K_1K_q)$ is substituted for the first term in equation (5.11). K_q is that part of the approach section conveyance corresponding to the projected width b.

The friction loss computed from equation (5.11) is only an approximation of the actual loss because of the rapid change in velocity from section 1 to 3. Satisfactory results cannot be obtained by the contraction method if the term h_f is large relative to the difference in head Δh.

The contraction method assumes tranquil flow at section 3. The Froude number is an index of the state of flow. In rectangular channels the flow is tranquil if the Froude number is less than 1. In irregular sections the Froude number computed as $V/\sqrt{(gy_3)}$, where y_3 is the average depth at section 3, is not an exact index of the stage of flow. If the computed Froude number exceeds 0.8, the computed discharge may not be reliable.

5.7.3 Classification of contractions

The geometry of a constriction affects the discharge characteristics of the flow. Most bridge openings can be classified as one of four types representing the distinctive features of their major geometric characteristics. The

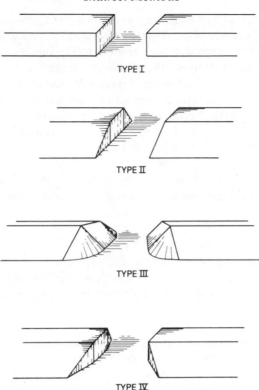

TYPE I

TYPE II

TYPE III

TYPE IV

Figure 5.3   Constriction geometries classified

coefficient C was defined by laboratory studies (Kindsvater and Carter, 1955) in terms of various geometric and hydraulic ratios, for the four types of bridge-opening shapes shown in Figure 5.3.

Type I: A type I contraction, as shown in Figure 5.3, has vertical embankments and vertical abutments with or without wingwalls. The entrance rounding or the wingwall angle, the angularity of the contraction with respect to the direction of flow, and the Froude number affect the discharge coefficient.

Type II: A type II contraction, as shown in Figure 5.3, has sloping embankments and vertical abutments. The depth of water at the abutments and the angularity of the contraction with respect to the direction of flow affect the discharge coefficient.

Type III: A type III contraction, as shown in Figure 5.3, has sloping embankments and sloping spillthrough abutments. The entrance geometry, and the angularity of the contraction with respect to the direction of flow, affect the discharge coefficient.

Type IV: A type IV contraction, as shown in Figure 5.3, has sloping embankments, vertical abutments, and wingwalls. The wingwall angle, the angularity of the contraction with respect to the direction of flow and, for some embankment slopes, the Froude number affect the discharge coefficient. Note that the addition of wingwalls does not necessarily make a type IV contraction. A type I contraction may have wingwalls. If the flow passes around a vertical edge at the upstream corner of the wingwall, the contraction is type I; if the flow passes around a sloping edge at the top of the wingwall, the contraction is type IV.

Besides the geometric influence of these four basic bridge types on the discharge coefficient, the effects of several other bridge situations have been defined.

Spur dikes

Spur dikes are added to some bridge abutments to modify the flow pattern and reduce scour at the abutments. The effect of elliptical and straight dikes on the discharge coefficient has been defined by laboratory study.

Dual bridges

The construction of divided highways has introduced dual bridges. For the special case where the abutments are continuous between the two bridges the geometry may still be classified as one of the standard types. Discharge coefficients have not been defined for dual bridges without continuous abutments. However, a detailed survey of high-water marks will provide a basis for judging the degree to which the divided abutments may have functioned as a continuous abutment. If the L/b ratio, measured from the upstream side of the upstream embankment to downstream side of the downstream embankment, is greater than 2.0, the abutments probably have about the same effect on the discharge coefficient as a single continuous abutment.

Abutments parallel to flow

The base discharge coefficients for all four types of contraction were determined for abutments perpendicular to the embankment, and then adjusted for angularity or the skew of the embankment. Many of the newer bridges have embankments at an angle to the channel, but abutments parallel to the flow. The discharge coefficient is the same for both conditions if the angle is less than 20 degrees. The effect of this change in geometry on the discharge coefficient for angles greater than 20 degrees has not been adequately defined and this geometry should be used with caution in computing peak discharge.

Non-standard types

Some bridges do not fit any of the four types described. Bridges with type I abutments set on type III embankments, or type IV wingwalls with vertical upstream ends for part of their height, or unique construction will require engineering judgement to select the type to be used. When there is a choice between two types, the discharge coefficient can be computed for each type; if the difference is less than 5 per cent, either type can be selected; if the difference is over 5 per cent, the two coefficients may be averaged.

Arch bridges often approximate a type I contraction; but if much of the arch is submerged, the reliability of the computed discharge is diminished by using type I coefficients.

Combination sites

Floods often flow over the road near a bridge in addition to flowing through the bridge opening. This is not a desirable condition for computing peak discharge, but if such a site must be used a combination of the contraction method and a computation of flow over the embankment may be adopted. Benson (1968), Hulsing (1967), and Kindsvater (1964) describe methods and coefficients applicable to road-overflow computations.

Multiple-opening contractions

Roadway crossings on large streams may include more than one bridge. Procedures for computing peak discharge through multiple-opening contractions have been defined by Davidian *et al.* (1962). In general, the same procedures used for single openings are applicable, but some of the geometric ratios and terms in the discharge equation are computed in a different manner.

5.7.4 Discharge coefficients

The discharge equation, equation (5.10), contains a coefficient, C, which represents the combination of (1) a coefficient of contraction, (2) a coefficient accounting for eddy losses caused by the contraction, and (3) the velocity-head coefficient (α_3) for the contracted section.

Dimensional analysis of the factors influencing the flow pattern through a bridge shows that the discharge coefficient can be expressed as a function of certain governing geometric and fluid parameters; or briefly, in functional notation,

$$C = f \text{ (degree of channel contraction, geometry of constriction, Froude number)} \tag{5.12}$$

Of all the factors influencing the coefficient of discharge, the degree of channel-contraction (m) and the length–width ratio of the constriction (L/b) were found the most significant, and they were common to all types of bridge openings. The Froude number has the least general significance. The factors comprising the term 'geometry of constriction' in equation (5.12) are not common to all bridge-opening types. They express, where applicable, such geometric features as: entrance corner rounding; wingwall and chamfer lengths; wingwall angles; pier and pile dimensions; measures of submergence; eccentricity of bridge opening; embankment slopes, dike dimensions; and angularity between plane of contraction and stream direction.

Because of their significance, ratios m and L/b have been selected as primary variables for determining the discharge coefficient for each type of opening, and adjustments to the coefficient are made for the effects of the other geometric terms and the Froude number where applicable.

5.7.5   Office procedures

After the high-water elevation on each bank has been determined from plots of the high-water marks, section properties for the approach and constriction cross-sections are determined. The various dimensionless parameters descriptive of the bridge geometry are tabulated in order to define the discharge coefficient, C.

The channel-contraction ratio m is one of the primary variables on which C depends. It is a measure of that part of the total flow which is required to enter the contracted stream from the sides—from the lateral regions upstream from the embankments. Thus m is defined by the equation

$$m = \frac{Q-q}{Q} = 1 - \frac{q}{Q} \qquad (5.13a)$$

where Q is the total discharge, and q is the discharge that could pass through the opening without undergoing contraction.

This concept has particular advantage in connection with the bridge-waterway problems in that the degree of channel contraction for irregular, natural channels can be computed as a ratio of hydraulic conveyances. It is assumed that the total discharge is distributed across the approach section in proportion to the distribution of conveyance, and that the energy slope is approximately constant across the section. If K_q is the conveyance of the subsection occupied by q, and K_1 is the total conveyance of the approach section (see Figure 5.4), then even for a considerable variation of depth, alignment, and roughness in the approach channel, a procedure for evaluating m is

$$m = (1 - K_q/K_1) \qquad (5.13b)$$

This value of m, and the ratio of L/b, are used to determine the base coefficient of discharge, which is then further adjusted for the effects of the various secondary parameters that are applicable to the particular bridge type.

If the bridge geometry is not one of the standard types shown in Figure 5.3, or if there are unusual flow conditions that were not tested in the laboratory studies to define C, the discharge formula cannot be applied directly. The

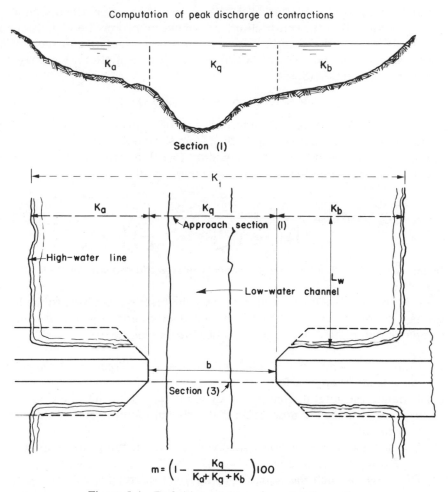

Figure 5.4   Definition of channel-contraction ratio

discharge coefficient for a bridge opening that cannot be classified exactly as any of the four types described, may be estimated by the engineer using his knowledge of the relative effects of the factors that influence the flow pattern and by a reasonable weighting of these factors. The most influential variables generally are m and L/b; therefore, a reasonable estimate of the adjustment factors will give results that are within the range of accuracy expected.

Some combinations of the adjustment factors applied to a base coefficient may yield a value of C greater than 1.00. Because this is impossible, a value of $C = 1.00$ is taken as the maximum under all circumstances.

Equation (5.10), the general discharge equation, requires a trial-and-error solution. A direct solution is obtained if Q/A_1 is substituted for V_1, and if the total friction loss, h_f, is computed in components. If the loss between sections

1 and 3 is taken to be the sum of the losses between section 1 and section 2 (the upstream end of the constriction), and between sections 2 and 3, through the constriction, and if section 3 is taken to be representative of the entire constriction (i.e. section 2 is equal to section 3), then the total loss may be expressed by equation (5.14):

$$h_f = L_w(Q^2/K_1K_3) + L(Q/K_3)^2 \tag{5.14}$$

where L_w is the length from the approach section to the upstream face of the bridge, and L is the length of the bridge, from section 2 to section 3.

With the above substitutions in equation (5.10), a direct solution for the general discharge equation becomes, in imperial units:

$$Q = 8.02 C A_3 \sqrt{\frac{\Delta h}{\left[1 - \alpha_1 C^2 \left(\frac{A_3}{A_1}\right)^2 + 2gC^2 \left(\frac{A_3}{K_3}\right)^2 \left(L + \frac{L_w K_3}{K_1}\right)\right]}} \tag{5.15}$$

In this form, the effects of the approach velocity and the friction loss can be identified.

Floods often cause flow over a road near the bridge in addition to the flow through the bridge, and approach section properties and most ratios must be computed differently. The following procedure is suggested:

(1) Compute flow over the road using the entire area of section 1 to calculate the velocity of approach.
(2) Estimate the total discharge and divide section 1 so that the total conveyance is divided in proportion to the discharges through the bridge and over the road.
(3) Use just that part of section 1 supplying flow to the bridge to compute A_1, L_w, K_a, K_q, K_b, α_1, and m.
(4) Discharge through the bridge plus that over the road should check estimated discharge within 1 per cent. If not, make new estimates until check is obtained.

When the flow over the road occurs on both sides of the bridge, divide the approach section into three parts.

If there is a series of independent, single-opening contractions, all of which freely conduct water from a common approach channel, the grouping is called a multiple-opening contraction. Independence of the openings is generally indicated when two or more pairs of abutments, and one or more interior embankments, exist. Structures in which piers or webs separate several openings between two abutments are considered single-opening contractions.

Discharge is determined by establishing in the approach channel pseudo-channel boundaries which divide the flow between openings. This procedure defines a separate approach channel for each individual opening. Take section

1—one opening width upstream from the embankment in the approach channel defined for each opening. Use the water-surface elevation at that point for h_1. As shown in Figure 5.5 the approach sections to the various openings will not be located on a continuous line across the valley unless the width of all openings is the same. The discharge is computed through each opening using identical procedures and discharge coefficients as for single-opening contractions.

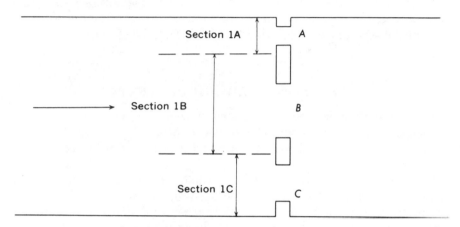

Figure 5.5   Location of approach section and flow division lines for a multiple-opening constriction

The upstream flow boundaries are located first by apportioning the length of each embankment between openings in direct proportion to the gross areas of the openings on either side, the larger length of embankment being assigned to the larger opening. Then, from the points on the embankment thus determined, lines are projected upstream parallel to the mean direction of flow. For computation, these lines are assumed to represent the fixed solid upstream boundaries of an equivalent single-opening contraction.

Once the multiple-opening constriction is divided into a series of equivalent single openings, the water-surface elevations must be determined for each opening.

The water-surface level, h_1, is determined at the location of section 1 for each opening.

Because high-water marks are commonly found only along the edge of the channel and the embankment, the value of h_1 for the central openings must usually be estimated from high-water marks on the upstream side of the interior embankments. Defining h_s as the maximum water-surface elevation along an interior embankment, the value of h_1 may be determined as

$$h_1 = h_s - \frac{\alpha_1 Q_1^2}{2gA_1^2} - \frac{Q_1^2 L_w}{3K_1^2} \qquad (5.16)$$

All quantities in the equation are for the pseudo single-opening channel. The procedure requires the use of an assumed discharge which must be later verified in the computation of discharge.

The downstream water level, h_3, is determined as for a single opening—the average of water-surface elevations on the downstream side of the embankment on each side of the bridge opening.

The values of h_1, h_3, and Δh will usually be different for each opening.

The discharge is computed separately for each opening using the same procedure and discharge coefficients as for single-opening contractions. The individual discharges are then added to obtain the total discharge.

5.8 FLOW THROUGH CULVERTS

A culvert can often be used as a convenient device for measurement of peak discharge by indirect methods. This is indeed fortunate because of the many difficulties involved in making current-meter measurements of flood flow on the smaller streams.

The flow-through-culvert method is similar to the contracted-opening method in that the change in water-surface profile in the reach between the approach and constricted sections largely reflects the effect of acceleration. Again, friction losses are generally of minor importance. A culvert constriction, however, is such that it may act as a control section, i.e. flow may pass through critical depth at the culvert entrance or outlet. The method, therefore, covers conditions of rapid as well as tranquil flow. The culvert could also flow under pressure.

The loss of energy near the entrance is related to the sudden contraction and subsequent expansion of the live stream within the barrel, and entrance geometry has an important influence on this loss. The important features that control the stage–discharge relationship at the approach section can be the occurrence of critical depth in the culvert, the elevation of the tailwater, the entrance or barrel geometry, or a combination of these.

The peak discharge through a culvert is determined by application of the continuity equation and the energy equation between the approach section and a section within the culvert barrel. The location of the downstream section depends on the state of flow in the culvert barrel. For example, if critical flow occurs at the culvert entrance, the headwater elevation is not a function of either the barrel friction loss or the tailwater elevation, and the terminal section is located at the upstream end of the culvert.

The procedures described below are based on the information obtained in laboratory investigations of the head-discharge characteristics of culverts, and information obtained in field studies of the flow through culverts at sites where the discharge was known. The laboratory studies have been reported by Bodhaine (1968).

5.8.1 Flow classification

The discharge characteristics of a culvert depend upon an evaluation of energy changes between an approach section upstream from the culvert and the control section. Depending upon the location of the control section and the relative height of headwater and tailwater, most culvert flow patterns may be classified in six types. These are described below and defined also in the sketches of Figure 5.6. Certain characteristics of each of these types of flow are shown in Table 5.2. From this information, the following general classification of flow types can be made:

(1) If h_4/D is equal to or less than 1.0 and $(h_1 - z)/D$ is less than 1.5, only types I, II, and III flow are possible.
(2) If h_4/D is greater than 1.0, only type IV flow is possible.
(3) If h_4/D is equal to or less than 1.0 and $(h_1 - z)/D$ is equal to or greater than 1.5, only types V and VI flow are possible.

Table 5.2   Characteristics of flow types (D = maximum vertical height of barrel and diameter of circular culverts)

Flow type	Battel flow	Location of terminal section	Kind of control	Culvert slope	$\dfrac{h_1 - z}{D}$	$\dfrac{h_4}{h_c}$	$\dfrac{h_4}{D}$
I	partly full	inlet	critical depth	steep	<1.5	<1.0	≤1.0
II	–do–	outlet	–do–	mild	<1.5	<1.0	≤1.0
III	–do–	–do–	backwater	–do–	<1.5	>1.0	≤1.0
IV	full	–do–	–do–	any	>1.0	—	>1.0
V	partly full	inlet	entrance geometry	–do–	≥1.5	—	≤1.0
VI	full	outlet	entrance and barrel geometry	–do–	≥1.5	—	≤1.0

Further identification of the type of flow requires a trial-and-error procedure which is described subsequently.

The discharge equation for each type of flow has been developed by application of the energy and continuity equations between the approach section and the terminal section. The discharge may be computed directly from these equations after the type of flow has been identified. Each of the six types of flow and its appropriate discharge equation is discussed below.

Type I: Critical depth at inlet—as indicated on Figure 5.6, flow passes through critical depth, d_c, near the culvert entrance. The culvert barrel flows part full. The headwater-diameter ratio, $(h_1 - z)/D$, is limited to a maximum

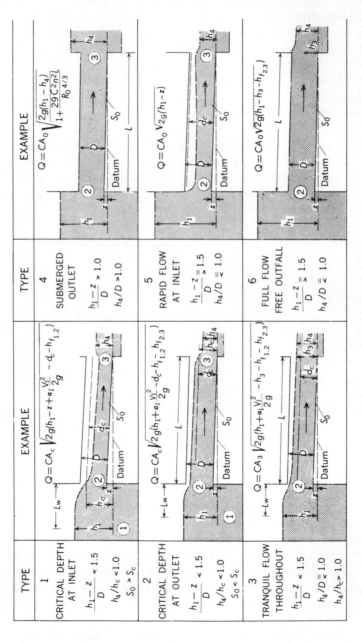

Figure 5.6   Classification of culvert flow

of 1.5. The slope of the culvert barrel, S_0, must be greater than the critical slope, S_c, and the tailwater elevation, h_4, must be less than the elevation of water surface at the control section, h_2. The discharge equation is

$$Q = CA_c \sqrt{\left[2g\left(h_1 - z + \frac{\alpha_1 V_1^2}{2g} - d_c - h_{fl \cdot 2} \right) \right]} \qquad (5.17)$$

where C is the discharge coefficient; A_c the flow area at the control section; V_1 the mean velocity in the approach section; α_1 the velocity-head coefficient at the approach section; $h_{fl \cdot 2}$ the head loss due to friction between the approach section and the inlet; $L_w(Q^2/K_1 K_c)$, and K is the conveyance = $(1.486/n)R_h^{2/3}A$, in imperial units or $(1/n)R_h^{2/3}A$ in metric units.

Other notation is evident from Figure 5.6, or has been previously explained. The discharge coefficient, C, is discussed in detail in Subsection 5.8.2.

Type II: Critical depth at outlet—flow passes through critical depth at the culvert outlet, and the culvert barrel flows part full. The headwater-diameter ratio does not exceed 1.5. Slope of the culvert is less than critical. Tailwater elevation does not exceed the elevation of the water surface at the control section, h_3. The discharge equation is

$$Q = CA_c \sqrt{\left[2g\left(h_1 + \frac{\alpha_1 V_1^2}{2g} - d_c - h_{fl \cdot 2} - h_{f2 \cdot 3} \right) \right]} \qquad (5.18)$$

where $h_{f2 \cdot 3}$ is the head loss due to friction in the culvert barrel, = $L(Q^2/K_2 K_3)$.

When backwater is the controlling factor in culvert flow, critical depth cannot occur and the upstream water-surface elevation for a given discharge is a function of the surface elevation of the tailwater. The culvert may flow partly full, in which case the headwater-diameter ratio is less than 1.5; or it may flow full, in which case both ends of the culvert are completely submerged and the headwater-diameter ratio may be of any value greater than 1.0. The two types of flow in this classification are III and IV.

Type III: Tranquil flow throughout—the culvert barrel flows part full, with the headwater-diameter ratio less than 1.5. The tailwater elevation does not submerge the culvert outlet, but does exceed the elevation of critical depth at the control or terminal section.

The lower limit of tailwater must be such that (1) if the culvert slope is steep enough that under free-fall conditions critical depth at the inlet would result from a given elevation of headwater, the tailwater must be at an elevation higher than the elevation of critical depth at the inlet; and (2) if the culvert slope is mild enough that under free-fall conditions critical depth at the outlet would result from a given elevation of headwater, then the tailwater must be at an elevation higher than the elevation of critical depth at the outlet. The discharge equation for this type of flow is

$$Q = CA_3 \sqrt{\left[2g\left(h_1 + \frac{\alpha_1 V_1^2}{2g} - h_3 - h_{f1 \cdot 2} - h_{f2 \cdot 3} \right) \right]} \qquad (5.19)$$

where h_3 is the depth of flow at the culvert outlet, and A_3 is the flow area at the culvert outlet.

Type IV: Submerged culvert—the tailwater elevation is high enough to submerge the culvert outlet. The culvert, therefore, is submerged and flows full. The headwater-diameter ratio can be anything greater than 1.0. No differentiation is made between low-head and high-head flow on this basis for type IV flow.

The discharge may be computed directly from the energy equation between sections 1 and 4. Thus,

$$h_1 + h_{v1} = h_4 + h_{v4} + h_{f1 \cdot 2} + h_e + h_{f2 \cdot 3} + h_{f3 \cdot 4} + (h_{v3} - h_{v4}),$$

where h_e is the head loss due to entrance contraction.

In deriving the discharge equation shown below, the velocity head at section 1 and the friction loss between sections 1 and 2 and between sections 3 and 4 have been neglected. Between sections 3 and 4 the energy loss due to sudden expansion is assumed to be $(h_{v3} - h_{v4})$. Thus,

$$h_1 = h_4 + h_e + h_{f2 \cdot 3} + h_{v3}$$

or in imperial units,

$$Q = CA_0 \sqrt{\left| \frac{2g(h_1 - h_4)}{1 + \dfrac{29 C^2 n^2 L}{R_0^{4/3}}} \right|} \qquad (5.20)$$

where A_0 is the area of culvert barrel, and R_0 is the hydraulic radius of culvert barrel.

High-head flow will occur if the tailwater is below the crown at the outlet and the headwater-diameter ratio is ≥ 1.5. This is an approximate criterion. The two types of flow under this category are V and VI.

As shown in Figure 5.6, partly full flow under a high head is classified as type V. The flow pattern is similar to that downstream from a sluice gate with rapid flow near the entrance. The occurrence of type V flow requires a relatively square entrance that will cause contraction of the area of live flow to less than the area of the culvert barrel. In addition, the combination of barrel length, roughness, and bed slope must be such that the contracted jet will not expand to the full area of the barrel. If the water surface of the expanding flow comes in contact with the top of the culvert, type VI flow will occur, because the passage of air to the culvert will be sealed off causing the culvert to flow full throughout its length. Under these conditions the headwater surface drops, indicating a more efficient use of the culvert barrel.

Within a certain range either type V or type VI flow may occur, depending upon factors that are very difficult to evaluate. For example, the wave pattern superimposed on the water-surface profile through the culvert can be important in determining full or part-full flow. However, within the range of geometries tested, the flow type generally can be predicted from a knowledge of entrance geometry and length, culvert slope, and roughness of the culvert barrel.

Type V: Rapid flow at inlet—the headwater–diameter ratio exceeds 1.5. The culvert entrance is such that flow is contracted in a manner similar to sluice- or unsubmerged orifice-type flow. The culvert barrel flows part full and at a depth less than critical depth. The tailwater elevation does not submerge the culvert outlet. The discharge equation is

$$Q = CA_0\sqrt{[2g(h_1 - z)]} \tag{5.21}$$

Type VI: Full flow, free outfall—the headwater–diameter ratio exceeds 1.5. The culvert barrel flows full under pressure. The tailwater elevation does not submerge culvert outlet. The discharge equation between sections 1 and 3, neglecting $V_1^2/2g$ and $h_{f1\cdot2}$, is

$$Q = CA_0\sqrt{[2g(h_1 - h_3 - h_{f2\cdot3})]} \tag{5.22}$$

A straightforward application of equation (5.22) is hampered by the necessity of determining h_3, which varies from a point below the centre of the outlet to its top even though the water surface is at the top of the culvert. This variation in piezometric head is a function of the Froude number. This difficulty has been circumvented by basing the data analysis upon dimensionless ratios of physical dimensions related to the Froude number. These functional relationships have been defined by laboratory experiment and their use is explained in a later section.

5.8.2 Discharge coefficients

Coefficients of discharge, C, for flow types I–VI were defined by laboratory study and are discussed extensively by Bodhaine (1968) and Benson (1968). They are applicable to both the standard formula and routing methods of computation of discharge. The coefficients vary from 0.39 to 0.98 and have been found to be a function of the degree of channel contraction and the geometry of the culvert entrance.

For some entrance geometries the discharge coefficient is obtained by multiplying a base coefficient by an adjustment factor to account for the effects of rounded entrances, levels, chamfers, or wingwalls. If this procedure results in a discharge coefficient greater than 0.98, a coefficient of 0.98 should be used as a limiting value in computing the discharge through the culvert.

The coefficients are applicable to both single- and multi-barrel culvert installations. If the width of the web between barrels in a multi-barrel installation is less than 0.1 of the width of a single barrel, the web should be disregarded in determining the effect of the entrance geometry.

Laboratory tests indicate that the discharge coefficient does not vary with the proximity of the culvert floor to the ground level at the entrance. Thus in types I, II, and III flow, the geometry of the sides determines the value of C; similarly, in types IV, V, and VI flow the value of C varies with the geometry of the top and sides. If the degree of rounding or bevelling is not the same on both sides, or on the sides and top, the effect of the rounding or wingwall must be obtained by averaging the coefficients determined for the sides, or for the sides and top depending on the type of flow.

The discharge coefficient does not vary with culvert skew.

Either Benson (1968) or Bodhaine (1968) should be used as a reference for selection of discharge coefficients and their use for varying types of culvert geometry, entrance conditions, and flow type.

5.8.3   Field data

High-water profiles must be defined both upstream and downstream from culverts, in general in the same manner as for bridge openings. At many culverts, crest-stage gages have been installed for obtaining data on peak stages. These alone are not sufficient to define headwater and tailwater elevations. Surveys of high-water profiles must be made at various stages until the relation between the crest-gage readings and the high-water profiles has been determined. If high-water marks are available within a culvert, they should be located and their elevations should be determined for use in checking computed elevations.

The location and elevation of floodmarks should be obtained along the embankment and upstream from the culvert. If there is a definite approach channel, marks should be obtained along the banks for a distance of at least two culvert widths. Where ponded conditions exist, the headwater elevation should be determined from marks along the banks or upstream from the opening where there is little or no velocity. In doubtful cases of ponding, conditions approximating ponding can be assumed if high-water marks along the embankments approach a level surface away from the culvert opening.

Tailwater elevations should be obtained along the embankment or the channel close to the outlet. They should not represent the elevation of the issuing jet, which may be higher than the tailwater pool. If marks cannot be found in the immediate vicinity of the culvert, the profile should be extended upstream to the outlet on the basis of the existing profiles.

An approach section usually is necessary, but if the area of the approach channel is estimated as equal to or greater than five times the area of the culvert barrel, zero approach velocity in the approach section may be assumed, and an approach section is not required. To avoid the possibility of the approach section being within the drawdown region, it should be located one culvert width upstream from the culvert entrance. Where wingwalls exist, it should be located upstream from the ends of the wingwalls equal to the width between the upstream end of the wingwalls. An approach section should never be closer than one culvert width. It should be at right angles to the channel.

One culvert width at a multiple-culvert installation may be considered as the sum of the widths at the individual culverts.

The selection of values of n, and any points of subdivision, are done as described for slope-area cross sections.

Complete details should be obtained of culvert dimensions (measured by steel tape), projections, wing-wall angles, size of fillets, degree of entrance rounding, types of entrance, culvert materials, and condition. Elevations of the culvert bottom, upstream and downstream, are necessary. Positions of the culvert, wingwalls, and other features must be located. The culvert length must be measured. The radius of rounding or degree of bevel of corrugated pipes should be measured in the field. These are critical dimensions that should not be chosen from a handbook and accepted blindly.

Photographs are a very important part of the field data. They should show culvert details and all pertinent conditions upstream and downstream from the culvert. A return trip to the site can be avoided if there is good photographic coverage of the general site and the construction details. Good lighting is particularly important because culvert entrances and exits are often dark, and details do not show up.

General views that show the relationship of the culvert to the approach channel, to crest-stage gages if they exist, and to the getaway conditions are useful. There should be at least one close-up of the culvert entrance to show entrance detail. A level rod standing at the entrance furnishes a permanent record of culvert height and is a good reference for other details. In cases of road overflow, a view showing the entire overflow section should be included.

Appropriate values of n, and for discharge coefficients, C, are discussed in detail in Benson (1968) and Bodhaine (1968). These references also describe procedures for special conditions, such as unusual entrances, skewed culverts, breaks in slope, shape, direction, or construction material part way through the culvert, culverts with natural bottoms or paved bottoms, multiple culverts, slope-area measurements within culverts, and sites associated with problems of road overflow and with storage effects of pondage upstream from a culvert.

5.9 FLOW OVER DAMS AND EMBANKMENTS

5.9.1 Theoretical basis

The hydraulics of channel controls and spillways has long been a basic tool of the hydraulician and designer. These principles may be adapted to indirect computation of discharge over such structures.

A weir, dam, or embankment generally forms a control section at which the discharge is related to the upstream water-surface elevation. The peak discharge at the control section can usually be determined on the basis of a field survey of high-water marks and the geometry of the particular structure. These methods are derived from investigations of the discharge characteristics of weirs, dams, and embankments as reported in the literature and from laboratory studies (see Chapter 3).

Many weirs do not fit any of the classes presented in Chapter 3 because of shape, channel conditions, or the hydraulics governing the particular flow condition. For example, a thin-plate weir may act like a broad-crested weir or a terminal sill because of fill in the approach channel, or a broad-crested weir may act like a thin-plate weir because the head is sufficient to make the nappe spring clear. It is thus necessary to consider many different aspects of the situation—shape of weir, channel geometry, ratio of head to crest length—before classifying the weir as to the appropriate discharge equation and coefficient.

Because friction loss between an upstream approach section and the control section is generally of minor importance, the selection of a roughness coefficient, n, at the approach section is not a critical factor. Rather, the reliability of the computed discharge over a weir depends primarily upon selection of the proper discharge coefficient. For some dams this coefficient must be estimated by comparison with calibrated weirs of similar shape, for others it must be computed from curves of general relations based on laboratory studies.

There is extensive information in the literature on sharp-crested weirs of various shapes, discharging either freely or submerged. The majority of weirs used for indirect measurements are broad- or round-crested dams, or highway embankments. Computations of discharge described below are, therefore, restricted to these latter types.

5.9.2 Broad-crested weirs

Standard broad-crested weirs are discussed in Chapter 3. The discharge equation for non-standard broad-crested weirs, in terms of the total energy head, H, is

$$Q = CbH^{3/2} \tag{5.23}$$

where Q is the discharge; C a coefficient of discharge having dimensions of square root of gravity acceleration; b the width of the weir normal to the flow, excluding width of piers; and H is the total energy head $(h + V_1^2/2g)$ referred to the crest of the weir and V_1 is the mean velocity at the approach section to the weir.

5.9.3 Round-crested weirs

Round-crested weirs are defined as weirs with crests that are smooth, single-curved (cylindrical) surfaces. The longitudinal profile of the crest usually is not a circular arc, as the name may imply; often it is a complex curve which cannot be described in simple geometric terms. The crests of overfall spill-ways of most modern high dams, which have the form of the lower surface of the nappe from a full-width, thin-plate weir, are special examples of round-crested weirs. These are commonly called ogee or design-head dams.

With respect to profile geometry, round-crested weirs are a class inter-mediate between thin-plate weirs and broad-crested weirs. The crest is long enough (in the direction of flow) to provide support for the nappe through the control section, but it is short enough that the flow over the crest is markedly curvilinear. The fixed lower boundary of the nappe is therefore subject to much variation in pressure; for any given head the pressure may vary considerably from point to point, and at any given point the pressure varies appreciably with the head.

The geometry and flow pattern for a simple round-crested weir are illus-trated in Figure 5.7. The intersection of the curvilinear crest and the vertical (or inclined) face of the weir is termed the 'spring point', and the height, P, of the weir is measured to this point. The head on the weir is referenced to the highest point on the crest.

The boundary geometry of round-crested weirs, specifically the height, P, and the inclination of the upstream face, uniquely determines the flow pattern over the crest. Any given head, then, has a unique flow pattern with unique velocity and pressure distributions, all of which determine the discharge coefficient.

The discharge equation for round-crested weirs is based on the piezometric head, h (see Figure 5.7):

$$Q = Cbh^{3/2} \tag{5.24}$$

where Q is the discharge; C the coefficient of discharge, having dimensions of square root of gravity acceleration; b the length of the crest normal to the flow excluding piers; and h is the static head referred to the crest of the dam.

The degree of submergence of a round-crested weir is defined by the ratio h_t/H, as illustrated in Figure 5.7. Total head, H, is the sum of the static head and the velocity head at the approach section,

$$H = h + h_v = h + V_1^2/2g$$

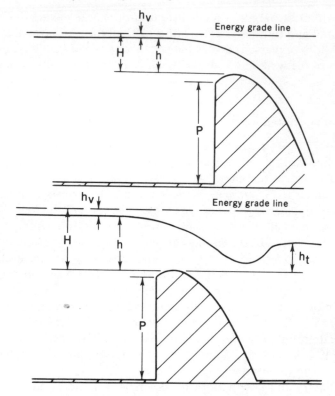

Figure 5.7　Definition of a nappe-fitting design-head
dam with free flow and with submerged flow

If the degree of submergence is greater than 0.6, the computed discharge may not be reliable and other methods of determining the peak discharge should be investigated.

5.9.4　Highway embankments

It is sometimes necessary to compute flow over highway embankments in combination with flow through bridge openings. The general procedure for dividing the flow and establishing the approach section and water-surface ·elevations for this type of indirect measurement was explained in the description of the contracted-opening method, in Section 5.7.4. Detailed explanations of the procedures are given by Benson (1968) and Hulsing (1967).

The geometry and flow pattern for a highway embankment are illustrated in Figure 5.8. Under free-flow conditions, critical depths occur near the crown line. The head is referred to the elevation of the crown and the length, L, in direction of flow is the distance between the top points of the upstream and downstream embankment faces. The height of the embankment has no influence on the discharge coefficient.

The discharge equation for flow over roadways is referred to the total head, H, and is

$$Q = CbH^{3/2} \tag{5.25}$$

where Q is the discharge; C the coefficient of discharge having dimensions of square root of gravity acceleration; b the length of the flow section along the road normal to the direction of flow; and H is the total head $= h + V_1^2/2g$.

Because of shallow depths over the road and very flat longitudinal slope (normal to flow) of the roadway it is often difficult to determine the length of the flow section, b, to be used in equation (5.25). It is thus convenient to assume that the elevation of the water surface at the crown line of the roadway (where b is measured) is equal to 5/6 of the maximum value of H for the section.

The degree of submergence of a highway embankment is defined by the ratio h_t/H, as illustrated in Figure 5.8. The effect of submergence is smaller than 0.9, and it increases so rapidly for higher values of ratio h_t/H that computed discharges may not be reliable. In some indirect measurements, however, the portion of the total flow which passes over the road, as compared with that which went through the bridge, may be small and thus a greater error can be tolerated in this computation.

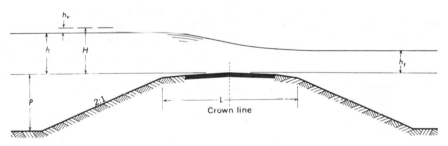

Figure 5.8 Definition of flow over a highway embankment

5.9.5 Discharge coefficients

Discharge coefficients for free discharge are a function of the geometry of the channel and the weir. The geometry is usually expressed as dimensionless ratios involving static head, weir height, length, and width, channel width, slopes of upstream and downstream faces of the weir, radius of rounding, side contractions, and velocity head at the approach cross-section. There may be additional effects due to abutments and piers, to gates and flashboards, to transversely non-level crests, and to submergence. Excellent descriptions and tabulations of information on discharge coefficients for sharp-crested, broad-crested, round-crested, and embankment-shaped weirs are provided in Hulsing (1967) and Benson (1968).

There are many weirs of unusual shape that cannot be classified as sharp-, broad-, or round-crested weirs. Information obtained from tests on weirs of similar shape may be used to determine the discharge coefficient. The best source of information on weirs of unusual shape is Horton (1907).

5.9.6   Field data

In order to be used as a site for indirect measurement of discharge, a weir, dam, or embankment must have adequately defined heads, h, on the upstream side (and also on the downstream side if the structure is submerged). It must also have a discharge coefficient, C. It must not have a velocity of approach that is high enough to make the velocity head a substantial part of the total head, H, because the reliability of the measurement would thereby be reduced.

The structure should fit within the verified range of parameters for which discharge coefficients have been verified so that C can be selected with confidence.

Thus, if the total head on the dam can be accurately determined and if the discharge coefficient is within the range for which adequate definition is available, then a reliable discharge can be computed for the dam.

For a dam measurement, the survey of physical features should be complete. There is tendency to use construction plans which are frequently available for such structures. It has too often been found that dams and other structures are not built as originally planned. All details should be measured in the field, including the length of dams, crest elevations, crest shapes, and approach cross-sections. Dams sometimes settle after construction, and pools are frequently filled in.

The exact shape of a dam crest longitudinally to the direction of flow is extremely important in determining the value of C. Measurements should be made to ± 5 mm at enough points along the profile to define the precise shape. Where the dam changes shape along its length, this must be done for each portion of the dam. In the case of roadway embankments, the widths of the roadway and shoulder and its bank slopes are required.

In the case of dams, there is frequently available from power plant operators a headwater elevation determined at a single gaging point. This is not sufficient for determining the flow over the dam. High-water profiles should be run, along both banks, as in any other type of measurement, to define the approach elevations. High-water marks should be surveyed on the downstream side of the dam to prove or disprove the presence of submergence, and to define it if present.

For highway embankments, marks must be surveyed both upstream and downstream to define effective elevations and submergence along the length of the fill.

Crest elevations should be taken along the length of the dam or roadway. Where necessary, approach cross-sections should be surveyed. If pool depths are sufficiently large, this may not be necessary. In such cases, evidence of the pool depths should be obtained.

Other features that frequently complicate dam computations, such as the presence of flashboards, gates, bypass channels, etc., should be measured, and the conditions of flashboards, gate openings, and possibly debris at the time of the peak must be ascertained from operators.

Pictures of the conditions of the dam crest or flow conditions at the time of the peak are very desirable. Detailed photographs of every pertinent feature of the dam should be obtained, with close-ups, and a scale reference.

5.10 CRITICAL-DEPTH METHOD

5.10.1 Theoretical basis

The solution of discharge by the critical-depth method requires the establishment of an approach cross-section, upstream from the suspected drop-off point, where the water-surface elevation is known from high-water profiles. An elevation of critical depth at the drop-off section is assumed. For this elevation the mean depth, d_m, can be computed as A/T, where A is the cross-sectional area, and T is the top width. The discharge is then computed as

$$Q = A\sqrt{(gd_m)} \tag{5.26}$$

In terms of Figure 5.1, the critical depth at the drop-off section establishes the downstream water-surface elevation, h_2; the elevation at the approach section, h_1, is established by high-water marks. With discharge determined from equation (5.26), the energy grade line at the downstream cross-section is established. The friction loss, h_f, in the length of reach, L, between the cross-sections is computed by using only the approach section conveyance, K_1:

$$h_f = LQ^2/K_1^2 \tag{5.27}$$

Deduction of the velocity head at the approach section, $(h_v)_1$, should give the known water-surface elevation there, h_1. If not, another elevation of critical depth is assumed downstream, its corresponding discharge is computed by means of equation (5.26), and the trial-and-error computation process is repeated until there is a match between the computed and known water-surface elevation at the approach section.

The method is reliable only where there is a sharp break in the channel bottom, with mild upstream, and steep continuing downstream slopes, in a channel of fairly uniform and nearly rectangular shape. These conditions are somewhat similar to those occurring where the pool behind the dam has filled in to the crest, and the procedures for computation are essentially the same. At times the length of the steep-sloping portion is not sufficient, so that there is certainty of free getaway and no backwater. In such a case a computation of this sort will give the maximum possible discharge, perhaps to be checked against some other computation of the peak.

This method should be used with caution as it is difficult to determine if flow actually occurred at critical depth through the control section.

5.10.2   Field data

Surveys of sites for the determination of flood flows by the critical-depth method are similar to those for slope-area reaches, with special attention to the cross-section of the control at which it is assumed that flow had taken place at critical depths.

An approach cross-section should be taken far enough upstream to be above the beginning of drawdown towards the drop-off section. This distance may probably be safely taken as at least $2\frac{1}{2}$ times the mean depth of the stream.

High-water profiles are taken upstream and downstream extending beyond the two cross-section points. There should be no question of free getaway downstream.

It is useful to obtain the elevation of points along the stream bottom both upstream and downstream from the break in slope. This will allow the comparison between critical slope and bed slopes.

Field data are plotted and data sheets are prepared in a manner similar to that used in the slope-area method.

5.11   CONVEYANCE-SLOPE METHOD

The conveyance-slope method is dissimilar to the foregoing indirect methods of measuring discharge, in that it makes no use of the high-water profile of a flood. Rather, it is a method for extending the upper part of the rating curve or interpolating the curve between high measurements which are far apart in stage.

5.11.1   Theoretical basis

Frequently, it is desirable to shape a rating curve between the highest current-meter measurement and a slope-area or other indirect measurement of a flood peak, or to extend a curve a considerable distance above the highest discharge measurement. Of several methods that can be used to shape or to extend a rating curve, the best method is the one that gives most consideration to the individual factors that determine the stage–discharge relationship. The conveyance-slope method gives greater consideration to these factors than do most other methods.

The discharge for any stage is expressed by the Manning formula, equation (5.1a) or (5.1b), which can be condensed further, by means of equations (5.3) and (5.5), to

$$Q = KS^{1/2} \tag{5.28}$$

All factors that determine the conveyance, K, $(1.486/n)AR_h^{2/3}$ in imperial units, or $(1/n)AR_h^{2/3}$ in metric units, can be obtained from simple field measurements with the exception of the roughness coefficient n. This coefficient must be based on judgement and experience, but tables and photographs that are readily available should restrict selection of values by different persons to a relatively narrow range. Some excellent references to the choice of n are mentioned in Subsection 5.5.1.

The values of n should always be selected by the most experienced person available after field observation of the channel.

The conveyance for a number of elevations should be computed and plotted to define a curve of gage height versus conveyance. The cross-section used should be an average for the stream and not one at an unusually large or contracted section. It should be located at the site of interest.

The energy slope for stages at which discharge measurements have been made can be computed by dividing the discharge by the corresponding conveyance and squaring the result, or $S = (Q/K)^2$. These points should be plotted to define a curve of gage height versus energy slope, S. This curve must be extended to the desired gage height, but usually the slope for flood stages tends to approach a constant so that an extension may be accurately made. For very high floods it is likely that the slope will approximate the general slope of the stream bed and may be determined from a low-water profile or from a topographic map. For some streams the slope will increase with increasing stages, while for others the slope will decrease with increasing stages.

The missing portions of the stage–discharge rating curve, either extrapolations beyond the highest measurements, or interpolations between high measurements where better definition is required, can be filled in by application of equation (5.28). From the stage-conveyance and stage-slope curves developed for the site, a value of conveyance and a value of slope is selected for any given stage and the corresponding discharge is computed.

5.11.2 Field data

The field data to be obtained should be a cross-section which is representative of the stream. It is usually most practicable to take this at the gage, if that is a representative section, because the stage is known there at the time of each measurement. The cross-section must be taken to as high a stage as the curve is to be extended. Values of n must be assigned to each portion of this channel. If other cross-sections are used than at the gage, some method must be available to obtain the stage there corresponding to the gage height.

5.12 STEP-BACKWATER METHOD

A very convenient method of extending, filling-in, or even synthesizing a complete stage–discharge rating curve is to make use of the characteristics of gradually-varied-flow profiles. Excellent descriptions of various methods of solution of the dynamic equation of gradually varied flow are given by Chow (1959) and Woodward and Posey (1941). Of these, the standard-step method is one of the most commonly used, particularly because it can be readily programmed for solution by high-speed electronic computers.

Although extensive channel-geometry data are required for application of the standard-step, or step-backwater method as it is commonly called, no high-water data are needed.

The method is used primarily in reaches of subcritical flow immediately downstream from the gaging station.

5.12.1 Theoretical basis

The water-surface profiles resulting from non-uniform flow conditions are called backwater curves. Various types of such profiles for gradually varied flows are described by Chow (1959), and Woodward and Posey (1941). Of the many backwater curves possible, those for subcritical flows on mild slopes are generally of most concern, and of these, the characteristics of the M1 and M2 curves (Figure 5.9) are utilized for the extension of stage–discharge ratings.

In steady, uniform subcritical flows, the normal-depth profile is higher in elevation than the critical-depth profile. The control is downstream. If the

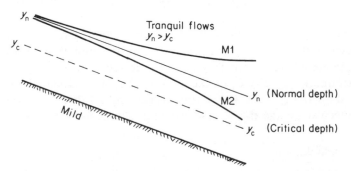

Figure 5.9 Subcritical-flow water-surface profiles M1 and M2 on mild slope

downstream control is a dam, bridge, channel constriction, or a flooding main stem of a river, the resulting water-surface profile represents a transitory, gradually varied flow condition, as shown by the M1 curve in Figure 5.9.

If the downstream control is a hydraulic drop, associated with falls, riffles, or any condition where the downstream bed slope is larger than the upstream bed slope, forcing the water-surface to pass through critical depth, the resulting backwater profile is depicted by the M2 curve in Figure 5.9.

It is characteristic of the tranquil-flow M1 and M2 backwater curves that one can start at any point on either of them, and by solving the energy equation, determine the elevation of the water surface at another point farther upstream. The intervening geometry and roughness, as well as the discharge, would have to be known. As may be seen on Figure 5.9, if the procedure were carried far enough upstream in increments, this step-by-step computed profile would converge asymptotically towards the normal-depth line. This characteristic of the backwater curves is utilized in determining normal depth in a channel.

In a given reach for a given discharge, therefore, all M1 and M2 curves for that discharge will converge to the normal-depth line if the computations are carried far enough upstream. One may begin computations at the downstream end of a reach, and compute several different water-surface profiles for a given discharge. Those profiles that start with an elevation higher than normal depth would be M1 curves; those starting with elevations lower than normal depth would be M2 curves. When computed values of water-surface elevations along any two profiles converge mathematically, it is generally assumed that the normal-depth profile has been reached. Computations along either profile continued further upstream would be identical, regardless of whether the profiles had been two M1s, two M2s, or one of each, at the start of computations at the downstream end of the reach. These continued computations would define the 'average' normal profile because the non-uniformities in the channel geometry and roughness would be translated to a series of minor, transitional backwater curves. Such continued computations, therefore, would be determining the water surface to be expected in this channel for the given discharge. The computed profile would define a locus of normal depths at each cross-section.

When computations along any two profiles that had been started with different elevations downstream (i.e. they were on different M curves) begin to give identical elevations at points upstream, it may be assumed that the normal profile will be computed thereafter; and when the normal depth is computed at the gaging station, that elevation together with the discharge used for the computations will provide a point on the stage–discharge rating.

In the step-backwater method, a total reach such as in Figure 5.9 is divided into a series of sub-reaches, downstream from the gaging station. Each sub-reach would then be represented by Figure 5.1, and the energy equation for open-channel flow, equation (5.2), is solved successively for each. All of the criteria for the selection of the reach, and its subsequent subdivision, and all of the criteria considered in the solution of equation (5.2), which were described in Section 5.6 for solution of indirect measurements by the slope-area method, are applicable for each sub-reach of the step-backwater reach.

With reference to Figure 5.1, the individual steps in the solution of the energy equation for subcritical flow are listed below. Definitions of various terms are provided in Subsection 5.6.1.

(1) The working discharge, for which the water-surface profile is to be determined, is chosen.

(2) All necessary channel geometry and roughness information are available in the lateral, longitudinal, and vertical directions, including cross-sectional shapes, subdivisions, and sub-reach lengths.

(3) The water-surface elevation, h_2, at the downstream end is chosen.

(4) For the value of h_2 chosen in step 3, the corresponding area, conveyance, velocity-head, and α values are computed for the downstream section.

(5) A water-surface elevation, h_1, is assumed for the upstream cross-section.

(6) For the value of h_1 chosen in step 5, the corresponding area, conveyance, velocity-head, and α values are computed for the upstream cross-section.

(7) The friction loss between sections 1 and 2 is computed: $(h_f)_{1-2} = LQ^2/K_1K_2$.

(8) The coefficient k is determined; k is 0.5 if Δh_v is positive, and zero if Δh_v is negative.

(9) The energy equation, equation (5.2), is solved.

(10) If the energy equation is not balanced within an acceptable predetermined tolerance, step 5 is repeated with a new value of h_1 chosen for the upstream water-surface elevation. If the equation is acceptably balanced, the next operation is step 12.

(11) Steps 5–10 are repeated until the energy equation is satisfactorily balanced.

(12) The sub-reach water-surface profile is computed, and the solution moves one step, or sub-reach, further upstream. The value of h_1 at the upstream end of the first subreach is now equivalent to the value of h_2 at the downstream end of the new sub-reach. This operation is equivalent to step 3, above.

(13) Steps 4–12 are repeated, sub-reach by sub-reach, until the water-surface profile throughout the entire reach has been computed.

If the first value of h_2 in step 3 for the most downstream cross-section were above the normal depth line, the profile computed would follow an M1 curve; if h_2 were started originally at an elevation below normal depth, the computed profile would follow an M2 curve. In order to determine the normal depth line in a channel, the procedure is to choose any two or more starting values of h_2 at the most downstream cross-section, and, for the same discharge, compute the resultant profiles until these profiles all converge further upstream, and thereafter give identical values of water-surface elevation at succeeding cross-sections.

Because of the trial-and-error nature of the solution of the energy equation, the manual determination of water-surface profiles is extremely tedious. Modern high-speed computers can determine several entire profiles in long, multi-reach channels, in only a few seconds.

Starting elevations for profile computations at the most downstream cross-section should, for each discharge, preferably be between normal depth and three-quarters of that depth for M2 curves, and between normal depth and $1\frac{1}{4}$ times that depth for M1 curves. The M1 curves, inasmuch as they are higher than the elevation of the normal depth profile, should start at elevations that are within definable banks at each cross-section. The general practice is to start with an M2 curve higher than critical depth, and an M2 or M1 curve that is within banks but fairly close to normal depth.

5.12.2 Field data

A transit-stadia survey for the entire stream channel is required for each study site. The surveys are run using the same basic techniques described by Benson (1968) and Dalrymple and Benson (1967) for indirect discharge measurements. Gage datum must be established by levels throughout the length of the reach.

The total length of reach to be surveyed, so that convergence of computed backwater curves will be ensured, is an important consideration where dense foliage and other problems make surveying a very difficult and costly operation. The length of reach needed depends on the slope, the roughness, and the mean depth for the largest discharge for which the normal-depth profile is desired. Because the length depends on the depth, and the depth itself is the unknown which must ultimately be determined, the total reach length must be computed by estimating the normal depth. A convenient rule-of-thumb for determining total reach length, L, is

$$L = 0.4(Y_n/S) \tag{5.29}$$

in which Y_n is the normal depth and S is the slope. In this expression, L is the length of reach required for an M2 curve starting at an elevation representing a depth equal to $3/4 Y_n$ in a rectangular channel, to converge within 3 per cent of the normal-depth profile.

The total reach should be fairly uniform in shape, area, conveyance, and slope, and the cross-sections should be located such that the individual sub-reaches can be represented by Figure 5.1. The various criteria for cross-section location, discussed in Subsection 5.6.4 for slope-area measurements, also apply for step-backwater reaches. In addition, the following considerations will influence the selections of cross-sections:

(1) The total reach length should be divided into at least 10–12 sub-reaches in order to establish convergence.
(2) No sub-reach should be larger than about 75–100 times the mean depth for the largest discharge to be considered, or about two times the width, at most. This is a maximum limit, and applies only if any other considerations as to cross-section locations do not particularly apply.
(3) The fall in a sub-reach should be equal to or greater than the larger of 0.15 m, or the velocity head, unless the bed slope is so flat as to require defaulting to the second criterion above.

5.12.3   Use of step-backwater method for indirect discharge measurement

A useful application of the step-backwater technique can be made in the determination of discharge by indirect means in a long slope-area reach. The reach may be ideal in every respect for a slope-area measurement, having a uniform shape and roughness, and steep sides, but it may lack high-water marks except for an excellent mark or two at the upstream end. A stage-discharge relation can easily be established at the upstream end where the high-water marks are located.

Cross-sections (at least 8–10) can be located through the reach, and two or more M2 profiles can be computed for each of a series of assumed discharges about the magnitude of the expected discharge. The reach should be long enough for that slope, geometry, and roughness of channel, for several M2 curves for each discharge to converge. In this manner, a stage–discharge

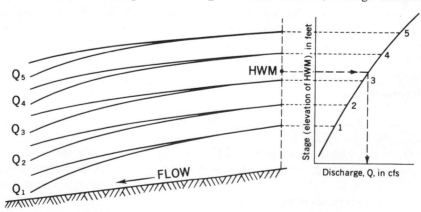

Figure 5.10   Definition of a rating curve at the upper end of a long reach by means of the step-backwater method, using convergence of M2 curves

rating is established for the cross-section at which the high-water mark is located, as in Figure 5.10.

The discharge corresponding to the elevation of the high-water mark, as determined from a well-defined rating as in Figure 5.10, should be every bit as reliable as a slope-area measurement made in that reach with good high-water marks to define the water-surface elevation.

5.13   SIMPLIFIED SLOPE-AREA METHOD

The basic slope-area method of computing discharge by indirect means after the passage of a flood was described in Section 5.6. A reach of uniform channel is selected on which the flood profile on both banks can be defined from high-water marks. Surveys are required of these high-water marks and of channel cross-sections, and estimates of the roughness coefficient are needed for the Manning equation—equations (5.1a), (5.1b), or (5.5).

Although judgement is required in selecting the stream reach and in interpreting the profiles from high-water marks, the major subjectivity is in selection of the roughness coefficient, n. The roughness coefficient of a natural channel is due to bed roughness, bank irregularity, effect of vegetation (if any), depth of water, channel slope, and perhaps other factors. No objective way of evaluating and combining these factors into one coefficient is available. Furthermore, a verified value of a roughness coefficient is affected by the inaccuracies in each of the other variables in the Manning equation. Inaccuracies may arise because of a poorly defined water-surface profile or non-representative cross-sections. A poorly defined profile may be caused by poor high-water marks, bank irregularity at the high-water line, or a changing cross-sectional area and shape throughout the reach. In rough cross-sections the detail to which the cross-section is surveyed will affect the computed area and thus the computed roughness coefficient.

Some measurable channel characteristic related to the roughness coefficient might permit an objective selection of the latter. Channel (or water-surface) slope is one such characteristic. Riggs (1976) has analysed n-verification data reported by Barnes (1967). He finds that these natural channel data show a relation between the roughness coefficient and the slope, although in practice the two parameters have been considered independent. By assuming that these two variables are closely enough related that only one need be used in computing discharge, Riggs has proposed a simplified method described in the next subsection.

5.13.1   Theoretical basis

The existence of a relation between channel roughness and water-surface slope in natural channels suggests that discharge might be computed without using a roughness coefficient and without defining the relation of roughness coefficient to slope. With the assumption that the slope will replace n and that the hydraulic radius, R_h, is closely related to cross-sectional area, A, the Manning equation is reduced to the model

$$Q = aA^b S^e \qquad (5.30)$$

if S is redefined as water-surface slope.

In an analysis of the data from Barnes' (1967) report, Riggs (1976) determined that the relation with slope in equation (5.30) is not linear. His analysis results in equation (5.31a) or (5.31b), which is the basis for the simplified method:

$$\log Q = 0.366 + 1.33 \log A + 0.05 \log S - 0.056(\log S)^2 \qquad (5.31a)$$

where Q is in ft^3/s, A is in ft^2, and S is dimensionless, or

$$\log Q = 0.191 + 1.33 \log A + 0.05 \log S - 0.056 (\log S)^2 \qquad (5.31b)$$

where Q is in m^3/s, and A is in m^2.

The standard error for this equation is about 20 per cent. Riggs further checked the suitability of equation (5.31a) with tests on 44 slope-area verifications that were not used in its derivation. The results were comparably good. The advantages of this method, according to Riggs, are:

(1) No subjective estimate of a roughness coefficient is required; both cross-sectional area and slope can be measured.
(2) A water-surface profile based on high-water marks throughout the reach is desirable, but well-defined water-surface elevations at the ends of the reach may be adequate.
(3) Computation of discharge is simple; and one answer is obtained. There is little room for subjective judgements either by the originator or by the reviewer.
(4) Two or more measurements at different stages in the same reach will give consistent answers.
(5) The method appears to give results comparable in accuracy with those from the slope-area method for natural streams.

5.13.2 Field data

The reach should be selected according to the same criteria used for a conventional slope-area measurement, except that a uniform cross-sectional area throughout is preferable to a contracting reach. Ideally, the reach should be four or five channel widths long in order that the water-surface slope can be measured accurately. The method has been verified only for nearly full natural channels without substantial overbank flow and without backwater from constrictions or a flooding tributary. This, it is recommended only for such channels at this time.

The criteria are not as restrictive as they may seem. Necessity often forces the use of the best available reach, which may have undesirable characteristics both for the simplified method and for the conventional slope-area method. Thus, a reach which does not meet the criteria is used when there is no better way to estimate the flood peak. The field procedures are:

(1) To determine peak discharge of a flood from high-water marks: (a) define slope of water surface from high-water marks as a straight line throughout the reach; (b) measure three cross-sections so located that their average defines the average cross-sectional area in the reach; (c) apply equation.
(2) To determine discharge corresponding to a specific stage, either arbitrarily selected or from a few high-water marks: (a) define water-surface slope as bed slope or slope of water surface at existing stream stage; (b) project defined water-surface slope through specified stage; (c) survey three cross-sections and average the areas; (d) apply equation.

5.14 OTHER METHODS

As mentioned in Section 5.3, there are other methods available for the indirect measurement of discharge that are considerably more independent of channel geometry, of channel roughness, of discharge coefficients for control structures, or of the principles of open-channel hydraulics than are the methods described above. Those other methods, though less 'direct', and of lesser reliability, must, nevertheless, occasionally be resorted to because of the unavailability of one of the more standard types of indirect-measurement site, or the lack of opportunity to use them.

Some of these other methods are being continually refined with more sophisticated application of the principles of statistics, mathematics, and hydrology. A general description of several of these methods is given below, but the state-of-the-art is steadily being upgraded.

5.14.1 Flood-routing procedures

The storage in the reaches of stream channels is used extensively as an index of the timing and shape of flood waves at successive points along a river. The principal uses of the technique are:

(a) to route hypothetical floods through river systems to determine the effects from proposed flood-control projects;
(b) to forecast river stages during high-water periods;
(c) to schedule the operations of hydroelectric power systems according to the predicted progress of a flood wave; and
(d) To compute and evaluate streamflow records.

The flood-routing techniques can also be used at gaging sites at which the stage of the flood wave is known from high-water marks, but not the discharge. The number of direct observations of discharge during major floods in a river basin is generally limited by the short duration of the flood and the inaccessibility of certain stream sites. Through the use of flood-routing techniques, all observations of discharge and other hydrologic events in a river basin may be combined and used to evaluate the discharge hydrograph, or the peak discharge, at another site.

Flood waves are subject to two principal kinds of movement—uniformly progressive flow and reservoir action. A uniformly progressive flow designates downstream movement of a flood wave without a change in shape, which would occur only under ideal conditions in a prismatic channel in which the stage and discharge are uniquely defined at all places. Reservoir action refers to the modification of a flood wave by reservoir pondage. Flood-wave movement in natural-channel systems is probably intermediate between the two ideal conditions cited, one or the other predominating in a particular place. However, the actual behaviour of the wave is sometimes obscured by the effects of local tributary inflow.

Many different methods are used to route flood waves through river reaches. All these methods are based on the law of continuity—the volume of water that is discharged from a reach during any interval must equal the volume of inflow during the interval plus or minus any increment in stored water during the period. In equation form it becomes

$$\bar{O} = \bar{I} - \frac{\Delta S}{\Delta t} \tag{5.32}$$

where $\bar{O}$ is the mean outflow during routing period Δt; $\bar{I}$ the mean inflow during routing period Δt; and ΔS is the net change in storage during routing period Δt.

Equation (5.32) is general. A modification frequently used is

$$\frac{\Delta t(O_1 + O_2)}{2} = \frac{\Delta t(I_1 + I_2)}{2} - (S_2 - S_1) \tag{5.33}$$

where O, I, S, and Δt are as before, and the subscripts identify the beginning and ending of routing period Δt. The assumption that mean discharge is equal to the simple arithmetic average of the flows at the end points of the interval can be justified if the period is equal to, or less than, the time of travel through the reach and no abrupt changes in flow occur during the routing period.

Stage-storage method

The simplest method of flood routing defines the storage in terms of the mean gage height in the reach. Thus, a gage-height record for both ends of the reach must be available if the flood is to be routed. The necessary stage–storage relation is generally defined on the basis of past flood-discharge records; although in certain reaches it may be defined from topographic data. Aerial photographs of rivers in flood are also valuable in defining this relation. Because storage is directly related to stage, $\bar{A}\Delta h$ may be substituted in equation (5.32) for ΔS:

$$\bar{O} = \bar{I} - \frac{\Delta S}{\Delta t} \tag{5.32}$$

$$\bar{O} = \bar{I} - \frac{\bar{A}\Delta h}{\Delta t} \tag{5.34}$$

where $\bar{A}$ is the average area of water surface in the reach during time Δt, and Δh is the average change in water-surface elevation in the reach in time Δt.

In equation (5.34), the value of $\bar{I}$ is obtained from the given inflow hydro-graph and from the value of $\Delta h/\Delta t$ which is obtained from the stage record. The water-surface area at a given stage is the slope of the stage–storage curve and may be easily computed from a stage–storage table, since it is the 'first difference'. Hence, the slope or first difference is a function of the mean stage in the reach. The outflow may be computed from the mean stage, the rate of change of stage, and the inflow.

Study of previous floods will help define a stage–storage relation for the reach.

Discharge-storage method

Most methods of flood routing now in use define the storage volume in terms of the inflow and outflow rather than the stage. One of the most widely used is known as the Muskingum method, which makes use of both the inflow and the outflow rates to express storage as a function of the weighted mean flow through the reach. Detailed explanations of this basic method are given by Chow (1959), the US Corps of Engineers (1960), and most hydraulics text-books.

5.14.2 Hydrologic comparisons

One of the techniques of reconstructing missing periods of discharge records at a gage is the use of hydrographic comparison with records at one or two nearby stations. Flood peaks in particular, especially those caused by wide-spread rainfalls rather than by cloudburst activity of more limited areal extent, often have a high degree of correlation.

Hydrographic comparisons of concurrent peak flows that are caused by various storm events covering a range of magnitudes can be made between two, or among several, stations in order to establish the mutual relationships of these peaks. These relations can provide a good means for estimating missing discharges. The relations, whether they are simple-linear or multiple-linear regressions, can be determined by well-established statistical procedures, either mathematicalty or by graphical analyses. Many textbooks describe these procedures; among them, Riggs (1968), in particular, provides the background material needed for understanding and applying statistical procedures most useful in hydrology.

5.14.3 Unit-hydrograph method

The basic unit-hydrograph method has been accepted as one of the best methods available for determining the time distribution of the precipitation excess. It may be used to obtain a peak-flow value on the basis of rainfall information for a given storm over the basin.

The unit hydrograph may be defined as the hydrograph for a given basin, of a unit volume of surface runoff that is produced by precipitation of uniform intensity occurring in a unit of time. The unit volume is taken as 1 in of precipitation excess over the entire basin.

Unit hydrographs are derived for a given site by study of past storm events. Reliable information on the areal extent, and depth and distribution of rainfall over the basin is needed, as well as the stage and discharge hydrographs resulting from those storms. After the unit hydrograph is determined, the discharge hydrographs for storms of other durations and with other depths of rainfall excess may be constructed in order to determine the peak flows. The method is well described in many hydrology textbooks.

There are several techniques for developing synthetic unit hydrographs. They are the methods proposed by Clark (1945), Snyder (1938, 1939), Mitchell (1948), and the US Soil Conservation Service (1957), and are described in detail in these original references as well as in many textbooks. Although these methods are useful in the synthesis of a unit hydrograph, their use could affect the reliability of a computed peak discharge if the shape of the synthesized unit hydrograph is not verified with actual flood data.

In addition to the requirement that the unit hydrograph should be based on actual flood hydrographs, there are a few other limitations in its use. It is assumed for the unit hydrograph that storms occur with uniform intensity over the entire basin; therefore, storm orientation is not considered to be important. Actual storms, however, move across a basin and their direction of movement does affect the nature of the runoff. The larger the drainage basin, the less the chance that any storm will satisfy the condition of uniform intensity. Unit hydrographs are therefore considered to be applicable only to basins having an area of less than 1000 square miles (about 2600 km^2) and preferably smaller if accuracy is important. Furthermore, large basins have substantial variations in soil conditions to consider.

For the purpose of indirectly determining the peak flow by use of this method, one must have available a well-defined unit hydrograph for a small basin, a good measure of the depth and distribution of rainfall over the basin, and high-water marks at the gage site.

REFERENCES

Barnes, H. H., Jr., 1967. Roughness characteristics of natural channels, US Geological Survey Water-Supply Paper 1849.

Benson, M. A., 1968. Measurement of peak discharge by indirect methods, WMO No. 225.TP.119, Technical Note No. 90, Geneva, Switzerland.

Bodhaine, G. L., 1968. Measurement of peak discharge at culverts by indirect methods, US Geological Survey Techniques of Water-Resources Investigations, Book 3, chap. A3.

Chow, V. T., 1959. *Open-Channel Hydraulics*, McGraw-Hill, New York, pp. 101–123.

Clark, C. O., 1945. Storage and the unit hydrograph, *Am. Soc. of Civil Engineers Trans.*, **110**, 1419–1446.

Corbet, D. M. *et al.*, 1945. Stream-gaging procedure, US Geological Survey Water-Supply Paper 888.

Dalrymple, T. and Benson, M. A., 1967. Measurement of peak discharge by the slope-area method, US Geological Survey Techniques of Water-Resources Investigations, Book 3, chap. A2.

Davidian, J., Carrigan, P. H., Jr. and Shen, J., 1962. Flow through openings in width constrictions, US Geological Survey Water-Supply Paper 1369-D.

Horton, R. E., 1907. Weir experiments, coefficients, and formulas, US Geological Survey Water-Supply Paper 200.

Hulsing, H., 1967. Measurement of peak discharge by indirect methods, US Geological Survey Techniques of Water-Resources Investigations, Book 3, chap. A5.

Kindsvater, C. E. and Carter, R. W., 1955. Tranquil flow through open-channel constrictions, American Society of Civil Engineers Transactions, **120**.

Kindsvater, C. E., 1964. Discharge characteristics of embankment-shaped weirs, US Geological Survey Water-Supply Paper 1616-A.

Limerinos, J. T., 1970. Determination of the Manning coefficient from measured bed roughness in natural channels, US Geological Survey Water-Supply Paper 1898-B.

Matthai, H. F., 1967. Measurement of peak discharge at width contractions by indirect methods, US Geological Survey Techniques of Water-Resources Investigations, Book 3, chap. A4.

Mitchell, W. D., 1948. Unit hydrographs in Illinois, Division of Waterways, State of Illinois, USA.

Riggs, H. C., 1968. Some statistical tools in hydrology, US Geological Survey Techniques of Water-Resources Investigations, Book 4, chap. A1.

Riggs, H. C., 1976. A simplified slope-area method for estimating flood discharges in natural channels, *US Geological Survey Journal of Research*, **4**, (3), 285–291.

Snyder, F. F., 1938. Synthetic unit graphs, *Am. Geophys. Union Trans.*, part 1, 447–454.

Snyder, F. F., 1939. A conception of runoff phenomena, *Am. Geophys. Union Trans.*, part 4, 725–738.

US Corps of Engineers (Department of the Army), 1960. *Routing of Floods through River Channels*, Engineering Manual 1110-2-1408.

US Soil Conservation Service, Department of Agriculture, 1957. *Engineering Handbook*, Sect. 4, Hydrology Supplement A.

Woodward, S. M. and Posey, C. J., 1941. *Hydraulics of Steady Flow in Open-Channels*, Wiley, New York.

CHAPTER 6

Aspects of Unsteady Flow and Variable Backwater

R. J. MANDER

6.1 INTRODUCTION

The principle of flow measurement in open channels relies upon the premise that the total energy state of the channel reservoir implied by the bulk parameters being measured exists as a factually steady and virtually uniform flow* throughout the time required to complete the overall measurement.

---

* In a long, wide, uniform channel of even grade, uniform flow would become established for a given constant discharge when the gravity and frictional forces acting along the grade achieve a balance. When this occurs the energy and water surface slopes are parallel to the bed slope, velocity and depth are constant along the channel. The flow is termed non-uniform when the forces are not in balance.

Since any cross-section of flow has the property of being the real transient mean of the cross-section in the reach which its total energy state defines, such a requirement of steady and uniform flows means that the fluid volume in storage in that reach shall not been seen to change either as a reach volume or as a volume distribution within the reach during the overall period of measurement. Thus in the case of a direct velocity-area measurement at a single cross-section, and of dilution methods also, steady uniform flow is always implied. When the reach length is given a quite arbitrary definition, as in the case of float measurements, and also of slope–area methods involving friction formulae such as those due to Chezy and Manning for example, steady and uniform flow is assumed.

However, such ideal steady-state conditions are not found in nature and all open-channel measurements are made always under conditions of changing channel storage although, in the practical terms of a measurement, in many circumstances the overall period of measurement may coincide with a reach energy state which may be accepted as steady. Generally, however, the accuracy of any single measurement of discharge will be affected by the flow conditions prevailing at the time of the measurement.

In a reach of channel free from intermediate inflows and abstrations, volumetric and distributive changes in reach storage will occur when the inflow rate changes or when backwater conditions change within the reach due to causes other than a change in inflow rate. It follows that a change in the volume and distribution of a reach storage can result from a change in inflow rate coincidental with a change in backwater not caused solely by the change in inflow rate itself.

Within the time scale of any single measurement of discharge it is therefore important to be able to recognize the presence of changes in channel storage, to have an adequate measure of their real effect upon depth of flow and velocity with respect to time at the cross-section concerned and, if possible and justifiable, to make an appropriate adjustment to a measured discharge in order to align it as closely as possible with that for the equivalent steady and uniform flow.

6.2 THE CONTROL REACH

When, by any method, a discharge measurement is made at or with reference to a single cross-section of flow, the parameters measured relate to a reach of channel and not simply to the measuring cross-section in isolation.

In a continuum of channel flow any wetted cross-section is, as a transient, the representative mean of a reach whose transient length is defined by the energy slope and mean depth of flow prevailing at the given cross-section, and whose transient volume of fluid storage is determined as the product of the wetted areas in the measuring cross-section and the reach length so defined. Irrespective of whether or not the flow is steady and the channel uniform, the instantaneous existence of the properties of the wetted cross-section thus imply an equivalent steady and uniform displacement of the actual transient storage in the reach, and so relate to a finite and characteristic time for such displacement.

As a result, the transient mean velocity in the measuring cross-section is likewise the mean velocity in the reach so defined and may therefore be valued correctly only if it is possible for the mean velocity in the measuring cross-section to be computed over a continuous period of time equal to the transient characteristic time. It follows that only when both stage and discharge at the measuring cross-section remain steady for a period of time not less that the characteristic time such that the characteristic time may be seen as constant, that the mean velocity in the measuring cross-section may be determined correctly over any interval of time within the characteristic time.

6.3 STAGE–DISCHARGE RELATIONSHIP

Although single measurements of discharge in open channels can be made by a variety of methods and serve a variety of purposes, the usual objective in the broadest sense is to establish or confirm a relationship between steady stage* and steady discharge at a selected channel cross-section in order to obtain a continuous statement of discharge from a continuous or frequent measurement of stage alone. As a result, in addition to the accuracy of the individual measurements from which it derives, the order of stability of a stage–discharge relationship over very long periods of time is to be recognized as a fundamental aspect of the calibration and operational processes and needs to be so treated (see Chapter 1). A relationship between depth of flow, in terms of stage, and discharge in open channels is usually an empirical statement deriving from a quantity of time-coincident measurements of stage and discharge at the location concerned although, in the case of purpose-built control structures inserted into a channel, or of velocity measurement by the ultrasonic method, frequently only confirmation of a model or theoretical relationship and of its continued stability with respect to time will be the requirement, but it may not prove always to be a practical undertaking. Such a relationship is controlled by the physical properties of the reach defined by the measuring cross-section, whereby the downstream extent controls the backwater condition and the upstream extent the approach velocity condition, such as to determine the reach mean wetted area and corresponding mean velocity, respectively, at the measuring cross-section. A stage–discharge relationship is therefore fixed if the control reach storage and its distribution

---

* Stage is defined as water surface level referred to a fixed site datum. It may be expressed either as a local arbitrary value or in terms of the national sea level datum. In the case of an artificial control, stage is normally referred to the lowest point on the flow boundary of the structure and is termed the 'head'. A change in stage with time is thus the same thing as a change in depth with time at the location of the gauge.

are fixed for any given steady discharge. If changes in reach storage with respect to any discharge are small relatively and without trend the stage–discharge relationship is said to be stable. Otherwise it is unstable.

6.4 STABILITY OF THE STAGE–DISCHARGE RELATIONSHIP

In the hydrometric sense backwater is simply channel storage expressed as depth of flow and so, at any given cross-section of channel, there will exist a neutral or normal depth which defines the condition of minimum or normal backwater, and hence also normal energy and water surface slopes*, at any given steady discharge. This applies equally to section controls, which may be weirs or flumes, when the normal backwater condition relates to what is termed the modular range of flow, so that for abnormal backwater conditions the flow would be non-modular. Thus, if at any steady discharge the volume and distribution of storage in the reach represented by the measuring cross-section are modified by changes in the control properties of the reach such as could be produced, for example, in the short term by high runoff rates from downstream tributaries or from artificial regulation, in the seasonal term by weed growth or by ice, and in the long term by changes in the geometry and configuration of the channel, the normal stage–discharge relationship established for the measuring cross-section would not then apply because of the variability between depth of flow, the energy slope, and the water surface slope, at any given discharge. Similarly, if the discharge is unsteady, then at any instant the energy and water surface slopes will differ from those which relate to the corresponding but steady discharge under conditions of normal backwater, the complexity of the problem being increased if both discharge and backwater are changing during a period of discharge measurement.

By definition, a stable channel is one wherein the physical form and frictional properties of the bed and sides remain constant with respect to time. Many natural channels are effectively stable, or may be made so by the construction of a section control of fixed geometry such as a weir or flume. A channel is unstable when those properties vary with respect to time such that the channel itself is mobile. But in any reach of channel, stable or unstable, any transient natural or artificial phenomenon whose effect is to exert influence upon the relationship between stage and discharge at any location within the reach will be a factor in the overall control which determines the stage–discharge relationship at any time. Thus a stable channel may exist as an unstable reach.

In all cases, however, instability in a stage–discharge relationship, whether in the short or in the long term, arises from variable conditions of backwater causing variable slopes for a given discharge at the measuring cross-section

---

* In hydraulics the energy slope for uniform flow is termed the normal slope and should not be confused with the established hydrometric usage of the term which means the slope at minimum (normal) backwater.

concerned and thus includes the effects of changing discharge and of changes in the mean velocity of approach at a given stage due to changed backwater conditions upstream from the measuring cross-section. But whereas in all circumstances of variable backwater at a given discharge a stable channel will provide a stable normal stage–discharge relationship and therefore also a stable stage–slope–discharge relationship when the energy slope in the reach concerned is included as a parameter in the discharge equation, such will not be the case for an unstable channel.

When determining the relationship between stage and discharge at a measuring cross-section the two general considerations are therefore:

(1) The accuracy with which the individual measurements can be made to represent steady and uniform conditions of flow at normal backwater.
(2) The stability of the relationship with respect to time.

These two considerations will now be discussed.

6.5 STAGE–SLOPE–DISCHARGE RELATIONSHIPS

The general equation for unsteady flow in open channels may be written as

$$-S_e = \frac{\partial z}{\partial x} + \frac{\partial \cdot \alpha \bar{V}^2}{\partial x \, 2g} + \frac{1 \cdot \partial \bar{V}}{g \, \partial t} \tag{6.1}$$

in which, at an instant of time, S_e is the slope of the energy line; $\partial z/\partial x = S_w$ the slope of the water surface; $\partial/\partial x (\alpha \bar{V}^2/2g)$ the slope increment due to velocity head; and $1/g(\partial \bar{V}/\partial t)$ is the slope increment due to acceleration head.

Upon the assumption that the acceleration term is always small relatively and may be neglected:

$$-S_e = -S_w + \frac{d}{dx} \cdot \frac{\alpha \bar{V}^2}{2g} \tag{6.2}$$

The coefficient α in the velocity-head increment is frequently assumed to be unity but in some circumstances can approach twice unity. When high velocities are involved it is desirable that values of α should be determined from the velocity distributions in the cross-sections defining the measuring reach.

When slope is included as a parameter in the stage–discharge relationship, stage is required to be measured usually at the two locations defining a reach whose length is sufficient to ensure that the measured loss of head, termed the fall, can be resolved without significant error, but close enough to ensure that no measurable gains or losses to flow occur between them. Usually the upstream gauge will be the stage references at the measuring cross-section. Both gauges should be set to the same datum.

Mathematical relationships involving slope, stage, and discharge at a reference cross-section at which discharge is measured or to which a discharge measured at a more suitable location can be legitimately referred are, like the principle of direct measurement itself, based upon the ideal of steady and uniform flow. But because in natural channels the flow is non-uniform and invariably unsteady, observed transient bulk values of the parameters must be substituted for the uniform values. Also, the arbitrary reach length used to determine the water surface and energy slopes, and its usually biased location with respect to the reference cross-section, will be such as to effectively prevent its coincidence, in characteristic terms, with the implied length and related time associated with the measured mean velocity in the reference cross-section. With those reservations in mind the mean velocity in a reach may be written as

$$\bar{V} = KR_h^\beta S_e^\gamma \qquad (6.3)$$

where $\bar{V}$ is the reach mean velocity, R_h the mean hydraulic radius, S_e the energy slope, β and γ are exponents, and K is a friction term. Inserting the wetted area of the reach mean cross-section into equation (6.3) gives

$$Q = AKR_h^\beta S_e^\gamma \qquad (6.4)$$

The friction term K may be expected to vary slightly as a function of slope, but if constant for a given stage then at that stage

$$\frac{Q}{S_e^\gamma} = AKR_h^\beta = \text{const} \qquad (6.5)$$

If therefore, at a given stage, two discharges Q_1 and Q_2 occur at different times with corresponding energy slopes S_{e_1} and S_{e_2}, then

$$\frac{Q_1}{Q_2} = \frac{S_{e_1}^\gamma}{S_{e_2}^\gamma} \qquad (6.6)$$

The energy slopes in equation (6.6) may be treated as functions of the falls in the reach concerned such that

$$\frac{Q_1}{Q_2} = \frac{F_1^\gamma}{F_2^\gamma} \qquad (6.7)$$

If the discharge Q_2 in equation (6.7) is that corresponding to defined unit fall then $F_2 = 1$ and

$$Q_2 = \frac{Q_1}{F_1^\gamma} \qquad (6.8)$$

When the velocity head increment in equation (6.2) is significant it should be added to the measured falls.

It is assumed frequently that the exponent γ equals 0.5, but in practice it may be expected to vary somewhat about that value depending upon the degree of non-uniformity of the reach and upon the type of flow.

A detailed description of the applications of equations (6.6) to (6.8) to practical situations lies outside the scope of this chapter, but it is appropriate to set out the general principles of established procedures.

6.6 VARIABLE SLOPE DUE TO VARIABLE BACKWATER

For reaches subject to variable backwater the adjustment of measured discharges to represent the equivalent minimum backwater condition relies upon the principle that if all discharge measurements could be made during either the same (constant) condition of fall or the condition of minimum (normal) fall, so as to define the relationship of equations (6.8), then a discharge measured at a given stage at any other condition of fall may be adjusted relative to the datum fall by means of the ratios of discharge and fall in accordance with equation (6.7).

If all available measurements of stage and discharge are plotted in the usual way, and against each plotted point is written the observed value of fall at the time of measurement, there is then implied a family of contour curves of equal fall which may be reduced to an equivalent single-valued relationship by means of the 'constant fall' or 'normal fall' methods of analysis, depending upon the control characteristics of the reach concerned.

The constant fall method is applicable when a reach is relatively uniform and backwater conditions affect the whole reach such that no definite stage–discharge relationship will be apparent within the plotted data. A curve of some suitable constant fall is therefore drawn for which a first approximation may be obtained by the use of equation (6.8) with the exponent equal to 0.5. This curve is then adjusted by inspection to obtain the best empirical value for the exponent when, in accordance with equation (6.7),

$$\frac{Q_m}{Q_c} = \left[\frac{F_m}{F_c}\right]^{\gamma} \tag{6.9}$$

in which the subscripts denote the measured and constant fall values.

When a reach is sufficiently non-uniform as to contain a control feature which prevents some backwater conditions from affecting the whole reach, there will occur a break in the energy slope and a quite definite stage–discharge relationship will be apparent, logically to the right of the plotted data with stage as ordinate, and which defines a condition of minimum backwater at the reference gauge at the head of the reach. For such a situation the constant fall method is unsuitable and a datum curve must be constructed which defines the minimum backwater condition at each stage. The fall represented by such a curve is therefore the normal fall and, in accordance with equation (6.7),

$$\frac{Q_m}{Q_n} = \left[\frac{F_m}{F_n}\right]^\gamma \qquad\qquad (6.10)$$

in which the subscripts denote the measured and normal fall values.

The next stage in the graphical process is to compute and plot the pairs of ratios of discharge and fall as in equations (6.9) or (6.10), as the case may be, and construct the mean curve through the data. When this has been done the measured discharges can be adjusted to the corresponding constant fall or normal fall discharge and replotted to give the required single-valued stage–discharge relationship. It may be necessary, depending upon the scatter of the computed data, to recompute the discharge and fall ratios after the curve of best fit has been drawn, replot them and then redraft their curve of relationship. In some non-uniform reaches several successive trials may prove necessary before satisfactory curves of relation are achieved. Figure (6.1) and (6.2) illustrate the processes described.

In the case of artificial controls subject to variable backwater the principles and methods of treatment are, of course, essentially the same, although discharge under conditions when normal backwater is exceeded is termed non-modular. Frequently, however, as in the case of the triangular profile Crump weir for example (Burgess and White, 1966), the discharge equations for both modular and non-modular conditions will have been determined from laboratory model tests, and field measurements are needed only to check the model calibrations.

6.7 VARIABLE SLOPE DUE TO UNSTEADY FLOW

When a flood wave passes through a reach implied and represented by a measuring cross-section the effect of the wavefront when in the upstream section of the reach is to increase the velocity of approach at the measuring cross-section. When the flood peak passes into the downstream section of the reach the rear of the wave increases the backwater condition and so reduces the velocity at a given discharge at the measuring cross-section. The result is that on rising discharges the stage at the measuring cross-section generally is lower at a given discharge than when discharges are falling, so that when stages are plotted against corresponding discharges measured during the passage of the wave, a looped curve is obtained as illustrated in Figure 6.3.

The problem is often complicated by the effects of accumulation and release of overbank storage, but may be treated nevertheless in a similar way to that of variable backwater alone by adjusting the discharges measured under conditions of changing discharge relative to measurements made during conditions of steady discharge.

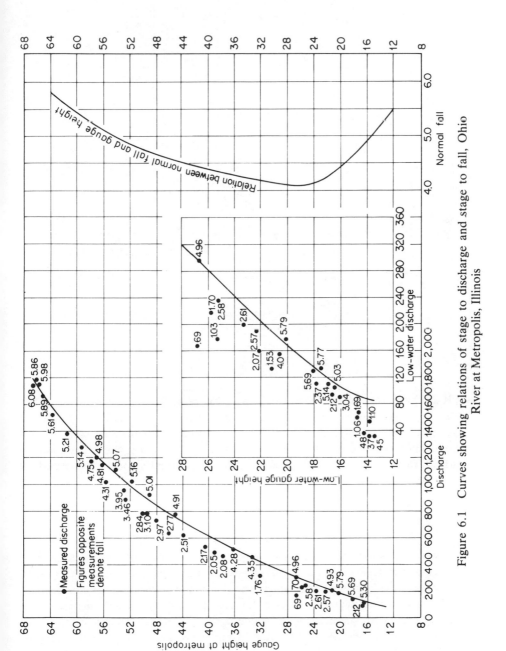

Figure 6.1   Curves showing relations of stage to discharge and stage to fall, Ohio River at Metropolis, Illinois

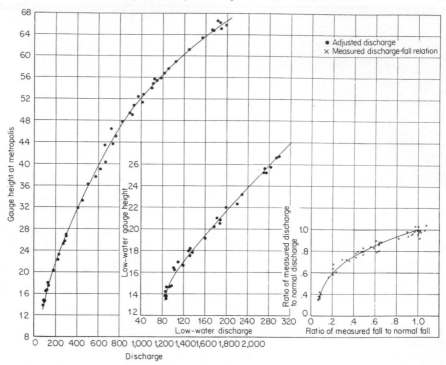

Figure 6.2   Curves showing relations of stage to normal discharge and discharge ratios to fall ratios, Ohio River at Metropolis, Illinois

If discharges are measured at or can be expressed correctly in terms of the reference cross-section at which stage is measured, if overbank storage is not significant, and if the celerity of the flood wave is assumed or can be estimated, then from consideration of the water surface slope, the wave celerity, and the rate of change of stage during the measurement period (Corbett, *et al.*, 1945),

$$\frac{Q_s}{Q_m} = \frac{S_{ws}^{1/2}}{\left[S_{ws} + \dfrac{1}{V} \cdot \dfrac{dh}{dt}\right]^{1/2}} \tag{6.11}$$

in which Q_m is the measured discharge; Q_s the equivalent steady discharge; S_{ws} the water surface slope at the steady discharge; and dh/dt is the rate of change of stage, positive or negative, and V is the celerity of the flood wave.

The use of equation (6.10) is limited to stations which provide two measurements of stage some distance apart. When stage measurement is restricted to the gauge at the measuring cross-section the relationship is modified as

$$\frac{Q_c}{Q_m} = \left[1 - \frac{1}{VS_e} \cdot \frac{dh}{dt}\right]^{1/2} \tag{6.12}$$

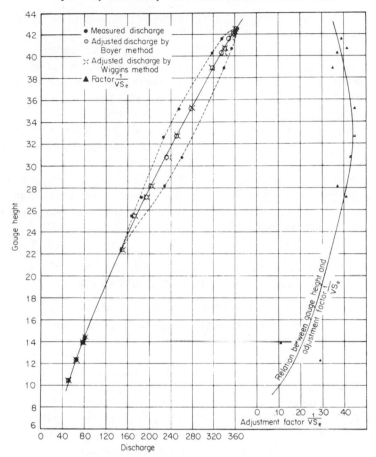

Figure 6.3   Adjustment of discharge measurements for changing discharge, Ohio River at Wheeling, West Virginia, during the period of 14–27 March, 1905

in which the energy gradient, S_e, is estimated by means of a standard formula, such as that due to Manning, which is

$$\bar{V} = \frac{1}{n} R_h^{2/3} S_e^{1/2} \tag{6.13}$$

in which n is the estimated roughness coefficient for the channel reach concerned.

The celerity V of the flood wave may be taken as $1.4\,\bar{V}$ for most natural non-uniform channels at high stages. However, empirical solutions are available for estimating the wave celerity in given situations, and also for computing the adjustment factor to be applied to discharges measured during unsteady flow conditions, without direct knowledge of wave celerity or slope, provided sufficient existing measured stage–discharge data are available for analysis (Boyer, 1939).

In reaches where artificial regulation controls the depth of flow for purposes of navigation the water surface slopes can be extremely small, especially below median discharge, and conventional methods of discharge evaluation which require water surface slope as a parameter are impracticable. A possible solution then is the generally complex one of calibrating the regulating structure which controls the reach, but the presence of river traffic and the effects of locking operations coupled with frequent changes in the weir geometry introduce special problems and, in any case, calibration of the lower range of discharges would necessarily be a theoretical assumption when reach velocities are likely to be well below the lower limit of calibration of current meters, and unmeasureable leakage past the weir gate seals is large relative to the total discharge.

6.8 VARIABLE SLOPE DUE TO UNSTEADY FLOW COMBINED WITH VARIABLE BACKWATER

The effects of a given backwater condition in conjunction with unsteady flow are usually contained in the slope determined at stations equipped for measurement of water surface slope. But if the backwater as well as the discharge is changing during the period of a measurement the problem becomes too complex for any form of general solution to be stated. Practical individual solutions are possible, however, if the quantity of discharge measurements available is sufficient to define completely the effects of changing backwater during measurements. But generally, for regulated reaches, and for reaches affected by variable changing backwater, the appropriate solution is the use of the ultrasonic method of continuous measurement of mean velocity at the measuring cross-section.

6.9 UNSTABLE (SHIFTING) CONTROLS

In unstable channels the channel geometry and friction properties and hence the control characteristics vary continuously as a function of time and so does the stage–discharge relationship at any given point on the channel. Such variations in control are evident particularly during and after flood periods, and during periods of weed growth and decay in channels susceptible to weed growth but may be due also to channel improvements or modifications to existing control structures. In many such cases acceptable estimates of discharge during periods when no actual measurements are available may be achieved as follows: all discharges measured during a water year are plotted against their corresponding stages and each point labelled by date order. The lie of the points is then examined for shifts in control with reference to their chronological order and, provided a sufficiency of measured values exists, smooth curves may be drawn for those periods when very little or no shift is apparent, and from which discharges in correct chronological context within that water year may then be estimated. Where weed growth is a cause of instability the general effect is seasonal. During the growing season the weeds progressively increase the friction and decrease the discharge capacity of the channel with the resultant backwater effect demanding progressively higher stages to pass any given discharge. As the weeds decay the reverse effect is observed. However, after periods of flood the effect of weed growth is usually changed radically, and when flood discharges occur during the growing season abrupt shifts in control can result but which reduce quite quickly as new growth becomes established.

6.10 THE THAMES AT TEDDINGTON

In open-channel hydrometry in particular, rarely are field problems so easily solved as might appear to be the case from simple consideration of theoretical principles and usually much general and local experience is required if they are to be seen in their proper perspective, and well enough understood to allow realistic appraisal of their importance, in terms of both magnitude and of time, as factors affecting a required result. It is appropriate therefore to review briefly the main aspects of a real, and important, hydrometric problem of typical complexity in which the essential requirement is an effectively continuous and reliable measurement of discharge deriving from the calibration of a large system of compounded weirs and gates designed primarily for head regulation. Our concern extends beyond the present-day discrete measurement to include the relative accuracies, or consistency, of measurement sustained over a long period of years which include significant changes in climate and in land and water use, and during which major structural changes affecting the control characteristics of the weir system have taken place.

The Thames to Teddington drains an area of 9951 km^2 which receives a long-term mean annual rainfall of 734 mm and provides a corresponding mean annual runoff of 247 mm or 33 per cent. The catchment contains a large groundwater storage potential, mainly in fissured limestones, the discharge from which is estimated to account for almost 50 per cent of the mean annual runoff. From the source at Thames Head, near Cirencester in Gloucestershire, the average gradient of the main channel is 1/2186 over the distance of 236 km to Teddington, of which 199 km are maintained navigable for small craft of draught not exceeding 1.98 m at Teddington, decreasing to 0.91 m at Lechlade, the upstream limit of regulation, by means of 44 locks and associated weir systems. Although Teddington defines effectively the tidal limit, the final weir is 5.13 km further downstream, at Richmond.

Except when discharges lie outside the limits within which headwater level at each weir can be controlled by varying the weir geometry, a standard headwater level (SHWL) is specified as the zero of the headgauge and, within the limits of practical operations, is maintained to serve the various requirements of navigation, land drainage, mill rights, and water use generally, including licensed abstractions. With certain exceptions, the design of Thames weirs is intended to provide a uniform level of all gate crests and overfalls when a weir is fully closed in, and ideally this would be SHWL at a location concerned. But actual headwater maintained normally needs to exceed SHWL in order to provide the required degree of operational flexibility and so, in practice, the designed crest level of weirs is usually slightly above SHWL.

As an aid to navigation, and hence also as an indirect aid to maintenance of SHWL at each weir, tailgauges are sited immediately downstream from each lock or lock system, gauge zero being at the cill level of the lock concerned. In the case of Teddington, the headgauge zero (SHWL) is at +4.38 m, and the tailgauge zero is at −1.08 m, referred to mean sea level Newlyn.

For the majority of Thames weirs records of headwater and tailwater levels observed at fixed times each day exist from January 1892, and for Teddington since January 1895. Earlier Teddington records dating from 1883, and from which the available monthly summaries of discharge for the period 1883–1894 were computed, appear no longer to exist.

The reach controlled by Teddington Weir is limited to 7.74 km by the weir at Molesey. At SHWL the reach averages 2.57 m deep by 76.20 m wide, and at zero discharge SHWL at Teddington is theoretically the same as standard tailwater level (STWL) at Molesey. The recorded extreme minimum discharge of zero occurred during the 1976 drought, with evidence of transient negative reach mean velocities, and the recorded extreme maximum discharge of 1066 cumecs occurred during the 1894 flood, when reach mean velocities probably exceeded 4 m s^{-1}. The recorded extremes of headwater relative to SHWL are −0.102 and +1.829 m.

Subject to a defined conditional relaxation, it is a statutory requirement that the measured discharge at Teddington shall not be reduced below 8.95 cumecs by specified upstream abstractions. The abstractions concerned, which serve the London Metropolitan area, the Greater London area and beyond into Hertfordshire, Essex, and Surrey, and are effected wholly within the 30 km reach between the weirs at Romney and Molesey, normally total about 18 cumecs, but can presently be maximized at 65 cumecs, subject to an average daily rate not exceeding 21.10 cumecs in the calendar year, and to an average rate not exceeding 17.40 cumecs during the period 1 May to 31 October. No significant proportion of that total abstraction returns to the river above Teddington Weir and so the natural discharge at Teddington is taken to be the measured discharge plus the related total abstraction. Since in a normal year the order of natural discharge during the low-flow period September–October is but 28 cumecs and in a dry year can be less than 14 cumecs, including an effluent content currently estimated as 5.40 cumecs, considerable emphasis is placed upon accurate measurement of the lower range of flow.

In order to improve the quality of the discharge measurements and dispense with the use of Teddington Weir as a hydrometric device, a single-path ultrasonic gauging station was commissioned in December 1974 at Kingston, in the Teddington reach. Although still in process of calibration, ultrasonic results now available permit valid comparison to be made with discharges calculated from the Teddington Weir relationships.

6.10.1 The weir system

The weir history at Teddington dates from 1811 but the hydrometric record commenced in 1883 when headwater and tailwater readings as a daily routine were first established. Since then the layout of the weir system and locks has conformed generally to that existing today, although major improvements in hydraulic design have been introduced over the years. Notable among these was the contruction in 1923 of a weir section specifically to measure low flows, this being achieved by adapting 70 ft length of the existing weir section A as a sharp-crested design, and capable of vertical adjustment over a range of 0.36 m primarily to make use of the tidal effects on storage in the head reach. In 1931 this thin-plate weir was reconstructed on a new line adjacent to the left bank and at the same time two additional deep-sill roller-type sluices were added and the Molesey–Teddington reach deepened to improve its flood discharge capacity. Finally, in 1950, the remaining sections of rymer-type structure dating from 1883 were replaced by radial-type gates. (The rymer weir is an elementary form of variable geometry weir consisting of fixed horizontal beams against which are supported vertical timber posts to form a series of rectangular openings. The openings may be partially or totally closed by means of timber gates fixed to the end of long poles (the combined gate and pole is referred to as a 'paddle') which may be inserted and removed by hand. A rymer weir is not intended to be a watertight structure.)

Figure 6.4(c) outlines the weir system as it exists today, each distinct weir section being lettered to facilitate historical comparison, and should be compared with the correspondingly lettered photograph in Figure 6.5 and also with the aerial view in Figure 6.6, in which the three locks and boatslide appear in the foreground. The individual weirs are shown in greater detail in Figure 6.7. In Figure 6.6 the housing for the tailwater recorder is clearly visible on the small lock island, next to the downstream gate of the centre lock.

Figure 6.4(a) outlines the weir system as it was in 1883 with the various sections lettered to conform with Figure 6.4(c), and should be compared with the aerial view in Figure 6.8, in which the earlier location of the headwater recorder is to be seen between weir sections A and B in the centre foreground.

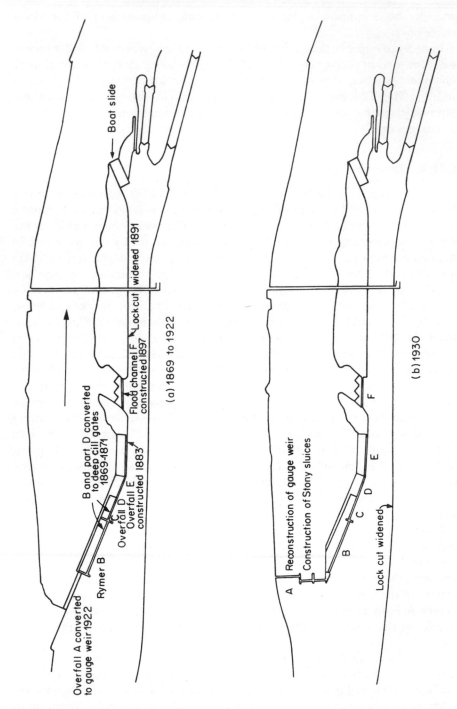

(a) 1869 to 1922

Boat slide

Overfall A converted
to gauge weir 1922

Rymer B

B and part D converted
to deep cill gates
1869-1871

Overfall D
Overfall E
constructed 1883

Flood channel F — Lock cut widened 1891
constructed 1897

(b) 1930

A   Reconstruction of gauge weir

Construction of Stony sluices

B   C   D   E   F

Lock cut widened

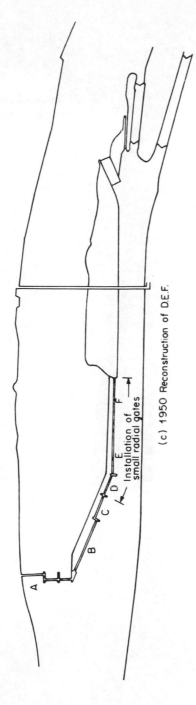

Figure 6.4   Teddington Weir—main stages of development: (a) 1869–1922; (b) 1930; (c) 1950

Figure 6.5   View of Teddington Weir to-day from downstream

Figure 6.6 Aerial view of Teddington Weir to-day from downstream

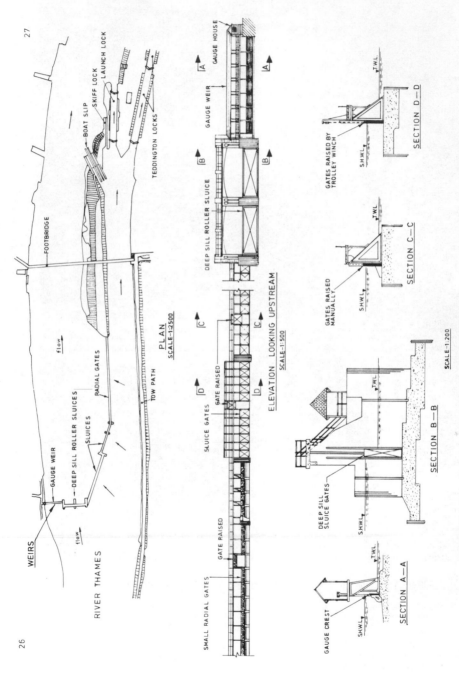

Figure 6.7   Teddington Weir—details of individual gates, sluices and thin-plate weir (gauge weir)

Figure 6.8   View of Teddington Weir from upstream, taken in 1920

Figure 6.4(b) shows the main intermediate stage of development of the weir following reconstruction of section A in 1930.

In the context of the homogeneity of the long record of discharge, changes in gate types, in gate and cill dimensions, in cill and crest levels, and the dates when such changes occurred are obviously essential information, and it is relevant to note the control geometry of the weir system as it was in 1883 and the major changes which have taken place since.

The history of the weir system can be summarized suitably by representing four periods of near stability in structural terms: 1883–1922; 1923–1930; 1931–1950; and 1951 to date in schematic form as in Figures 6.9(a) to 6.9(d) respectively, all cill levels, crest levels and effective cill lengths being given. It should be noted nevertheless that structural changes did occur within those defined periods and included the construction at section F of a flood channel during 1897 and, throughout 1941–1950, the sealing-off below SHWL of wartime bomb damage to section D, E and F (see Figure 6.4).

The design levels of cills and gate crests used in discharge calculations during any period were not necessarily the factual levels. For example, check levelling in 1928 and 1929 recorded errors of up to 0.046 m compared with levels in use up to that time. It was concluded then that since 1924 the overestimate of discharge for a fully drawn weir was +5.5 per cent and the revalued levels were introduced for discharge calculations from 1st March,

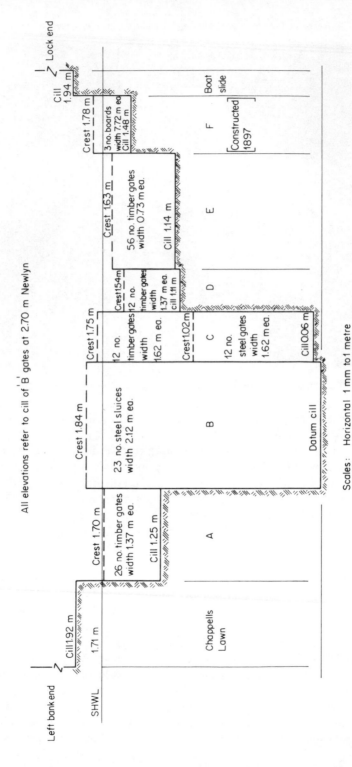

Figure 6.9a Teddington Weir 1883–1922

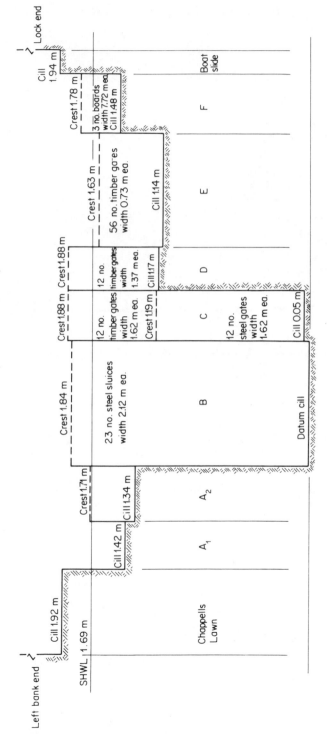

All elevations refer to cill of 'B' gates at 2.70 m Newlyn

Scales: Horizontal 1 mm to 1 metre
Vertical 50 mm to 1 metre

Figure 6.9b Teddington Weir 1923–1930

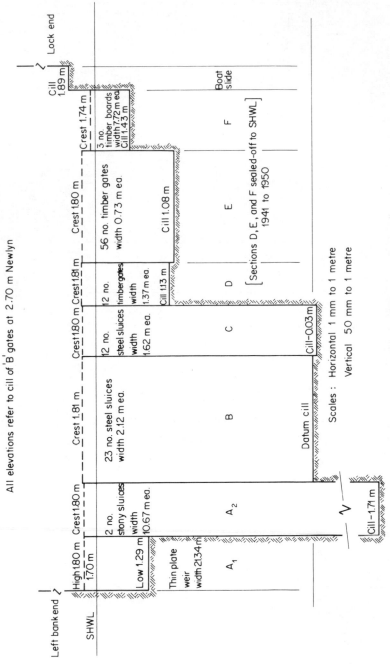

Figure 6.9c Teddington Weir 1931–1950

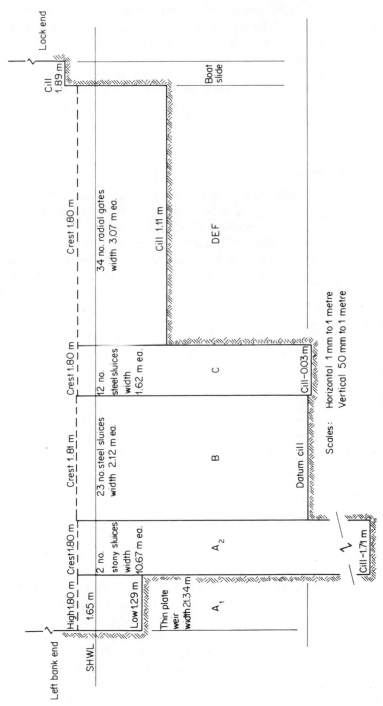

Figure 6.9d   Teddington Weir 1951 onwards

1929. Partly for such reasons, but also because of a lack of complete continuity and detail in the record of the weir geometry in time, small errors in retrospective valuation of levels and effective dimensions of the various weir sections are unavoidable.

Exact knowledge of an individual gate setting at any time is, of course, essential for calculating total discharges which depend, in effect, upon individual gate ratings, and so a log of all weir and lock operations is maintained for this and other purposes.

6.10.2 Stage measurements

Since 1895, headwater and tailwater levels at Teddington have been read from staff gauges located at the head and tail of the lock system, and have been logged at fixed times between 0900 and 1800 hours and at times of high and low water daily. The lock headgauge carries side-by-side calibrations both of inches and tenths of a foot, but the lock tailgauge is calibrated in inches only. Headwater and tailwater levels are also recorded continuously by float-operated chart-type instruments, evidently at least since 1891, although the exact dates of the original installations have not been determined. Prior to 1931, the headwater recorder was located between weir sections A and B (see Figures 6.8), when it was moved to its present site at the left bank immediately upstream from the sharp-crested weir. The tailwater recorder is at the river side of the small lock island and in line with the downstream end of the centre lock. Additional staff gauges in the vicinity of the recorders are used for check purposes.

It is important to note that throughout the record period headwater and tailwater levels at all points on the weir system have, in effect, been inferred by observation at a single fixed location in each case.

6.10.3 Discharge measurements

Standard formulae dating from 1883 in respect of each discharging gate or gate crest are applied when the total discharge is less than about 85 cumecs. It is on record that coefficients and dimensions in the formulae have been revised from time to time, generally supported by current meter measurements, to reflect the affects of changes in gate and cill ratings and of dimensions necessitated by changes in the weir structure or re-evaluation of cill and crest levels.

When discharge exceeds 85 cumecs, estimates are presently obtained from one of two tailwater stage–discharge curves, the applied relationship depending upon the backwater condition created by Richmond Weir. Except for tidal releases, Richmond Weir is effectively closed for upland discharges below 227 cumecs.

Because of tidal backwater, valid tailwater levels occur only twice each day at times of low tide, and linear interpolation between the related twice-daily discharges given by the appropriate tailwater curve is used to derive an average discharge for the day commencing at 0900 hours. An assumed daily mean discharge is thus the basic element of the Teddington Weir record. To a calculated daily discharge is added an allowance for unmeasured discharge resulting from locking and leakage combined. Until 1921 the allowance appears to have been 1.05 cumecs for all weir gates closed. From 1921 to 1932 it was 1.42 cumecs, from 1933 to 1950 it was 0.57 cumec, and since 1950 it has been 0.42 cumec.

6.11 FACTORS AFFECTING ACCURACY OF DISCHARGE ESTIMATES

6.11.1 Leakage

With a large number and complexity of gate types it is impossible at any time, currently or retrospectively, to estimate reliably the leakage past gate seals, under gate bottoms, and under gate cills. At low flows with all gates closed the potential leakage is a maximum both in absolute terms and as a proportion of total discharge. Incomplete closure of weir gates due to debris does occur, and gradual deteriortation of gate cills leads to piping discharge which can be large relatively. Debris collects also on the crests of closed gates when the head on the crest is small, and may at any time effectively destroy the head–discharge relationship of the thin-plate weir if not immediately cleared, as is well illustrated in Figure 6.10. The result of worn gate cills coupled with incomplete closure of gates due to cill obstruction is evident in Figure 6.11, which shows section B of the weir in 1972 when all B gates were supposedly closed and headwater was at or near the designed uniform crest level of the gates. When the rymer type of weir was in use, leakage per unit area of gate will have been much greater than that occurring at the present time.

6.11.2 Locking

Whilst there is always some leakage through closed lock gates, frequencies of locking vary with season and with time of day. Maximum daily frequencies occur during the summer but tend to continue into the months of September and October when discharges are at their seasonal lows, and the disturbing effects of locking are then greatest in terms of discharge measurement.

The surge initiated by a locking, at either end of the reach, undergoes repeated reflexion with gradual attentuation at the reach ends, and at a location within the reach can be evident, especially at low discharges, as a velocity pulsation whose fundamental period and relative magnitude depend, at such discharges, primarily upon the reach length. A pulsation waveform becomes complex when lockings are sufficiently close in time.

Figure 6.10   Teddington Weir, showing thin-plate weir
obstructed by debris, 1972

At low discharges, when headwater level is being held at or near gate crest level, the effects of locking can be observed as a periodic surge of discharge over the gates crests, frequently from zero to a fully aerated state, with an intermediate period when the nappe clings to the faces of the gates, as illustrated in Figure 6.12. If the thin-plate gauging weir is accommodating the nominal total discharge, possibly with the help of a small opening at one roller sluice, there are at least 180 m of supposedly uniform crest over which such surge discharge may occur.

The effect upon observed headwater level at the weir when locking occurs is significant particularly during the summer-months when river traffic is at its

Figure 6.11 Teddington Weir, showing leakage under
bucks, 1972

peak. In terms of a graphical record, oscillations can of course be damped by
severely restricting the area of the intake to the float well. In the case of
Teddington, however, true headwater at any time is assumed to be at the
estimated centroid of the undamped oscillatory range.

6.11.3 Variable backwater

Tailwater level at Teddington is influenced by variable backwater from tides
and also, at low discharges, from the effects of the control variability of
Richmond Weir. When discharge calculation is made in terms of the
Teddington Weir geometry any effect due to backwater on the modularity of

Figure 6.12   Eynsham Weir—radial gates showing unmeasurable crest discharge
and leakage

weir gates and cills is allowed for by the implicit inclusion of tailwater level in
the discharge equations. However, although Teddington Weir does not
relinquish the final element of control on headwater level until the discharge
reaches some 350 cumecs (nominal bank-full and normal flood is about
260 cumecs) the present calibration requires reversion to the tailwater curves
at a mere 85 cumecs.

At tidal low water, Richmond has to pass only the prevailing upland
discharge at the required retention level, the resultant backwater is treated as
normal and, for discharges in excess of 85 cumecs, tailwater level at Teddington is valid for application to the appropriate tailwater rating curve.

When, during a flood tide, a tailwater level is attained at Richmond which
requires that weir to be fully drawn, tailwater level at Teddington then rises
rapidly and, if it exceeds significantly the level of 3.46 m, Newlyn (Trinity
High Water) headwater level at Teddington is raised increasingly and is
reflected in an increased tailwater level at Molesey.

The effect of tidal backwater on Teddington tailwater and headwater for
tides of various amplitudes at the low upland discharge of 72 cumecs, including the well-documented high tide augmented by surge which occurred on
1st February, 1953, is illustrated in Figure 6.13. On that day, although the

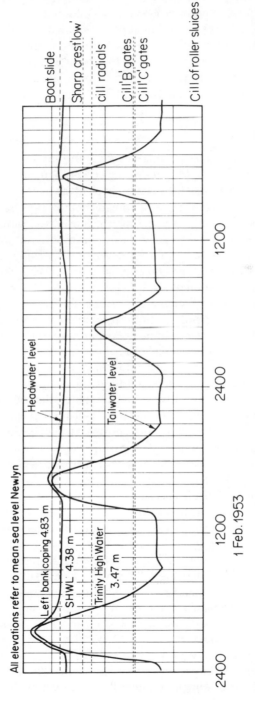

Figure 6.13 River Thames at Teddington—effects of tidal backwater on headwater and tailwater levels (includes tide of 1 February, 1953)

predicted tide itself was not exceptionally high, the combined effect of the tide and of the North Sea surge upon headwater at Teddington was to raise if from 4.51 m at preceding low tide to 5.42 m at peak tide, with a related increase of tailwater level at Molesey from 4.70 to 5.49 m. It has been estimated that if the upland discharge on 1 February, 1953 had been 525 cumecs (tenth-order flood), headwater level at Teddington would have attained 5.95 m, with a corresponding tailwater level at Molesey of 7.47 m.

6.11.4 Weir procedures

When a weir comprises a large number of gates of different kinds and dimensions, the number of permutations and combinations of gate openings which will maintain the same headwater level at a given discharge can be considerable. At the same time, a supposition that the discharge of any number of gates, located randomly among a group of the same kind, will be the known discharge of one type-gate multiplied by the number of identical gates open may well prove to be in obvious error, especially so when partial drawing of gates is permitted, for it takes no account of the relative changes in hydraulic interference between gates.

For the field calibration of such weirs, when the practical limitations of available methods of independent measurement of discharge restrict calibration to selected discharges within the operating range of the weir, it becomes an important part of the calibration to identify standard sequences of opening and closing the weir gates and to ensure that these be maintained as routine procedures. Usually, such standard procedures will depend primarily upon the structural and hydraulic designs of the weir and upon the operational and environmental requirements, but in some circumstances the hydrometric interest may be important enough to justify procedures particularly favourable to discharge measurement.

At Teddington no such standard procedures are in use and there is no evidence that they may have existed in earlier times, although certain general operational policies have been laid down, and also changed, from time to time. In hydrometric terms probably the most important of these was a major change in policy involving the use of the two roller sluices which, from the time of their construction in 1930, until 1958, were used only for flood discharges, that is they were opened only when, on a rising river, the remaining sections of the weir had become fully drawn. As a result, before either of the roller sluices was discharging, the undeflected discharge mainly from sections B and C of the weir (of the order of 300 cumecs) caused erosion problems on the opposite bank, and so in 1958 it was laid down that in future the two roller sluices would discharge at all times when the 15 cumecs capacity of the sharp-crested weir was exceeded.

At the present time, therefore, discharge assessments by weir formulae are limited effectively to the ratings of the roller sluices and the sharp-crested weir (with some radial gates open to provide the necessary degree of fine control), for at about 1 m combined opening of the roller sluices the total discharge requires that the tailwater rating curves to be introduced. However, the roller sluice discharge is approximated, in linear terms, as being equivalent to discharge through section B of the weir, the rule being that 1 m of combined opening is equivalent to six fully drawn and freely discharging gates of section B. In such cases, therefore, the effects of variable backwater upon the discharge through the roller sluices are assumed to be non-existent and, bearing in mind the location of the tailwater recorder, the changed tailwater condition which could obtain for a given backwater condition and given discharge due to the change configuration of flow immediately downstream from the roller sluices is ignored, albeit knowingly.

6.11.5   Methods of discharge evaluation

Prior to 1970 the headwater and tailwater charts provided records from instruments of such design that friction effectively damped out any transient changes in water levels so that periods of several hours of apparently constant water levels could be defined and, when weir formulae were applied, the mean discharge for the day commencing at 0900 hours was computed as the weighted mean of the resulting irregular periods of varied discharge during that 24 hours. However, since 1973, in view of the more sensitive instruments installed, discharges appropriate to the use of weir formulae have been evaluated every hour.

Since 1950 and at the present time, discharges in excess of about 85 cumecs derive from the tailwater rating curves, by which the effect of tidal backwater on tailwater levels limits valid tailwater readings to two per day, and so the mean discharge is, in effect, represented by the mean of two evaluations during any 24 hours, based upon the continuous record from the tailwater recorder. However, it seems clear from the records that prior to 1950 the weir formulae were applied until the weir was fully drawn, and also possibly thereafter, which means that some lack of consistency, and also relative change in accuracy of discharge evaluations before and after 1950, is to be expected.

6.12   ASPECTS OF DATA VALIDATION

The following discussion on data validation is to demonstrate simple examples of applied processes only and is not intended to present formal conclusions regarding the integrity of the long record of Teddington discharges.

6.12.1 Historical discharges

Little if any profit is likely to accrue from any primitive attempt to value the historical record of lower order discharges at Teddington, and it is sufficient to say here that it is on record that the coefficients in the various stage–discharge equations were established and checked, as occasions demanded, from 1883 onwards by current meter measurements. It is reasonable, however, to review the higher order discharges.

With increasing discharge the weir gates at Teddington are drawn progressively in order to maintain SHWL until, at about 350 cumecs when the weir system is fully drawn, the weir geometry is constant for a continued increase in discharge. However, well before that stage is reached the reduced relative importance, in control terms, of the weir gates yet to be drawn is such that above about 230 cumecs a high degree of correlation between depth of flow in the head reach and discharge is to be expected. The two main factors which can disturb that correlation are the relative phasing of inflows from the river Mole, a major tributary capable of high rates of runoff which enters the Thames immediately below Molesey Weir, and very high tides.

Although a subjective adjustment in respect of tidal backwater may be made to any headwater level so affected, and the actual stability of the correlation theoretically improved by including the water surface slope in the head reach as a parameter, certain features of the historical record are more easily observed if the water surface slope is omitted, and if the correlation is stated directly in terms of the available unadjusted but consistent record of water levels obtained from the tailwater gauge at Molesey Weir.

Dividing the full record into the two periods 1883–1931 and 1932–1976, in order to exclude from the earlier period the effects of the major structural changes and channel improvement in 1931, which also provides broadly equal-sized samples, and using available published data supplemented by low-order floods to define adequately the practical lower limits, and without consideration of possible tidal effect upon certain tailwater levels at Molesey, the two resulting correlations defined by eye-fitted curves appear as in Figures 6.14a and 6.14b, respectively. On that evidence the following points could be inferred:

(1) Apart from the relative shift in datum the mean curve for each of the two periods defines essentially the same relationship.
(2) The variance of the calculated discharges appears significantly greater for the earlier period.
(3) An effect of the Thames Improvement Scheme has been to lower the tailwater level at Molesey Weir by a maximum of about 0.436 m at discharges exceeding 450 cumecs.
(4) The extreme flood peak estimate of 1076 cumecs in 1894 appears to be consistent with the remaining data in the sample to 1931.

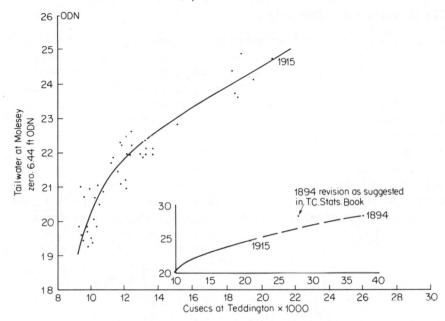

Figure 6.14(a)   Teddington Weir; 24-hour flood discharges related to tailwater at Molesey Lock; period 1883–1931

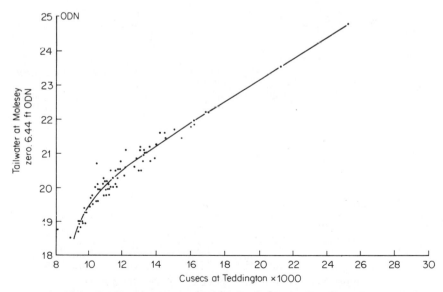

Figure 6.14(b)   Teddington Weir; 24-hour flood discharges related to tailwater at Molesey Lock; period 1932–1976

6.12.2 Present-day discharges

In the case of the analysis demonstrated in respect of historical discharges greater than 230 cumecs, no suitable and completely independent reference was available by means of which the absolute, as distinct from merely the relative, accuracies of those discharges could be assessed. However, in respect of the full range of contemporary discharges derived from the Teddington Weir relationships the ultrasonic gauging station in operation since December 1974 at Kingston, in the Teddington reach, can provide such a reference. At the same time, a prior and essential requirement is acceptable proof of the overall accuracy of the ultrasonic instrumentation itself and so a fundamental part of the calibration procedure for that installation consists of comparison of the vertical distribution of mean velocity measured concurrently by both ultrasonic and traditional current meter methods. Although completion of the

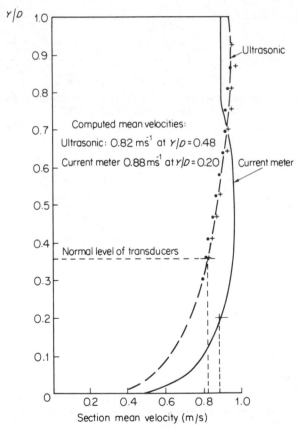

Figure 6.15   River Thames at Kingston—comparison of ultrasonic and current meter velocity profiles; December 1976. Y is depth above bed level; D is total depth

calibration phase of the ultrasonic station has been delayed by the long sequence of low discharges experienced since commissioning, available results permit adequate comparison of ultrasonic and current meter measurements, and those are listed in Table 6.1 as evaluated discharges.

An example of the mean velocity distribution compared in each case is given in Figure 6.15. The current meter result is obtained from arithmetic integration of the segments represented by each vertical at which mean velocity was in turn measured using five current meters on a single rod, the mean velocity at each vertical then being determined subsequently by graphical integration. The ultrasonic result is determined by moving the single pair of transducers in the vertical plane and recording velocities at selected points in the vertical, then using graphical integration to determine the section mean velocity. The two sets of measurements are, of course, concurrent, and the relative shapes of the two distributions are typical of all the pairs of measurements in Table 6.1.

Table 6.1   River Thames at Kingston—Comparison of Current Meter and Ultrasonic Discharges

Date	Current meter measurement (cumecs)	Ultrasonic measurement (cumecs)	Difference of ultrasonic from current meter result (%)
8/7/75	14.95	14.29	−4.41
9/7/75	10.40	10.62	+1.02
10/7/75	15.81	15.20	− 3.86
2/12/75	80.55	81.59	+1.01
3/12/75	142.58	151.30	+6.12
8/12/76	191.50	192.30	+0.42
9/12/76	182.70	181.00	−0.93
18/1/77	200.20	209.10	+4.45
19/1/77	155.40	160.70	+3.41
1/2/77	140.90	145.50	+3.26

On the basis that the ultrasonic measurements are the accurate reference since December 1974, a sample graphical comparison of daily mean ultrasonic discharges and those based upon the Teddington Weir relationships for the two periods October 1975 to June 1976 and October 1976 to March 1977 are given in Figure 6.16(a), for the discharge range 0–80 cumecs, and in Fig. 6.16(b), for the discharge range 80–310 cumecs. From simple inspection of those results it would be concluded that:

(1) From 0 to 13 cumecs the Teddington estimates are low by an average of about 20 per cent. Note: Since the defined maximum capacity of the

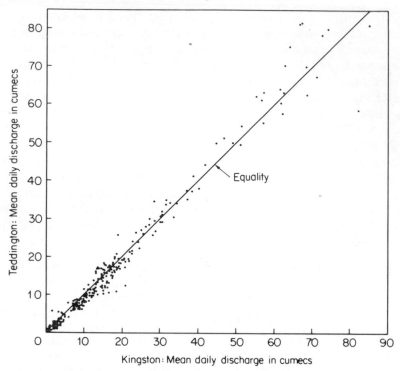

Figure 6.16(a)   River Thames—correlation between discharges derived from Teddington Weir calibration and ultrasonic measurements Kingston; data periods October 1975 to June 1976, October 1976 to March 1977 inclusive. See also Figure 6.16(b) for discharges >80 cumecs

thin-plate gauging weir is near 15 cumecs, it seems clear that the unmeasured discharge is much greater than the 0.42 cumec allowance for locking and leakage. In June 1970, with the weir fully closed-in apart from the thin-plate weir and one partial radial gate, the author estimated the leakage through the closed gates as not less than 1.7 cumecs for an office estimate of 15.70 cumecs total discharge. Subsequent to the bomb damage to weir sections D, E, and F in 1941 one of the rollers on the roller sluice was found to be fractured and it is possible that some leakage through the structure of section A of the weir may have developed.

(2) From 13 to 85 cumecs both sets of measurements are in good agreement on average. Note: For this range of discharge the Teddington estimates are based upon standard weir formulae.

(3) From 85 to 210 cumecs the Teddington estimates are high by an average of about 7 per cent. Note: It is at about 85 cumecs that the Teddington estimates change from weir formulae to tailwater curve derivations.

(4) From 210 to 310 cumecs (no comparative measurements above 310 cumecs are available) the Teddington estimates appear low by an

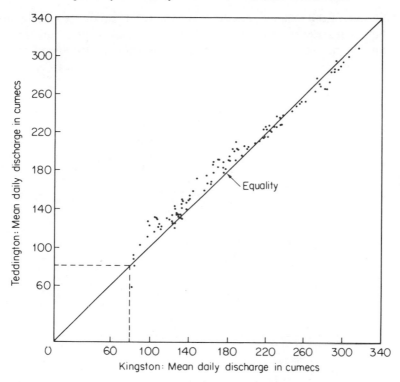

Figure 6.16(b)   River Thames—correlation between discharges derived from Teddington Weir calibration and ultrasonic measurements at Kingston; data periods October 1975 to June 1976, October 1976 to March 1977 inclusive. See also Figure 6.16(a) for discharges < 80 cumecs

average of about 1 per cent. Note: Above about 230 cumecs the degree of artificial control upon headwater by Teddington Weir becomes increasingly reduced until, at 350 cumecs, it is relinquished altogether. Also, it is in the region of 230 cumecs that Richmond Weir becomes fully drawn, and therefore that discharge reflects the changeover point between the two tailwater rating curves.

6.12.3   Runoff trends

It is essential to be reasonably sure that none of the many structural changes to which the weir system has been subjected, including change in location of the headgauge and deepening of the head reach, or changes in the method of operation of the weir and of evaluating discharges, have caused false trends to be introduced into the long record of accumulated runoff.

When the long-period mean of the annual volumes of runoff computed from the daily discharges is subtracted from each annual volume and the algebraic sum of the resulting residuals is computed and plotted on an annual basis, the result is as shown in Figure 6.17, in which the slope of any trend line represents the ratio of the mean annual runoff during that trend period, to the long-period mean.

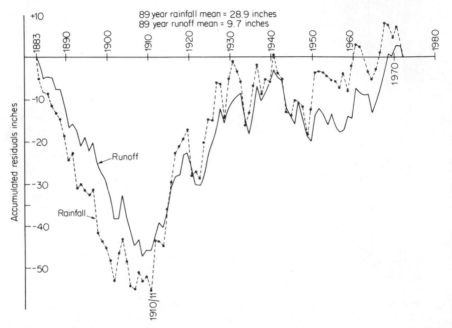

Figure 6.17   River Thames at Teddington—residual mass curve of rainfall and runoff about the corresponding 89-year means, based upon water years

Since it is possible in a relatively long record for any observed trends in annual runoff to be due essentially to climatic changes alone, the same procedure is carried out for annual areal rainfall on the Teddington catchment. The result can then be superimposed on Figure 6.17 to allow comparison with the annual runoff, in this case conveniently to the same absolute scale of residuals. The form of the result of the areal rainfall analysis is substantiated by similar individual analyses of the standard twelve rainfall stations used for computing the catchment areal amounts.

Allowing for the expected differences in phasing, the correspondence of the two sum curves throughout the data period would seem to justify the assumption that no major trend has been introduced into the runoff records as a result of the various changes made in and relating to the control structure at Teddington, all trends evident conforming to those exhibited by the mean annual rainfall.

This is not to say, of course, that the runoff record does not contain some persistent errors which, in a particular context, could be very important. For example, in the case of discharges below 15 cumecs, when the thin-plate weir is accommodating the total discharge, the mean velocity in the head reach is at or below the rating threshold of the impeller-type current meter and no calibration at such discharges has been possible prior to the commissioning of the Kingston ultrasonic station. As a result it would be a reasonable assumption on the basis of Figure 6.16(b) that, since 1931 certainly, all such discharges could have been seriously underestimated.

In terms of Figure 6.17 generally, the period of best average consistency between runoff and rainfall is from 1911 to 1949. However, in 1950 a sudden reduction in runoff relative to rainfall persists into the 1970s, and it would be difficult not to associate this with the possible effects of the reconstruction of weir sections D, E and F in 1950 combined with the change in policy regarding the use of the roller sluices following upon that reconstruction.

ACKNOWLEDGEMENT

The author is grateful to Thames Water for permission to use the information relating to the River Thames catchment, particularly that concerning Teddington Weir, and for permission to publish; the views expressed, however, are not necessarily the official views of Thames Water.

BIBLIOGRAPPHY

Andrews, F. M., 1962. Some aspects of the hydrology of the Thames Basin, *Proc. ICE*, **12**.

Allen, F. H., Price, W. A. and Inglis, Sir Claude C., 1955. Model experiments on the storm surge of 1953 in the Thames Estuary and the reduction of future surges. *Proc. ICE*, Part III Discussion 1955 Part III.

Boyer, M. C., 1939. Determining discharge at gauging stations affected by variable slope, *Civil Engineering*, **9**.

Burgess, J. S. and White, W. R., 1966. The triangular profile Crump weir: Two dimensional study of discharge characteristics, Hydraulics Research Station, Wallingford, UK.

Corbett, D., *et al.*, 1945. Stream gauging procedure, US Geological Survey Water Supply Paper 888.

Gilchrist, B. R., Flood routing. Chap. X. *Engineering Hydraulics*, Chapter X, Hunter Rouse (Ed.), Wiley, New York.

Grover, N. C., and Harrington, A. W., 1943. *Stream flow*. Wiley, New York.

Herschy, R. W. and Loosemore, W. R., 1974. The ultrasonic method of river flow measurement, *Water Research Centre and Department of the Environment*, Water Data Unit, Symposium on river gauging by ultrasonic and electromagnetic methods, University of Reading, UK.

ISO 1100, 1973. Establishment and operation of a gauging station and the stage discharge relationship, International Organization for Standardization, Geneva.

Johnstone, D. and Cross, W. P., 1949. *Elements of Applied Hydrology*, Ronald Press, New York.

Jones, B. E. 1916. A method of correcting river discharge for changing stage, US Geological Survey, Water Supply Paper 375.

King, H. W., and Brater, E. F., 1963. *Handbook of Hydraulics*, 5th edn, McGraw-Hill, New York.

Posey, C. J., 1949. *Gradually varied channel flow, Engineering Hydraulics*, Chapter IX, Hunter Rouse (Ed.), New York.

Steers, J. A., 1953. The East Coast floods Jan 31–Feb 1, 1953, *Geographical Journal*, **CXIX,** Part 3.

CHAPTER 7

New Methods

M. J. GREEN AND R. W. HERSCHY

7.1 INTRODUCTION

There are circumstances when the existing traditional techniques of stream gauging, detailed in earlier chapters, may be unsuitable. The velocity-area method, for example, requires a stable stage–discharge relation (Chapter 1) and this is not met in open channels where the water level is maintained at or near a constant level. Measuring structures (Chapter 3) are normally suitable for small rivers and only when a constriction in the river is acceptable.

The locations where existing techniques are unsuitable may be listed as:

(a) where no stable stage–discharge relation is possible;
(b) where afflux or backwater cannot be tolerated;
(c) wide rivers over 500 m wide;
(d) rivers where weed growth is predominant;
(e) rivers with moving beds;
(f) rivers where reverse flow is experienced.

This chapter deals with two new methods introduced in recent years to gauge open channels where one or more of the above difficulties is experienced.

7.2 THE ELECTROMAGNETIC METHOD

7.2.1 Introduction

One of the most difficult problems in connection with river gauging in the UK in recent years has been the profuse growth of weeds at gauging stations (see Chapter 9). This has no doubt been due to the increased use of farm fertilizers. Several attempts have been made to establish a method of gauging weedy rivers, but without success. The procedure therefore has been to develop a family of curves for different conditions of weed growth. Clearly this is a hit-and-miss approach and presupposes selecting the appropriate curve for the assumed condition unless the situation is carefully monitored on site. The results at best can only be approximate and at worst extremely questionable. It was for situations such as this that the electromagnetic method was developed in the UK.

As in the case of the ultrasonic method (Section 7.3) the civil engineer has called on the help of the electronics engineer, and the method requires a measure of sophisticated electronics particularly in the area of signal detection and processing. The electronics aspects of the system have been dealt with by Newman (1974, 1975) and will only be superficially covered here. The electromagnetic method requires an on-site empirical calibration usually by current meter, and a source of electrical power (mains supply) must be available. A feature of the system, as with the ultrasonic method, is its capability of measuring reversed flow. Silt or gravel deposited on the bed due to floods does not affect the system nor does the geometry of the river section. Cost alone determines the maximum width of river suitable, but generally this will be of the order of 50 m. Three stations are now in operation in the UK, the results from which are distinctly encouraging.

7.2.2 Principle of operation

The motion of water flowing in a river cuts the vertical component of the earth's magnetic field and an electromotive force (e.m.f) is induced in the water. This e.m.f. can be sensed by electrodes at each side of the river and is directly proportional to the average velocity of flow in the cross-section (Faraday, 1832). Unlike the ultrasonic method, therefore, which measures the velocity across a path, the electromagnetic system performs an integration over the entire cross-section. However, the e.m.f. induced by the earth's magnetic field is too small to be distinguished from other direct potentials due to electrical interference from mains, electrical motors and other ambient electrical noise present in some form or another in the ground. To induce a measurable potential therefore in the electrodes a vertical magnetic field is generated by means of a large coil buried beneath the river bed through which an electric current is driven (Figure 7.1).

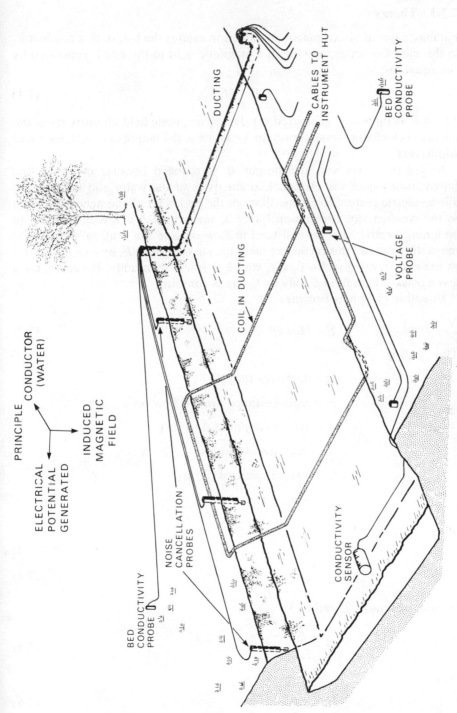

PRINCIPLE

CONDUCTOR (WATER)

ELECTRICAL
POTENTIAL
GENERATED

INDUCED
MAGNETIC
FIELD

DUCTING

CABLES TO
INSTRUMENT HUT

BED
CONDUCTIVITY
PROBE

VOLTAGE
PROBE

COIL IN DUCTING

NOISE
CANCELLATION
PROBES

BED
CONDUCTIVITY
PROBE

CONDUCTIVITY
SENSOR

Figure 7.1 Diagrammatic view of an electromagnetic gauging station

7.2.3 Theory

Faraday's law of electromagnetic induction relates the length of a conductor, in this case the water, moving in a magnetic field to the e.m.f. generated by the equation

$$E = Hvb \qquad (7.1)$$

where E is the e.m.f. generated (v); H the magnetic field intensity (t); v the average velocity in cross-section (m/s); and b is the length of conductor (river width) (m).

In practice, however, the output E is reduced because of signal loss (attenuation) since the electrical conductivity of the water and the bed will allow electric current to escape through the bed. This phenomenon is known as the conductivity attenuation factor δ, say, where δ is less than unity. In addition, electric currents will tend to flow outside the artificially produced magnetic field and this further reduces the signal output E, by a factor known as the end shorting factor β, say, where β is less than unity. However, for a given coil configuration β may be taken as constant.

Equation (7.1) now becomes

$$E = Hvb\,\delta\beta \qquad (7.2)$$

Thus, letting

$$Q = \text{discharge (m}^3\text{/s)}$$

$$A = \text{cross-sectional area of flow (m}^2\text{)}$$

$$h = \text{average depth of flow (m)}$$

$$I = \text{coil current (A)}$$

$$r_w = \text{water resistivity } (\Omega/\text{m})$$

$$r_b = \text{bed resistance } (\Omega)$$

and

$$Q = Av \qquad (7.3)$$

$$= bhv \qquad (7.4)$$

from equations (7.2) and (7.4)

$$Q = \frac{Eh}{H\delta\beta} \qquad (7.5)$$

Now

$$H = k_1 I \qquad (7.6)$$

where k_1 is the coil constant, and it can be shown (Herschy, 1975, 1976); (Newman, 1975) that

$$r_b = k_2 \frac{r_w \delta}{h} \tag{7:7}$$

from which

$$\delta = k_2 \frac{r_b h}{r_w} \tag{7.8}$$

Substituting for H and δ in equation (7.5) gives

$$Q = \frac{1}{k_1 k_2 \beta} \cdot \frac{Er_w}{Ir_b} \tag{7.9}$$

$$= K \frac{Er_w}{Ir_b} \tag{7.10}$$

where

$$K = \text{a dimensional constant} = \frac{1}{k_1 k_2 \beta} \tag{7.11}$$

and E is in microvolts (μV). The variables E, r_w, I, r_b and stage (h) are recorded on site, and the coefficient K is empirically calibrated by current meter at a few points in the range of measurement required.

In view of the non-uniformity of the magnetic field and the fact that H decreases with distance from the coil to the water surface, the empirical calibration may reveal either a straight line, or a curve, or a series of curves which may have a logarithmic transformation. That is,

$$Q = K_1 \frac{Er_w}{Ir_b} + c \tag{7.12}$$

$$- K_1 E_r + c \tag{7.13}$$

or

$$Q = K_2 \left(\frac{Er_w}{Ir_b} \right)^n \tag{7.14}$$

$$= K_2 E_r^n \tag{7.15}$$

where

$$E_r = \frac{Er_w}{Ir_b} \tag{7.16}$$

It will be noted that stage does not appear in the above equations. However, because the bed resistance includes both bed and water conductivity, r_b is also, in practice, a measure of stage.

Because of the smoothing effect of the current meter calibration, however, it has been found possible in practice to establish a relation between Q, E, h, and I as follows. From equations (7.5) and (7.6)

$$Q = K_3 \left(\frac{Eh}{I} \right)^n \tag{7.17}$$

$$= K_3 E_h^n, \text{ say} \tag{7.18}$$

where

$$E_h = \frac{Eh}{I} \tag{7.19}$$

It will be noted that δ in equation (7.5) has now been included in the dimensional constant K_3. Again the relation may be in the form of a straight line, or a curve, or series of curves which may have a logarithmic transformation.

It should be emphasized that although, in theory, equation (7.1) applies to a rectangular section, in practice it is found that the equation is adaptable to irregular sections. As stated above, this is mainly due to the method of calibration and to the random uncertainties in the current meter observations which may be of the order of ±7 per cent (95 per cent level). These uncertainties are probably significantly larger than the uncertainties due to the shape of the cross-section and to the conductivity attenuation. A calibration curve with a standard error of ±10 per cent (95 per cent level) is considered acceptable for river gauging purposes and this should be the criterion to be aimed at when designing an electromagnetic gauging station.

At noisy sites, where electrical interference is significant, four additional probes (noise cancellation probes) may be required. These are placed upstream and downstream of the coil and the signal from them deducted from the signal produced by the voltage probes.

7.2.4 Design and operation

Figures 7.2 and 7.3 show the layout for an electromagnetic station having sloping voltage probes. The station is on the river Rother at Fittleworth in Sussex. The bank-full width of the river (w) is about 25 m and this dimension is used to position the ducts for the coil and probes, the former being placed 25 m (w) apart. The voltage probes are located at the centre, and the noise cancellation probes each placed 12.5 m ($\frac{1}{2}w$) from the cable duct as shown. Finally the bed resistance (conductivity) probes are each located 25 m (w) back from the top of the banks.

On completion of the ducting the cables were threaded through the ducts in the usual way. To ease insertion the cable was cut into four lengths so that an end could be drawn through each of the two ducts. Those parts of the cables

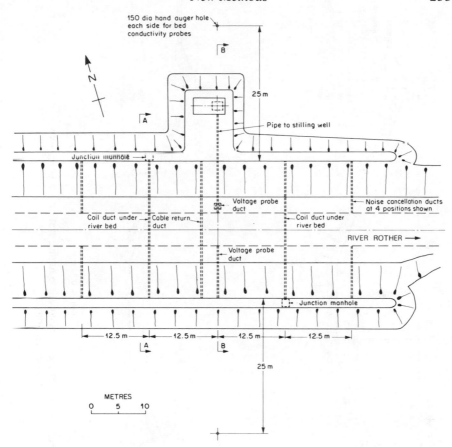

Figure 7.2   Layout of the Fittleworth electromagnetic gauging station showing plan
view

which were positioned in the banks were placed in a trench about 0.5 m deep.

At Fittleworth, although allowance has been made for two pairs of noise cancellation probes these have not been found necessary since the site noise is low.

All probes, including the bed conductivity probes, consist of 50 mm diameter stainless steel tubes and are taken down to the coil level. All are in 150 mm PVC ducts except the bed resistance probes. The ducts for the voltage probes and noise cancellation probes are perforated to allow the entry of water and the probes are centred in the ducts by means of insulated spacers.

The cross-sectional area of flow over which the voltage is measured is contained by that area between the voltage probes, the coil and the water surface. Therefore, any flow in the banks or bed within this area is included in

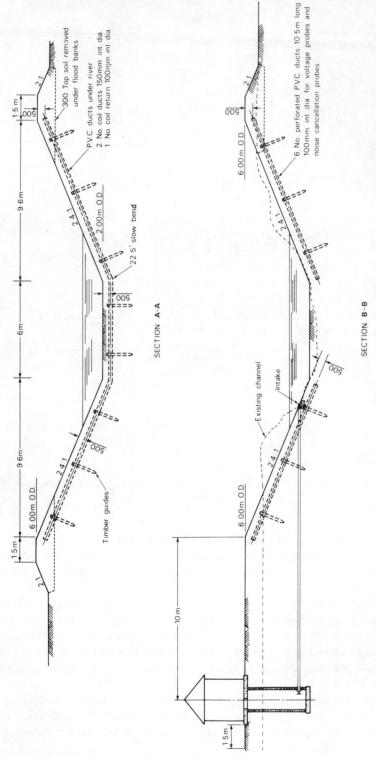

Figure 7.3   Layout of the Fittleworth electromagnetic gauging station showing view
of cross-section

the measurement. Silt or gravel deposits or weed growth do not therefore affect the measurement, and zero flow, negative flow and positive flow arc automatically integrated by the system. If the flow is negative a negative sign precedes the value of probe voltage in the output. Water conductivity is monitored by a standard sensor. The output is by teletype, which produces an 8-channel punched tape and a print-out of the variables E, r_b, I, r_w and stage at 15-minute intervals, the stage being recorded on a Fischer and Porter punched tape recorder interfaced with the processor unit.

The station was installed in October 1976 and the first calibration by current meter gave the following relations:

$$\text{using equation (7.13)} \quad Q = 0.280\,(E_r - 18.165)$$

(standard error of estimate ±13 per cent; standard error of mean ±5 per cent), or

$$\text{using equation (7.18)} \quad Q = 0.130\,(E_h + 1.576)^{2.676}$$

(standard error of estimate ±5 per cent; standard error of mean ±2 per cent). However, shift in the rating was later discovered and the station is being recalibrated (January 1978).

Advantage was taken of channel realignment works at Fittleworth to install the ducts in the dry (see Figure 7.4) and the cost of the site works were therefore reduced, the final cost coming to about £11 000.

Figure 7.4   View of Fittleworth electromagnetic gauging station under construction

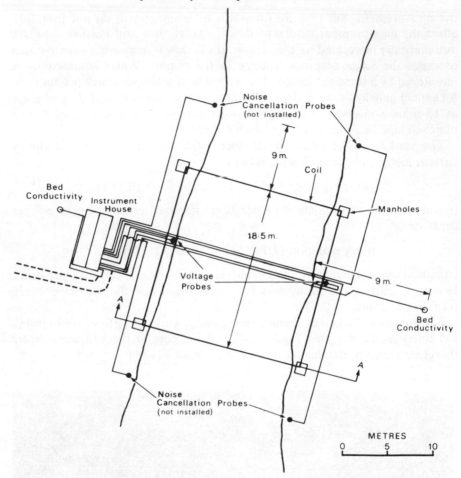

Figure 7.5 Layout of the Broadlands electromagnetic gauging station showing plan view

At Broadlands on the river Test, in Hampshire, however, the ducts had to be laid in a flow of about 20 m^3/s in velocities of over 1 m/s. Details of the station at Broadlands are given in Figures 7.5 and 7.6. Using a dragline, two trenches were excavated across the bed to a depth of about 0.5 m below the lowest point of the bed. Two rows of temporary piles were then placed at about 5 m intervals across the river on the downstream edge of each of the two trenches. Two 150 mm diameter PVC ducts were preformed to the shape of the bed (Figure 7.7) and each duct floated downstream to rest against each row of piles. The ducts were filled with water and submerged with the aid of divers (Figure 7.8). Bagged concrete was used to retain the ducts in position prior to back-filling. The voltage probes at Broadlands were driven into the

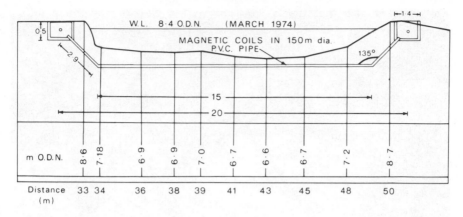

SECTION A-A

METRES

0 _____ 5

Figure 7.6 Layout of the Broadlands electromagnetic gauging station showing view
of cross-section

Figure 7.7 Broadlands electromagnetic gauging station showing installation of
coil ducts

Figure 7.8 Broadlands electromagnetic gauging station showing coil ducts being submerged

soft banks in a vertical position to a depth of about 2 m so that the base of the rods were level with the coil and the tops were just below ground level. Ambient electrical noise was, as at Fittleworth, low enough to dispense with the noise cancellation probes, although ducts were installed in case they should, at any time, be found necessary. A Fischer and Porter punched tape recorder was also installed to record stage. The cost of the civil engineering work including the brick build recorder house came to about £8000 and the station was put into operation in July 1976.

The present calibration (March 1978), by current meter, gave the following rating equations:

$$\text{using equation (7.15)} \quad Q = 1.147 \, (E_r - 16.08)^{0.60}$$

(standard error of estimate ±10 per cent; standard error of mean ±1.5 per cent), or

$$\text{using equation (7.18)} \quad Q = 2.352 \, (E_h - 0.92)^{0.94}$$

(standard error of estimate ±12 per cent; standard error of the mean ±2 per cent).

The coil at Fittleworth consists of four turns of 4-core aluminium 25 mm^2 cable (total area 400 mm^2) with aluminium armouring, giving a total of 16 turns. The same design was used at Broadlands but with three turns of 4-core cable, giving 12 turns. The coil at Fittleworth is excited with a current of about 24 A and at Broadlands about 19 A is used. Figure 7.9 shows a typical junction box at Broadlands, in which the lengths of coil cable were joined.

Figure 7.9 Broadlands electromagnetic gauging station showing view of coil junction box

The range of flows at Fittleworth is expected to be about 0.5–100 m^3/s and at Broadlands about 2–50 m^3/s.

The third electromagnetic gauging station at present operating in the UK is the experimental station on the river Rother at Princes Marsh, which has been described by Newman (1974, 1975) and Herschy (1976). The coil at Prince's Marsh is of diamond shape and has three turns of 4-core aluminium 16 mm^2 cable and is excited with a current of about 15 A. The station was sited about 20 m downstream of an existing Crump weir for comparison purposes. To compare the flow from two measuring devices in series on a river is a severe test. Factors to be considered include the uncertainty in the results from the existing station, the hydraulic features of the reach, and the lag in the recorded discharge between the stations. In general, however, the agreement between the Crump weir and the electromagnetic station is good.

The rating, unlike Fittleworth and Broadlands, is not linear and very much more complex. This is no doubt due to hydraulic conditions, particularly backwater, caused by downstream ponding effects. The station is now used mainly for research purposes, the final design of both Fittleworth and Broadlands being based on the design modifications made at Prince's Marsh.

7.2.5　Instrumentation

The success of the electromagnetic method depends on the accurate detection of small voltages in high ambient noise voltages. At Prince's Marsh, for example, the voltages being measured may be as low as $\frac{1}{10}$ of a microvolt (100 nanovolts) in noise as high as $\frac{1}{10}$ of a volt (100 millivolts).

Note:
$$
\begin{aligned}
1 \text{ volt} &= 1000 \text{ millivolts} \\
1 \text{ millivolt} &= 1000 \text{ microvolts} \\
1 \text{ microvolt} &= 1000 \text{ nanovolts}
\end{aligned}
$$

The instrumentation required is therefore of necessity, highly sophisticated and a detailed description is beyond the scope of this chapter. Ths subject, however, is dealt with by Newman (1974, 1975) and Herschy (1976).

Figure 7.10 shows a view of the hardware instrumentation. A rack contains the power supplied, the processor, a card bin containing the various amplifiers, phase sensitive detector, filters, analogue to digital converter, etc., and a twin-channel chart recorder. A second cabinet contains the coil power supply and the coil switching unit.

7.2.6　Conclusion

The results from the electromagnetic gauging stations now in operation are encouraging and it is anticipated that the method may have wide application in difficult situations where existing methods are unsuitable. The main use is

Figure 7.10   Electromagnetic gauging station showing the instrumentation

expected to be in rivers where weed growth is becoming an ever-increasing problem at many existing stations. The method, however, may be adapted to other difficult situations, for example where backwater is a problem. The system is expensive at the present time, the cost of the instrumentation at Fittleworth and Broadlands being estimated at about £25 000 for each station. However, it is anticipated that, from the knowledge gained so far and with the advent of microprocessors, the cost may be substantially reduced. The installation of the coil, at one time thought to be a difficult and expensive civil engineering task, was found at Broadlands to be neither difficult nor expensive and the whole operation was carried out quickly and effectively.

A most interesting and important feature of the system is that a relation is established between Q and E_r or E_h. It is possible that the main reason for this feature is the method of calibration by current meter, which is a crude instrument compared with the sophisticated electromagnetic signal detection hardware. However, a relation having a standard error of estimate of less than ±10 per cent is quite satisfactory for stream gauging purposes and there is good reason to suppose that this will be obtained when additional gaugings are available. One difficulty of an *in situ* calibration, however, is having to wait for a range of flows in order to complete the calibration, and a theoretical calibration is at present being investigated.

7.3 THE ULTRASONIC METHOD

The ultrasonic technique, sometimes known as the acoustic method, of stream gauging is now well established and to date there are many such stations in operation in a number of countries. The use of ultrasonics to measure discharge dates back to the 1950s (Swengel and Hess, 1955; Swengel *et al.*, 1955), and development has proceeded and complete systems are now commercially available.

7.3.1 Principle of operation

The ultrasonic method in itself is basically a velocity-area method, where the velocity at a known depth in the river is measured by recording the time it takes for a beam of acoustic pulses to cross the river. In operation, transducers that are capable of transmitting and receiving acoustic pulses are positioned on either side of the river, as in Figure 7.11. It is necessary to stagger the transducers along the river bank so that there is a time difference between pulses travelling downstream and those travelling upstream.

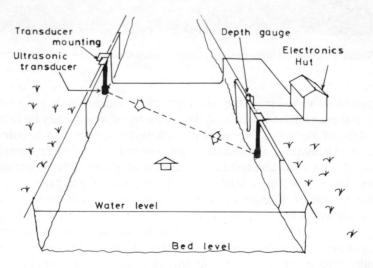

Figure 7.11   Schematic of an ultrasonic gauging station

There are a number of ultrasonic techniques available to measure the speed or line velocity and these may be summarized as follows:

The time- or phase-difference method

Two transducers simultaneously emit acoustic signals across the river, one upstream and the other in a downstream direction. The difference in the time of travel of the two signals or waves, or the phase difference, is a measure of the velocity of the river. This method may be affected by small variations of temperature.

A modification to this method is the 'leading-edge technique'. Here signals of short duration are transmitted simultaneously in either direction. The signal that travels downstream arrives first and initiates a time clock. The clock is switched off when the upstream pulse arrives at the opposite transducer. In addition, the total time of travel has to be measured. The leading-edge technique refers to the procedure in detecting the first negative or positive portion of the received waveform.

Phase-comparison time-sharing method

This is similar to the previous method, except that the paths in either direction are superimposed. This is achieved by using a pair of transducers which switch from transmit to receive mode.

Frequency-difference method

This is the method selected by one European manufacturer and in one mode is referred to as the 'sing around' technique. In practice a signal is emitted from the first transducer and received by the second. The received signal at the second transducer is used to trigger again the first transducer. This cycle is repeated for a short period at a known frequency. The procedure is reversed at another frequency and the difference between the two frequencies is again a measure of the velocity.

A novel adaption of this technique is the Harwell single-path system (Herschy and Loosemore, 1974; Loosemore, 1975) which incorporates the leading-edge detection technique. Here two voltage-controlled variable frequency oscillators are designed such that their output frequencies are proportional to the transit times, without the need for any electronic memory or store. By accumulating the beat frequency between these two oscillators for a given time, which is proportional to the depth of water, the output can be displayed directly in m^3/s.

The ultrasonic method is similar to other velocity-area methods where it is necessary to measure the river bed profile and the water level. Therefore ideally the river should be flowing in a stable, prismatic channel.

The speed of sound through the water is dependent on the elasticity and density of the water, i.e. pressure, temperature, and salinity. Any change in these parameters along the flight path will affect the results, i.e. produce an error or uncertainty. The errors associated with ultrasonic river gauging may be listed as refraction of the signal, attenuation of the signal, velocity errors, area errors, and integration errors.

Refraction is caused by the acoustic beam being refracted or bent so that the two opposite transducers do not 'see' each other; for example, if a temperature or salinity gradient occurs along the flight path. Another situation that could arise is in a shallow river where the transducers are too close to the water surface or the river bed. In this case the signal reflected off the interface may reach the opposite transducer before the direct line of site signal.

Attenutation is the result of the acoustic beam not reaching the opposite transducer. Ultrasonic systems are designed to tolerate a blockage of the signal for a short duration; the Harwell system will tolerate up to a 90 per cent loss. It has been shown that the signal is attenuated by:

(1) High loads of suspended solids. It was found that loads in excess of 2000 mg/l in a river 64 m wide with the ultrasonic equipment operating at 500 kHz resulted in loss of the signal.
(2) Animal and plant life. Weed growth has been shown not only to block the signal but to cause reflections as well, and errors of up to 36 per cent in discharge have been reported. Algal growth on the face of the transducers is not a problem; Figure 7.12 is a photograph of a transducer removed for inspection from a river after two years' operation. The transducer was still operating.
(3) Entrained gases. Air that becomes entrained in the water, for example downstream of a spillway or weir, will attenuate the signal (Smith, 1974).

Figure 7.12 Ultrasonic transducer removed for inspection after two years' operation

Velocity errors arise from three sources. First, in measuring the path length or distance between the two transducers. This error should be small <0.1 per cent with the accurate distance-measuring equipment now commercially available.

Secondly, the path angle (Fig. 7.13) will give rise to an error. It is easy to measure θ relative to the river bank provided the banks are parallel. The problem arises when they are not parallel or there is a skew or cross-flow component. It is the angle between the flight path and the mean direction of flow that is required and it is therefore important when selecting the site not to locate it near a bend or an obstruction that will give rise to a skew or cross-flow component.

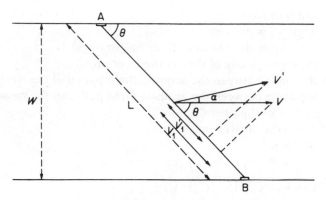

Figure 7.13   Principle of the ultrasonic method

Thirdly, there are delay errors in the transit times. The most common cause of this type of error is when there is dead or stationary water in the flight path. This could be the result of recessing the transducers into the river bank. Alternatively, if an acoustic window is placed over the transducers to protect them from damage, the time of travel through the 'window' may be appreciable. Errors associated with the electronic circuitry are negligible.

Area errors. The river cross-section or bed profile should be stable, otherwise frequent bed surveys are required.

Integration errors. Integration of the velocities varies according to the systems used.

7.3.2   Theory

Referring to Figure 7.13, the time taken for an acoustic pulse to travel from A to B is

$$t_1 = \frac{L}{C + V_1} \tag{7.20}$$

Similarly the time taken for a pulse to travel from B to A is

$$t_2 = \frac{L}{C - V_1} \tag{7.21}$$

where

L = acoustic path length (m)
C = velocity of sound in water (m/s)
V_1 = velocity component along the acoustic flight path (m/s)
t_1 = time taken for a pulse to travel from A to B (s)
t_2 = time taken for a pulse to travel from B to A (s)

and

W = width of river (m)
D = depth of water (m)
V = velocity of river at transducer height (m/s)
$\bar{v}$ = average velocity of flow in the river (m/s)
θ = the angle between the acoustic flight path and the river bank (°)
$\theta + \alpha$ = the angle between the acoustic flight path and the mean direction of flow of the river (°)
Q = discharge of river (m^3/s)

Let $t_1 - t_2 = \Delta t$, then

$$\Delta t = \frac{L}{C + V_1} - \frac{L}{C - V_1} = \frac{2LV_1}{C^2 - V_1^2} \qquad (7.22)$$

Assuming V_1^2 is small when compared with C^2

$$\Delta t = \frac{2LV_1}{C^2} \qquad (7.23)$$

But

$$V_1 = V \cos \theta, \text{ if } \alpha = 0 \qquad (7.24)$$

and

$$Q = WD\bar{v} \qquad (7.25)$$

then

$$Q = DL \sin \theta \bar{v}$$

$$Q = V_1 DL \tan \theta \qquad (7.26)$$

or

$$Q = \tfrac{1}{2}\Delta t C^2 D \tan \theta$$

7.3.3 Design and operation

There are two types of system: the single-path, where only one pair of transducers is in operation, and the multi-path, where more than one pair of transducers are used. The multi-path system is primarily required for rivers that exhibit a wide range in stage throughout the year.

Figure 7.14 Single-path ultrasonic gauging station on the River Avon near Bournemouth, showing the transducer carriage assembly and depth gauge

In the single-path system the transducers may be fixed permanently in the river or mounted on a sliding carriage arrangement (see Figure 7.14). In the fixed position it is necessary to calibrate the gauging station by an alternative method, the most common of which is by current meter. The calibration of the gauging station employs a velocity index which is directly proportional to V_1 (Smith, 1974), where

$$V_1 = C_1 I \qquad (7.27)$$

C_1 is a constant and I is the velocity index. From equation (7.24)

$$V = \frac{C_1}{\cos \theta} I \qquad (7.28)$$

The relationship between V and $\bar{v}$ will vary according to the stage because V, the velocity as measured by the ultrasonic system, is at a fixed position above the river bed, whereas $\bar{v}$, the mean velocity in the vertical velocity profile, will vary with depth. In an idealized logarithmic-shaped velocity profile, $\bar{v}$ will be 0.6 times the depth of water below the water surface. The ratio of $V/\bar{v}$ is required and can be expressed as

$$\bar{v} = C_2 V \qquad (7.29)$$

where C_2 is a function of stage. But

$$Q = DW\bar{v} \qquad (7.30)$$

Substituting equations (7.28) and (7.29) into (7.30) gives

$$Q = \frac{IC_1C_2DW}{\cos\theta} \tag{7.31}$$

Simplified, $Q = KIDW$ where $K = (C_1C_2)/\cos\theta$.

The system is calibrated by obtaining discharge measurements for values of area and velocity. The measured values of area are plotted against stage to obtain a stage–area relationship. The measured values of $\bar{v}$ are divided by concurrent values of I to obtain values of K. The values of K are plotted against stage to obtain an empirical relationship, as in Figure 7.15. The

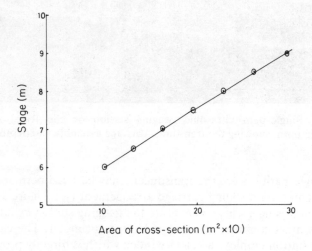

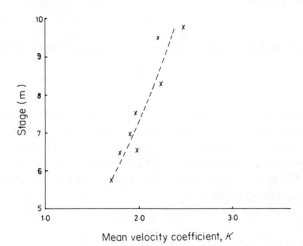

Figure 7.15   Single-path calibration procedure for
a fixed transducer system

discharge for any given value of I is calculated by inserting the value of stage into the graphical relationships to obtain area and K. The values of K, I and area are multiplied together to obtain discharge Q.

Alternative methods can be employed by substituting mathematical equations instead of graphical solutions, i.e. a second-order equation to relate area and stage, where

$$\text{area} = A_1 + A_2 H + A_3 H^2$$

A_1 A_2 and A_3 are constants, and H is the stage.

Similarly, to obtain the index K a second-order equation can be employed, where

$$K = \frac{\bar{v}}{I} = K_1 + K_2 H + K_3 H^2$$

K_1 K_2 and K_3 are constants.

In the movable transducer method, the transducers are designed to slide or move under manual control on either a vertical or inclined carriage assembly. This method is most appropriate where the channel is either rectangular or uniformly trapezoidal in cross-section, i.e. a canal. Figure 7.14 shows a single-path station with a movable transducer assembly. By moving the transducers over a range it is possible to measure V_1 at a number of positions in the vertical velocity profile (Herschy and Loosemore 1974; Green and Herchy, 1975). This enables the shape of the velocity profile to be found and an estimate of the optimum position of the transducers to be determined. It is not mechanically practical to arrange for the transducers to move according to the stage.

The advantage of the transducer carriage is that the ultrasonic system is now self-calibrating and check gaugings are only necessary to verify that the gauge is operating correctly. If, for different stages and discharges, the velocity profiles are measured (Figure 7.16a), a composite curve may be drawn enabling a discharge coefficient to be calculated (Figure 7.16b).

In practice the vertical velocity profiles may be logarithmic in shape, hence $\bar{v}$ is at 0.6 depth below the water surface. The transducers in normal operation are again fixed so that as the stage rises or falls the transducers will no longer be at their optimum position. Therefore a relationship between stage and a coefficient C' is established, where C' is the actual discharge divided by the recorded discharge, as in Figure 7.16b. If the water level in the river is maintained at or near a constant level then the coefficient C' may be small or not necessary. Output can be in graphical form, but it is normal for it to be punched on to 8-hole paper tape, thus enabling the data to be compatible with a computer for processing.

There is a need to measure flow where there is a wide range in stage, and in this instance the coefficient C' may differ appreciably from unity. Another situation is where a uniform vertical velocity profile does not develop and it is

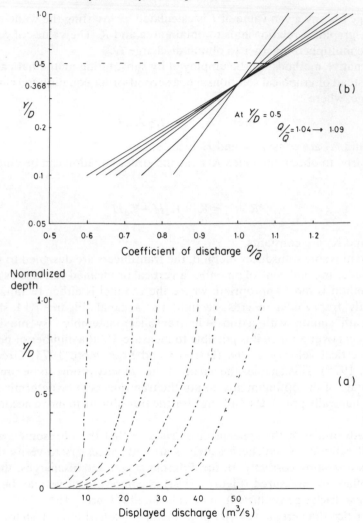

Figure 7.16 Single-path calibration procedure for a movable
transducer system. D is the total depth; Y is the depth of the
transducers from bed level

then not possible to determine C'. Here the multi-path system is applicable
and the number of pairs of transducers required is selected for the particular
range of stage and accuracy. Again, with these systems the time of flight of the
acoustic beam is measured, but a mini-computer or microprocessor is
required to undertake the operation of switching from one pair of transducers
to the next and storing the timing difference. A block diagram of a multi-path
system is seen in Figure 7.17 and a photograph of a multi-path station is seen
in Figure 7.18. The calculation of discharge is undertaken by integrating the

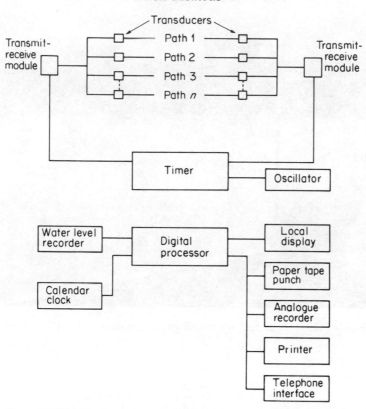

Figure 7.17 Block diagram for a multi-path ultrasonic gauging
station

velocity profile as measured by the ultrasonic system. This can be accomplished on or off line as required.

In operation the errors in equation (7.26) can be minimized by careful site selection. For instance, signal refraction may be diminished if the site is located above the tidal limit of the saline water in an estuary or avoiding sluggish water where in summer there may be differential heating. A simple guide when calculating whether the reflected signal will affect the direct signal, assuming that there must be a wavelength difference between the two signals (Lowell, 1977), is given by

$$H = \sqrt{\frac{LC}{2F}}$$

where H is the distance between the direct acoustic path and the air/water or water/river bed interface; L is again the path length; C is the velocity of sound in water; and F is the operating frequency (Hz) (Figure 7.19).

Figure 7.18   Multi-path ultrasonic gauging station on the River Severn

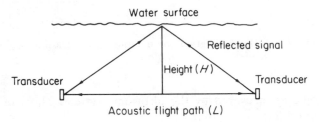

Figure 7.19   Simple method to determine signal interference

From this it can be seen that H will decrease for higher operating frequencies. Therefore, as it is ideal to measure velocities as close as possible to the water surface or river bed, the operating frequencies should be high, 500 kHz or above. The higher frequency also improves the timing accuracies and hence smaller errors in the line velocity measurement. Unfortunately the higher operating frequencies are more susceptible to signal attenuation and therefore for rivers that exhibit high loads of suspended solids a lower operating frequency, less than 500 kHz, is desirable.

As previously mentioned, ultrasonic river gauging systems have been designed to withstand a blockage of the signal for a short duration, i.e. the passage of a boat through the beam. The blockage caused by weeds can only be remedied by regular cutting, especially since weeds have been shown to produce reflections as well as complete blockage of the signal.

Table 7.1

Path angle, $\theta(°)$	Error in θ (%)		
	1°	2°	3°
30	±1.0	±2.0	±3.0
45	±1.7	±3.5	±5.2
60	±3.0	±6.0	±9.1

The error caused by either measuring θ incorrectly or having a skew or cross-flow component is given in Table 7.1. As can be seen, appreciable errors are involved for a small error in θ. It appears to be an advantage to keep θ to about 30 degrees but this would involve longer acoustic path lengths than if θ is 60 degrees. The longer the path length, again the more susceptible the signal is to attenuation and refraction. A compromise has to be sought between short path lengths and possible errors in θ, and longer path length and possible attenuation. A solution to the path angle is to introduce an extra pair of transducers to give a crossed-path system, as in Figure 7.20.

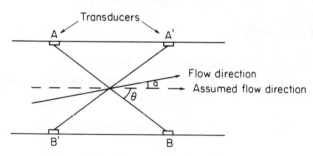

Figure 7.20   Cross-path ultrasonic system

Here the extra pair of transducers are installed at the same level but at an opposite angle. The cross-flow component α can then be determined and eliminated. For a multi-path system a single cross-path can be used to correct all the other paths or a corresponding number of crossed paths are required.

The calculation of discharge depends on the system in use, the fixed single path having to be calibrated and the movable single path depending on the estimate of the ratio of the measured velocity to the mean velocity. In the multi-path system, inherently more accurate than the single-path system, because of the additional measurements of line velocities, the greatest uncertainty is in estimating or assigning a velocity to a volume of water above the highest operating transducer and the water surface. Conversely, near the river bed a velocity has to be assigned to a volume of water not measured. As previously discussed the transducers cannot be placed too close to the water/air or water/river bed interface because of possible reflections interfering with the received direct signal. The integration of the line velocities is by a mean or mid-point technique, but at the surface it is based on the velocity gradient between the two uppermost transducers extrapolated to the water surface. Near the river bed the uncertainty is of lesser importance as the velocities are lower and it is normal to assume that the velocity is zero at the mean bed level and either fits a logarithmic curve or a straight line through the origin and the lowest velocity measurement. It has been shown that the single-path ultrasonic river gauging system with the movable transducer assembly for calibration purposes can have an uncertainty of ± 2.2 per cent (Green and Ellis, 1974), while for the multi-path system ± 1 per cent has been reported (Lowell, 1977).

7.3.4 Conclusion

The ultrasonic system of river gauging is now an accepted technique and its advantages over other types of river gauging can be summarized as follows:

(a) does not require a stage–discharge relationship;
(b) measures flow reversals;
(c) does not obstruct flow;
(d) high accuracy;
(e) can be self-calibrating (measures the velocity profile);
(f) cost of equipment is independent of the size of the river, and
(g) suitable for rivers up to 500 m wide.

Against these advantages can be set the following limitations of the ultrasonic method:

(a) The signal may be refracted by temperature and/or salinity gradients along the flight path;
(b) the signal may be attenuated by high loads of suspended solids, entrained gases, plant and marine growth; and
(c) the method is not recommended for wide shallow river channels or rivers that have an unstable bed.

ACKNOWLEDGEMENT

This material is published with the permission of the Director of the Water Data Unit, UK, and the Director of the Water Research Centre. UK.

BIBLIOGRAPHY

Faraday, M., 1832. *Phil. Trans. of the Royal Society*, p. 175; also included in *Experimental Researches in Electricity*, 1839, vol. 1, Taylor, London.

Green, M. J. and Ellis, J. C., 1974. Knapp mile ultrasonic gauging station on the River Avon, Christchurch, Dorset, Water Research Centre and Department of the Environment, Water Data Unit, Symposium on River Gauging by Ultrasonic and Electromagnetic Methods, University of Reading, UK.

Green, M. J. and Herschy, R. W., 1975. Site calibration of electromagnetic and ultrasonic river gauging stations, International Seminar on Modern Developments in Hydrometry, World Meteorological Organization, UNESCO, and IAHS, Padova.

Herschy, R. W., 1974. The ultrasonic method of river gauging, *Water Services*, **78**, 198–200.

Herschy, R. W., 1976. New methods of river gauging, in *Facets of Hydrology*, J. C. Rodda (Ed.), Wiley, London.

Herschy, R. W., 1977. The electromagnetic method of river gauging, *Water Services*, **81**, 544–548.

Herschy, R. W. and Loosemore, W. R., 1974. The ultrasonic method of river flow measurement, Water Research Centre and Department of the Environment, Water Data Unit, Symposium on River Gauging by Ultrasonic and Electromagnetic Methods, University of Reading, UK.

Herschy, R. W. and Newman, J. D., 1974. The electromagnetic method of river flow measurement, Water Research Centre and Department of the Environment, Water Data Unit, Symposium on River Gauging by Ultrasonic and Electromagnetic Methods, University of Reading, UK.

Holmes, H., Whirlow, D. K. and Wright, L. G., 1970. The LE (leading edge) flowmeter—a unique device for open discharge measurement, Proc. International Symposium on Hydrometry, Koblenz. Unesco/WMO/IAHS Pub. 99, pp. 432–443.

Loosemore, W. R., 1975. Ultrasonic river gauging, International Seminar on Modern Developments in Hydrometry, World Meteorological Organization, UNESCO, and IAHS, Padova.

Lowell, F. C., 1977. Designing open channel acoustic flowmeter for accuracy, Parts 1 and 2, *Water and Sewage Works* (July and August), 70–75 and 84–88.

Newman, J. D., 1974. Princes Marsh electromagnetic gauging station on the River Rother, Water Research Centre and Department of the Environment, Water Data Unit, Symposium on River Gauging by Ultraonic and Electromagnetic Methods, University of Reading, UK.

Newman, J. D., 1975. Electromagnetic stream gauging in the UK, International Seminar on Modern Developments in Hydrometry, World Meteorological Organization, UNESCO, and IAHS, Padova.

Smith, W., 1969. Feasibility study of the use of the acoustic velocity meter for measurement of net outflow from the Sacramento–Sai Joaquin Delta in California, US Geological Survey Water Supply Paper 1877, 54 pp.

Smith, W., 1971. Application of an acoustic streamflow measuring system on the Columbia River at The Dalles, Oregon, *Water Research Bulletin*, **7**, 1.

Smith, W., 1974. Experience in the United States of America with acoustic flowmeters, Water Research Centre and Department of the Environment, Water Data Unit, Symposium on River Gauging by Ultrasonic and Electromegnetic Methods, University of Reading, UK.

Smith, W., Hubbard, L. L. and Laenen, A., 1971. The acoustic streamflow measuring system on the Columbia River at The Dalles, Oregon, US Geological Survey Open-File Report, Portland, Oregon, 60 pp.

Smith, W. and Wires, H. O., 1967. The acoustic velocity meter—a report on system development and testing, US Geological Survey Open-File Report, Menlo Park, California, 43 pp.

Swengel, R. C. and Hess, W. B., 1955. Development of the ultrasonic method for measurement of fluid flow, Sixth Hydraulic Conference, University of Iowa.

Swengel, R. C., Hess, W. B. and Waldorf, S. K., 1955. Principles and application of the ultrasonic flowmeter, *Electrial Engineering*, **74**(4), 112–118.

Vanoni, V. A., 1941. Velocity distribution in open channels, *Civil Engineering*, **11**(6), 356–357.

CHAPTER 8

Flow Measuring Instruments

GEORGE F. SMOOT

8.1 INTRODUCTION

River flow measurement practices and instruments have evolved over the years to meet local stream characteristics. Ice-covered streams require different methods than do streams in warm regions. Large streams demand more complex equipment than do small streams. Sand-channel streams must be treated differently from those with stable channels. Additionally, the practices and equipment have evolved somewhat independently from country to country throughout the world, even for similar conditions.

Despite all these factors, there is a remarkable similarity in equipment regardless of the area in which it originated with only minor and usually insignificant differences. Staff gauges, for example, may be made of wood in one country, of cast iron in another, and of enamelled steel in a third. Their graduations may be spaced differently, but the basic operating principle and function of the gauges are the same. This chapter, describes the basic principles of the various instruments used in river flow measurement, rather than discussing their insignificant differences.

8.2   STAGE MEASURING AND RECORDING INSTRUMENTS

The stage of a stream is the elevation of its water surface with respect to an established datum plane. Its measurement is the most important of all measurements made in the course of flow measurement, because it is the base to which all other measurements are related. Readings of stage may be required as a single instantaneous measurement, as a short series of instantaneous measurements, or as a continuous or virtually continuous record.

Stage measuring instruments may be conveniently separated into two types, direct reading and indirect reading.

8.2.1   Direct-reading gauges

The significant feature of direct-reading gauges is that the stage measurement is made directly in units of length without any intervening influences.

Staff gauges

A staff gauge is a vertical or inclined staff with graduations. It is intended to be permanently installed, extending from the elevation of the lowest stage to be measured upwards to the elevation of the highest stage to be measured.

Certain characteristics are essential for staff gauges. These are:

(a) they must be accurately and clearly graduated with permanent markings;
(b) they should be constructed of durable material, particularly with respect to alternating wet and dry conditions and also with respect to wear or fading of the markings; and
(c) they should be constructed of material with a low coefficient of expansion with respect to both temperature and moisture.

Vertical-staff gauges are manufactured in convenient sizes, usually sections 1 m in length—Figure 8.1(a). They are designed to be mounted on a vertical plate or backing board which is securely anchored to a foundation extending below the ground surface to a level free of disturbance by frost.

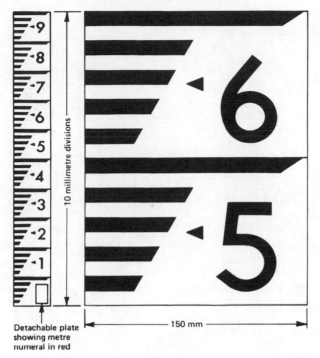

10 millimetre divisions

Detachable plate
showing metre
numeral in red

150 mm

Figure 8.1(a)   Typical details of vertical-staff gauge

Inclined-staff (ramp) gauges are sometimes used and sloped in a manner as to closely follow the contour of the streambank. The profile of the bank may be such that a single slope may suffice, but frequently it is necessary to install the gauge in several section, each with a different slope.

Inclined-staff gauges are usually calibrated *in situ* by means of a surveyor's level. A typical detail of a ramp (inclined) gauge is shown in Figures 8.1(b) and 8.1(c) (see also Chapter 1).

Crest-stage gauges are used to obtain a record of the peak level reached during a flood when other methods of recording levels cannot be used. Peak discharges may be calculated from the water levels of two gauges installed some distance apart in a stretch of channel, provided that the time lag between measurements is negligible.

These gauges are locally made to different designs. Basically they may be a tube about 50 mm internal diameter, down the centre of which runs a rod. The tube is perforated to permit rising water to enter, the perforations being located to prevent drawdown or velocity head from affecting the static water level. The top of the tube must be closed to prevent the entry of rain, but it should have an air vent to permit water to rise up the tube without significant delay. Powdered cork in the bottom of the tube floats on the surface of the

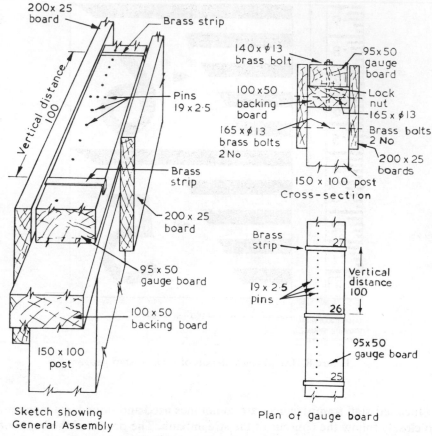

Sketch showing
General Assembly

Plan of gauge board

All dimensions given in millimetres.

Figure 8.1(b)   Typical details of inclined (ramp) gauge

floodwater being deposited on the centre tube as the water recedes. Alternatively the centre tube is coated with a paint whose colour is permanently affected by water. Figure 8.1(d) shows a typical crest-stage gauge.

Needle gauges

A needle gauge consists basically of a metal rod with a tapered point, a means of adjusting its vertical position, and a means of precisely determining the elevation of the tip of the point. Position is usually adjusted by a rack-and-pinion arrangement and determined from vernier and a scale graduated downwards from top to bottom. There are two basic types of needle gauge (Figure 8.2):

(a) the point gauge with a tip that approaches the free-water surface from above; and

Figure 8.1(c) Rampgauge showing use of stilling box

(b) the hook gauge with a tip which is immersed and approaches the free-water surface from below.

Application of the needle gauge consists of placing the gauge on a bracket or datum plate so that the needle is directly above the water, positioning the tip of the gauge near the free-water surface and carefully adjusting its position until the tip just touches the free surface, apparently trying

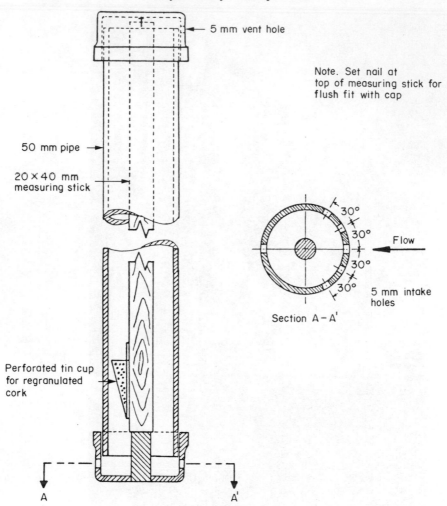

Figure 8.1(d) Typical details of a crest-stage gauge

to pierce its skin. A variety of electronic and optical devices have been incorporated with the basic needle gauge to facilitate the detection of the water surface.

Needle gauges have high measuring accuracy, but also have the disadvantage of a limited measuring range, the length of the movable scale being usually 1 m. This disadvantage can be overcome, however, by installing datum plates at different levels throughout the range in stage to be measured. Datum plates must be mounted on a secure foundation which extends below the frost line.

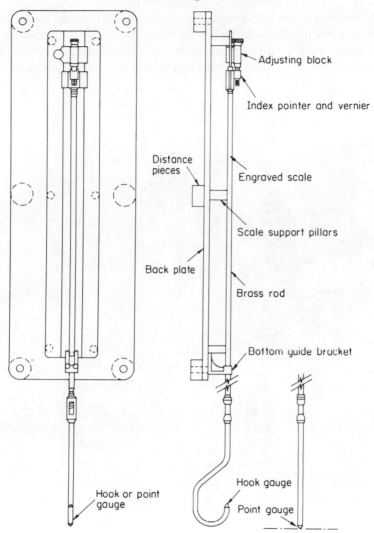

Adjusting block

Index pointer and vernier

Distance pieces

Engraved scale

Scale support pillars

Back plate

Brass rod

Bottom guide bracket

Hook or point gauge

Hook gauge

Point gauge

Figure 8.2   Arrangement of hook and point gauges

Wire-weight gauges

A wire-weight gauge (Figure 8.3) consists of a drum wound with light cable, a weight attached to the end of the cable, and a metering counter by which the amount of cable played out can be accurately determined. This mechanism is housed in a metal box which usually incorporates a check bar or plate. The elevation of this check bar is established and the calibration of the gauge can be checked occasionally by resting the bottom of the weight on the check bar.

Wire-weight gauges are designed to be mounted directly over the water surface on a bridge, dock, or similar structure. Readings are taken by lowering the weight until it just touches the water surface and observing the value indicated on the metering device. Their operating range is usually 25 m or more.

Figure 8.3   Wire-weight gauge

Float gauges

The typical float gauge (Figure 8.4) consists of a float, a graduated steel tape, a pulley, a counterweight, and an index pointer. The float pulley is grooved on the circumference to accommodate the tape and is mounted on a support. The tape is fastened to the upper side of the float and runs over the pulley. It is kept tight by a counterweight at the free end. The stage fluctuations, sensed by the float, position the tape with respect to the pointer.

A float gauge must be positioned directly above the water surface and should be housed in a stilling well to protect the float from water surface oscillations and the tape from the effects of wind.

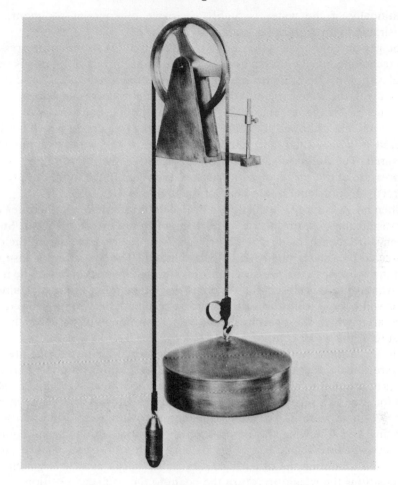

Figure 8.4 Float gauge

8.2.2 Indirect-reading gauges

Indirect-reading gauges include those devices which convert a pressure or acoustic signal to an output which is proportional to the water level.

Pressure gauges

The measurement of stage by pressure utilizes the principle that the hydrostatic pressure at a point in a column of liquid is proportional to the height of the column above that point. If the pressure sensor can be located at or below the point, the pressure can be transmitted directly. However, if the sensor is located above the point, the direct method is usually not satisfactory because gases entrained in the water accumulate in the line and can create air locks.

Additionally, if the water is highly corrosive or contains sediment, it is undesirable to bring it into contact with the sensor.

The gas-purge (bubbler) technique is widely used for transmitting pressure. This technique may be used when the elevation of the water column is below the elevation of the pressure sensor; and, since the sensor does not come in contact with the water, it is suitable for use in highly corrosive waters.

In the gas-purge technique a small quantity of non-corrosive gas or compressed air is allowed to bleed into a tube, the free end of which has been lowered into the water and fixed at the base of the water column to be measured. The pressure sensor which is located at the opposite end detects the pressure of the gas required to displace the water in the tube; this pressure is directly proportional to the head of liquid above the orifice.

When no gas supply is available, a pressure-bulb system is sometimes used to transmit the pressure of the water column to the pressure sensor. Such a system is frequently made up of a cylinder with one closed end and the open end sealed by a slack, highly flexible, diaphragm. The pressure is transmitted from the cylinder to the pressure sensor by interconnecting tubing. The major disadvantage of this closed-gas system is that ultimately an excessive amount of gas will escape from the system, with a resultant stretching of the diaphragm, after which the pressure in the system no longer represents the true pressure of the water column.

The servo-manometer (Figure 8.5) and the servo-beam balance are both pressure sensors that convert the pressure sensed to a rotational-shaft position proportional to the height of the column of water. The shaft action is used for driving a recorder and a water-level indicator. As the name implies, the servo-manometer is essentially a manometer with a servo-mechanism for detecting and following the liquid-level differential within the manometer. The servo-beam balance is a beam balance with a pressure balance on one side and a weight on the other. The servo-mechanism detects any imbalance and positions the weight to return the beam to the balanced position.

Both systems usually have incorporated either automatic or manual methods for correcting for changes in the density of water. Such changes occur with changes in temperature, dissolved-solid content, and suspended-solid content.

Acoustic-level gauges

Acoustic-level gauges operate on the same principle as echo-sounders. An acoustic pulse is generated and directed to the water surface from above. Here it is reflected, and a part of it returned to the point of origin. As the pulse is released, a timer is initiated and runs until the echo is received. The distance from the origin to the water surface is computed using the formula

$$h = \frac{t}{2} c$$

Figure 8.5   Servo-manometer

where h is the distance above the water; c is the velocity of sound in air; and t is the duration of travel time.

As can be seen from the above formula, h is directly proportional to the velocity of sound in air, which in turn varies with changes in both temperature and humidity. Hence, such gauges are not highly accurate. Most such gauges incorporate one of several methods for automatically correcting for changes in velocity of sound, which improves the overall accuracy but does not entirely eliminate errors from this source.

Figure 8.6   Weekly recorder

8.2.3   Recorders

Water-stage recorders are generally of the shaft-angular-input type. Such recorders can be classified as either analogue (Figures 8.6 and 8.7) or digital (Figures 8.8 and 8.9), depending upon the mode used in recording the rotational position of the input shaft. The analogue recorder produces a graphic record of the rise and fall of parameter values with respect to time, while the digital recorder punches coded parameter values on paper tape at preselected time intervals. Analogue recorders can further be classified into

Figure 8.7   Continuous recorder

two groups—those which record for a fixed period of time (daily, weekly, monthly, etc.) and those which record continuously.

Angular movement of the input shaft drives the stylus of an analogue recorder or the encoding discs of a digital recorder through a mechanical linkage. The driving force to overcome the friction generated by these moving parts is usually supplied by water displacing a float and transmitted by float line and counterweight to a drive pulley on the input shaft. Since this force is very low and can only be increased by increasing the size of the float, it is highly important that the friction is kept to a minimum.

Figure 8.8   Digital recorder

Hysteresis or lost motion in the mechanical linkage is another source of error of significance. If the input shaft is rotated in one direction until the stylus follows and then the direction of rotation is reversed, the total hysteresis is that amount of motion required to cause the recorder to follow in the reversed direction. In order to minimize errors from friction and hysteresis, it is necessary to have a well-designed and properly manufactured drive mechanism on a water-level recorder.

8.3   VELOCITY MEASURING INSTRUMENTS

The velocity of a body is defined as the rate of change of position of the body in reference to a specified direction, i.e. the derivative of the displacement (s) (vector) with respect to time (t) (scalar):

$$v = \frac{ds}{dt}$$

Figure 8.9   Digital recorder

In many problems related to river flow measurements, the direction of velocity is known, so the only additional information needed is the determination of its absolute value. Hence, most velocity measuring instruments described in this section are designed to provide only the magnitude of the velocity vector.

8.3.1   Floats

Man's first attempts at measuring velocity were undoubtedly by timing the travel of floating debris over measured distances. Velocity, as determined by use of floats, is neither a local velocity nor an instantaneous one, for one determines the mean value of the velocity in the time (t) over a measured distance (s):

$$v = \frac{s}{t}$$

Surface floats

Surface floats (Figure 8.10a) are the most primitive of all velocity measuring devices still in use. Wind acting upon that portion of the float above the water can significantly affect its course and speed causing gross errors in the velocity observations. Despite the drawbacks, surface floats can, for certain applications, prove highly useful.

Canister floats

Canister floats (Figure 8.10b) consist of a closed canister connected by a thin line to a small surface float with a flag. The canister dimensions and its immersion depth are chosen so that the float velocity is equal to the mean velocity in the vertical section parallel to the hydraulic axis of the river.

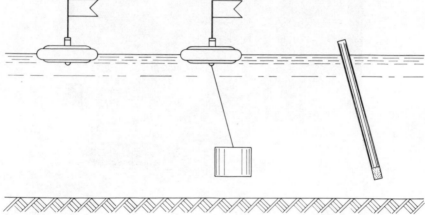

Figure 8.10   Floats: (a) surface float, (b) canister float, (c) rod float

Rod floats (see also Chapter 2)

Cylindrical rods weighted so that they float vertically in still water with only the tip protruding above the surface, are called rod floats (Figure 8.10c). They are used to measure the mean velocity in the vertical section along the travelled path. Rod lengths are selected so that they extend through as much of the stream depth as possible without the lower end touching or dragging the bed.

8.3.2   Drag-body current meters

Quite a wide assortment of velocity measuring devices have been developed employing various shapes and configurations of drag elements and readout tehchniques. All such devices are governed by the basic equation

$$F = \frac{C_{D}\rho A v^{2}}{2}$$

where F is the drag; ρ the density of fluid; A the projected area; v the velocity of fluid; and C_D is the coefficient of drag.

As can be seen, the drag is proportional to the square of the velocity. Consequently, all drag-body current meters are insensitive to extremely low velocities. The drag is also proportional to the coefficient of drag, a function of Reynolds' number. C_D changes if the surface of the drag element becomes fouled with aquatic growth or debris.

Drag-element current meters first appear in literature in the first part of the seventeenth century, and their development and use continues to the present. Therefore, variations of such meters are too numerous to cover here, and only a few of those in most common usage will be briefly described.

Two basic types of deflection vane have evolved—the vertical-axis and the horizontal-axis types. Figure 8.11 shows two variations of the first type. The vane on the left is designed to obtain a 'point' or local velocity, whereas the one on the right has been designed to integrate velocities throughout the greater portion of a vertical.

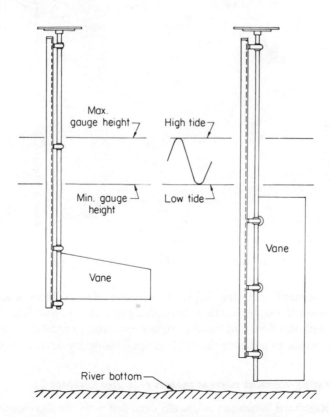

Figure 8.11   Two types of vertical-axis deflection vane

In these vane assemblies, a weight-and-pulley system is used to resist the force of the flow on the vane, and the deflection is indicated by a scale and pointer at the top.

Figure 8.12 shows one type of horizontal-axis deflection vane, sometimes referred to as a pendulum type. The weight at the bottom edge is used to resist the force of the flow on the vane and can be changed to provide different velocity ranges. A small pendulum is sealed inside the housing at the top of the vane assembly, which always hangs straight down. An electrical signal, proportional to the angle formed between the two pendulums (the amount of deflection), is generated.

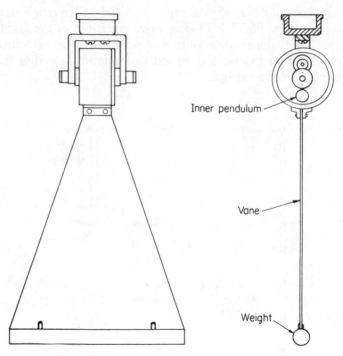

Figure 8.12   Pendulum-type deflection vane

The inclinometer (Figure 8.13) is a meter lighter than water and is anchored below the water surface through a swivel at its base. It is thus free to align itself with the flow and has a compass enclosed to provide flow direction as well as a device to indicate the angle of inclination for determining velocity.

8.3.3 Rotating-element current meters (see also Chapter 2)

Although Reinhard Woltman is usually credited with the development of the rotating-element current meter, Robert Hooke conceived a meter very

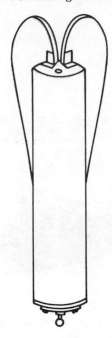

Figure 8.13   Inclinometer

similar in principle in 1663, considerably more than a century earlier. Since that time, rotating-element meters have received much attention and are still today the most widely used and the most reliable of all velocity measuring devices.

The principle of operation of a rotating-element current meter is based upon the proportionality between the local-flow velocity and the resulting angular velocity of the meter rotor. The relationship (calibration) between velocity and rotor speed is established experimentally. Calibration consists of towing the meter at various velocities in straight open tanks through sensibly still water and recording the revolutions of the rotor, the distance travelled, and the times of each. After calibration, velocity can be determined by counting the number of revolutions of the rotor over a specified interval of time.

Rotating-element current meters can be divided into two types: (1) vertical-axis meters with their rotating axis perpendicular to the direction of flow, and (2) horizontal-axis meters with their rotating axis coinciding with the direction of flow. Each type can be further divided: the vertical-axis type into (a) cup or bucket wheel with six cups comprising the rotor (Figure 8.14), and (b) vane with two S-shaped vanes comprising the rotor (Figure 8.15); the horizontal-axis type into (a) vane with two or more vanes comprising the rotor (Figure 8.16), and (b) helical screw with two or more blades comprising the

Figure 8.14   Price and Price–Pygmy–type current meters

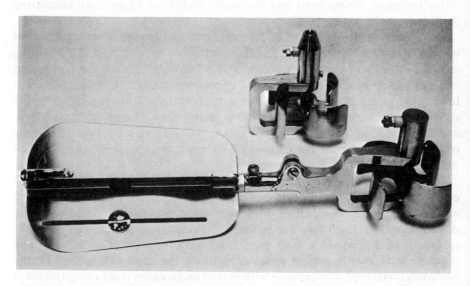

Figure 8.15   Vane (S-type) current meter

Figure 8.16   Hoff-type meter

rotor (Figures 8.17 and 8.18). Each type of these meters has its strong points and its weaknesses, and no one meter can be said to be superior to all others for all applications.

8.3.4   Heated-element meters

When an electrically heated element (wire, film, bead, etc.) is placed in a flowing fluid, heat will be transferred from the element of the fluid. The rate of transfer is a function of the velocity, temperature, viscosity, and thermoconductivity of the fluid, as well as the temperature and exposed surface area of the heated element. If all parameters except velocity are kept contant, then the heated element becomes a transducer for measuring the velocity of the fluid.

The convective heat transferred from a circular cylinder has had extensive theoretical and experimental study, with the potential-flow solution attributed to King. King's law relation may be expressed as

$$\frac{I^2 R_w}{R_w - R_a} = A + B\sqrt{v}$$

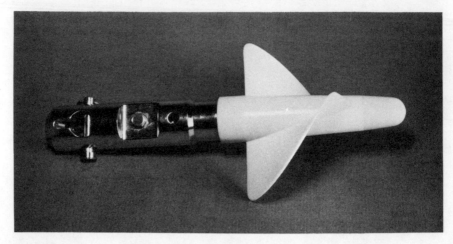

Figure 8.17   Braystoke current meter

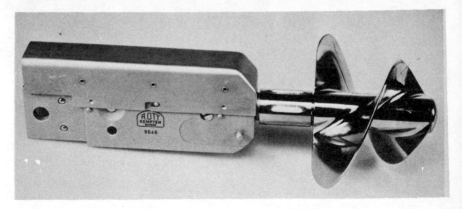

Figure 8.18   Ott-type current meter

where I is the electric current; R_w the electrical resistance of wire at wire temperature; R_a the electrical resistance of wire at fluid temperature; A and B are coefficients; and v is the velocity of flow.

It has been found that, at best, King's law holds only for continuum flow, and that the best experimental relationship between mean heat loss and mean velocity is

$$\frac{I^2 R_w}{R_w(R_w - R_a)} = A + Bv^n$$

Heater-element velocity sensors are used extensively in the study of turbulence because of their small size, high-frequency response, and high sensitivity to minute changes in velocity. Their use as general velocity-measuring devices has not proved very successful. Fouling of the element by foreign material and by minute gas bubbles drastically changes the heat transfer capability of the element and consequently its calibration.

8.3.5 Eddy-shedding meters

The eddy-shedding current meter is based upon the relationship of the count of the vortex street frequency formed behind a strut in fluid flow and the velocity of the flow. This shedding frequency is proportional to the magnitude of the free-stream velocity and is approximated by the relation

$$f = \frac{vS}{d}$$

where f is the shedding frequency; v the local velocity; S the Strouhal number, and d is the rod diameter.

The Strouhal number for a circular probe, which is the shape of most probes, can be approximated by

$$S = \frac{0.21}{C_d^{3/4}}$$

in the range of Reynolds' number from 10^3 to 10^6, with Reynolds' number $= (vd)/\nu$ where C_d is the coefficient of drag, and ν is the viscosity.

From the above it can be seen that the threshold velocity is fairly high, of the order of 0.03 to 0.05 m s^{-1}.

The natural turbulence in flowing water tends to complicate the uniform pattern of eddy shedding, making counting somewhat uncertain. Hence, this type of current meter has not been widely accepted for general use.

8.3.6 Electromagnetic current meters

The electromagnetic current meter is based upon Faraday's law of electromagnetic induction (Figure 8.19). As a liquid, in this case water, cuts lines of magnetic flux, an electromotive force is induced in the liquid. This e.m.f. is detected by a pair of electrodes on the surface of the meter. The magnitude of the induced e.m.f. is proportional to the strength of the magnetic field and velocity of the water.

In order to avoid thermoelectric effects, electrochemical effects, and d.c. detection problems, an altering-polarity magnetic field is generated. An alternating magnetic field can cause a quadrature voltage problem resulting in an offset in zero point, thereby indicating a flow under actual no-flow conditions. This effect, however, can be compensated for with proper design.

Some minor errors are introduced into readings of the electromagnetic meter as a result of hydrodynamic effects. The sensor is usually formed as a

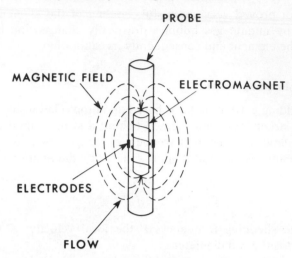

Figure 8.19 Operating principles of electromagnetic current meter

strut or cylindrical rod. Its very presence in the flowing water creates disturbances in the flow field, which in turn produces erroneous readings of velocity. The strength of the magnetic field is inversely proportional to the square of the distance from the magnet; therefore, the greatest potential is produced nearest the surface of the sensor where the greatest disturbance exists. It is of importance that the sensor be shaped to minimize any disturbances in the flow field. The total-flow electromagnetic method is described in Chapter 7.

8.3.7 Ultrasonic (acoustic) velocity meters

The acoustic-velocity meter is based upon the principle that the velocity of a sound pulse in moving water is the algebraic sum of the acoustic propogation rate (c) and the component of water velocity parallel to the acoustic path. Several acoustic-velocity meter systems have been developed using this same basic principle. Common to each is the measurement of water velocity by determination of the travel times of sound pulses moving in both directions between sonic transducers which are located so that the path between them is diagonal to the flow.

Three general approaches have been employed in designing acoustic-velocity meter systems. They are: (1) the total travel-time circuit, (2) the total sing-around circuit, and (3) the differential time circuit.

From Figure 8.20 the mathematical relations of the total travel-time system can be derived. The travel time of an acoustic pulse originating from a

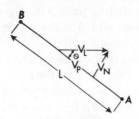

$$T_{AB} = \frac{L}{c - V_p} \qquad (1)$$

$$T_{BA} = \frac{L}{c + V_p} \qquad (2)$$

$$V_p = \frac{L}{2}\left(\frac{1}{T_{BA}} - \frac{1}{T_{AB}}\right) \qquad (3)$$

$$V_p = V_L \cos\theta \qquad (4)$$

$$V_L = \frac{L}{2\cos\theta}\left(\frac{1}{T_{BA}} - \frac{1}{T_{AB}}\right) \qquad (5)$$

c = velocity of sound in still water
L = the length of the acoustic path from A to B
T_{AB} = travel time from A to B
T_{BA} = travel time from B to A
V_L = the average velocity along the acoustic path A to B
θ = the angle of departure between the streamline of flow and the acoustic path
V_p = average water velocity parallel to the acoustic path

Figure 8.20 Operating principles of acoustic-velocity meter

transducer at point A and travelling in opposition to the flow of water along path AB can be expressed as

$$T_{AB} = \frac{L}{c - V_p}$$

Similarly, the travel time for a pulse travelling with the flow from B to A is

$$T_{BA} = \frac{L}{c + V_p}$$

Combining these two equations and solving for V_p, we have

$$V_p = \frac{L}{2}\left(\frac{1}{T_{BA}} - \frac{1}{T_{AB}}\right)$$

And as

$$V_p = V_L \cos\theta$$

then

$$V_L = \frac{L}{2\cos\theta}\left(\frac{1}{T_{BA}} - \frac{1}{T_{AB}}\right)$$

where c is the velocity of sound in still water; L the length of acoustic path AB; T_{AB} the travel time from A to B; T_{BA} the travel time from B to A; V_p the average water velocity parallel to path AB; V_L the average water velocity along the streamline of flow; and θ is the angle of departure between the streamline of flow and the acoustic path.

In this type of system, changes in the velocity of sound are automatically compensated for by the treatment of the data.

Travel times are measured sequentially for pulses originating at A and travelling against the flow, and then for pulses originating at B and travelling with the flow. Accuracy of a system of this type depends upon the precision with which the individual travel times can be measured. Errors in indicated velocity are a linear function of timing errors in either direction.

The sing-around circuit, sometimes referred to as the pulse-repetition frequency system, operates in the following manner. Cumulative measurements of travel times are made by using the received pulse at the far end of the acoustic path to immediately trigger a second transmittal pulse at the originating transducer. Arrival of the second pulse triggers the next transmission, and the system is allowed to continue operation in this repetitive pattern. Transit time of the acoustic pulse is resolved either by measurement of the total time for completion of a fixed number of cycles, or by reduction of the cycling rate to a continuous pulse-repetition frequency. Where a single pair of transducers is employed, such systems measure transit times for a given period in one direction along the acoustic path and then in the opposite direction, for a given period. Where two pairs of transducers are used to operate simultaneously, one pair in each direction, they are tuned to operate at different frequencies in order to avoid interference.

The quotients $1/T_{AB}$ and $1/T_{BA}$ in the above equation are the pulse-repetition frequencies for acoustic transmission in each direction that is measured in a sing-around circuit system.

The differential time-circuit system differs from the two circuits previously discussed in that the difference in the time of arrival of acoustic pulses, triggered simultaneously at each end of the acoustic path, is directly measured. When the two transducers transmit signals simultaneously toward each other, the flow of water in the path will increase the velocity of one and decrease the velocity of the other. The signal transmitted in the downstream direction arrives first and is used to start a time clock; the signal transmitted in the opposite direction arrives later in time and is used to stop the clock. The time increment recorded is thus the differential between the total travel times involved, a differential that is linearly proportional to the water velocity. In this system, the measure of average total travel time in each direction must also be measured and used to compensate for changes in the velocity of sound in water.

Acoustic-velocity meters have been manufactured with path lengths of less than 1 m to measure a local velocity to path lengths of several hundred metres to measure the mean velocity across a stream on the acoustic path. The application of the ultrasonic (acoustic) method is described in Chapter 7.

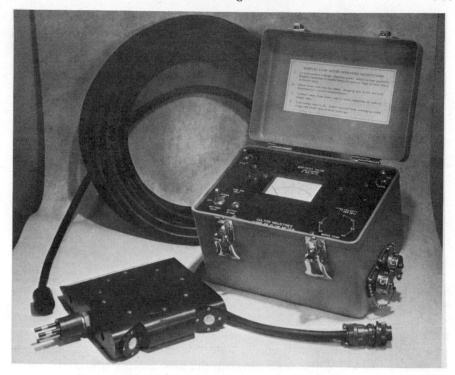

Figure 8.21   Doppler-shift velocity meter

8.3.8   Doppler-shift velocity meters

Another device utilizing acoustic energy to measure velocity is the Doppler-shift velocity meter (Figure 8.21). It is a continuous-tone acoustic device employing the Doppler effect. Its principle of operation, as depicted in Figure 8.22, utilizes a beam of acoustic energy which is projected into a non-homogeneous liquid. It is scattered by the suspended particulate matter in the liquid. Some of the energy is reflected to the receiver. This received signal, called volume reverberation, is found to shift in frequency owing to the Doppler effect, provided there is a net movement of the non-homogeneities with respect to the transmitter or receiver. If the scatterers are stationary with respect to the liquid, the observed Doppler shift is proportional to velocity of the liquid relative to the meter.

The frequency of the received signal is given by

$$f_R = \frac{c + V \cos T}{c - V \cos R} f_T$$

Solving for V, we have

$$V = \frac{c(f_R - f_T)}{f_T \cos T + f_R \cos R}$$

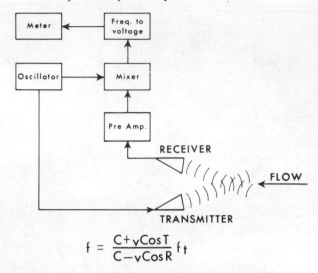

$$f = \frac{C + v\cos T}{C - v\cos R} f_t$$

Figure 8.22   Operating principles of the Doppler-shift
velocity meter

where c is the velocity of sound in water; V the velocity of the suspended particulate matter; T the angle between the flow and the transmitter; R the angle between the flow and the receiver; f_T the transmitted frequency; and f_R is the received frequency.

The acoustic beam is projected upstream and the observation is made at the point of intersection of the transmitter and receiver paths; thus, little or no disturbance is created in the flow at this point. These meters work well when the size and concentration of suspended particulate matter in the stream given adequate signal return. When concentrations are high, no signal is returned and when concentrations are low or particle size too small, there is either sporadic or no signal return.

8.3.9   Deep-water isotopic current analysers

The deep-water isotopic current analyser or nuclear current meter is in actuality a time-of-travel meter. It is a wheel-shaped device with an isotope release mechanism at its hub and 12 scintillation detectors spaced at 30 degree intervals around its periphery (Figure 8.23). An isotope is released by injection at the hub and detected at one or more of the detectors on the periphery. The isotope's travel over the fixed distance is timed and from this the water velocity is computed and the direction determined. Iodine-131 has been used, but any high specific activity gamma source with a short half-life would be satisfactory.

12 SCINTILLATION DETECTORS
SPACED AT 30° INTERVALS

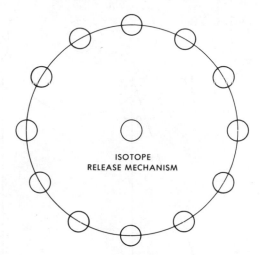

Figure 8.23 Deep-water isotopic current
analyser

This type of current meter is particularly useful in reservoirs and in large sluggish rivers where extremely low velocities of unknown direction are encountered.

8.3.10 Optical current meters

The optical current meter (Figure 8.24) is a stroboscopic device designed to measure surface velocities of water in open channels without immersing equipment in the stream. The meter is a lightweight, battery-powered unit, consisting of a low-powered telescope, a set of rotating mirrors, a variable-speed drive motor, and a tachometer. Velocity measurements are made from an observation point above the stream by looking down at the water surface through the telescope of the meter while gradually increasing the angular speed of the rotating mirrors. As synchronization is reached, the apparent motion of the water surface, seen through the eyepiece as a succession of frames or images, slows down and is finally stopped. The angular speed of the mirror wheel at this null point, and the vertical distance from the optical axis of the meter to the water surface, are the only factors needed to compute the velocity:

$$v_s = (2W)D \, s^2 \, \theta$$

where v_s is the surface velocity; D the vertical distance from the mirror to the

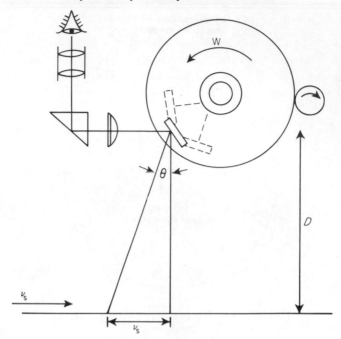

Figure 8.24   Operating principles of optical
current meter

water surface; W the mirror speed (rad s^{-1}); and θ is the angle subtended by
the mirror.

This meter, although measuring only surface velocities, is useful where
equipment cannot be placed in the stream because of heavy debris or
extremely high velocities. It has been successfully used to measure velocities
above 15 m s^{-1}.

8.4   SOUNDING AND SUSPENSION EQUIPMENT

Sounding (determination of depth) is usually done in connection with
measurement of discharge. Consequently, sounding equipment frequently
serves the dual purpose of measuring the depth and suspending the current
meter at the desired point in the stream.

Discharge measurements are made by wading, from ice cover, from
bridges, from cableways, and from boats; and many of the components that
make up a sounding and suspension system are used for more than one type
of measurement. For example, the same sounding weights, hanger bars, and
reels are used for bridge, cableway, and boat-type measurements. For the
application of sounding and suspension equipment see Chapter 1.

8.4.1 Wading rods

There are two basic types of wading rod, with many variations of each type. The top setting rod (Figure 8.25), sometimes referred to as the dry hand rod, consists of dual rods, a baseplate, and a calibrated handle. The main rod is graduated for measuring depth and the smaller rod used for positioning the current meter.

The rod is placed in the stream with the baseplate resting on the stream bed and the depth of water read on the main rod. When the setting of the smaller rod is adjusted to read the depth of water, the meter is positioned automatically to the 0.6 depth. When the depth of water is divided by two and the new

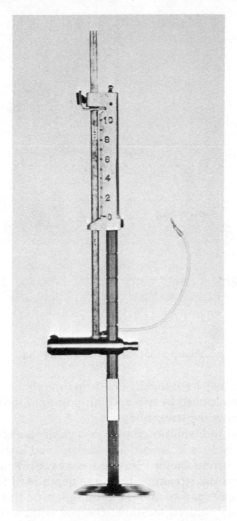

Figure 8.25 Top-setting wading rod

Figure 8.26   Sectional wading rod

value set, the meter will be at the 0.2 depth up from the stream bed. When the depth of water is multiplied by two and this position set, the meter will be at the 0.8 depth up from the stream bed.

The other type of rod typically consists of a graduated sectional rod (Figure 8.26) approximately $1\frac{1}{2}$ m in length, a baseplate, and a slide support to hold and position the current meter. This rod is placed in the stream with the baseplate resting on the stream bed and the depth read from the graduated rod. The rod is then removed and the slide support adjusted to position the current meter to the desired location. The rod is then re-inserted into the stream and the current meter reading is taken.

8.4.2 Handlines

Handlines (Figure 8.27) are sometimes used for making soundings and discharge measurements from bridges where only a light sounding weight, not exceeding 15 kg, is needed. The handline is made up of two separate cables that are electrically connected at a small reel. The upper or hand cable is a heavy two-conductor electric cable, whose thick rubber protective covering makes the cable comfortable to handle. the lower or sounding cable is a light reverse-lay steel cable with an insulated core conductor. A connector joins the lower cable with a hanger bar for mounting the current meter and sounding weight. Sounding cable, in excess of the length needed to sound the stream being measured, is stored on the reel. The sounding cable is tagged at convenient intervals with streamers of differently coloured binding tapes,

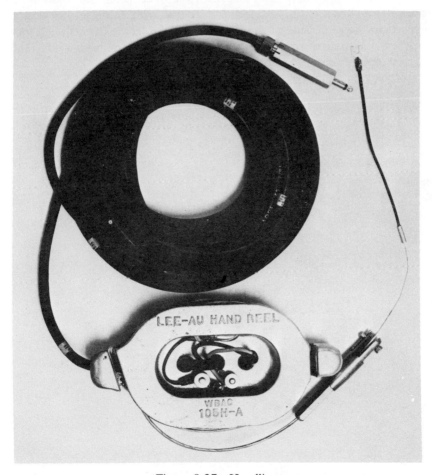

Figure 8.27 Handline

each coloured streamer being at a known distance above the current meter and sounding weight.

Although the use of a handline is limited to streams where light sounding weights can be used, they are particularly useful under certain conditions. One example is the measurement of discharge from truss bridges without cantilevered sidewalks.

8.4.3 Bridge frames

Bridge frames (Figure 8.28) serve a purpose similar to that of handlines, i.e. for making soundings and discharge measurements from bridges where only a light sounding weight, not exceeding 20 kg, is needed. The assembly consists of a tubular steel frame approximately 1 m in length and 0.3 in width, a seat for mounting a light-duty reel, and a sheave over which the sounding cable travels. The sheave end is extended over the bridge rail, and a steel angle fastened to the underside of the frame keeps the frame from sliding forward on the rail. A chain or rope to secure the assembly to the bridge is fastened to the rear of the frame.

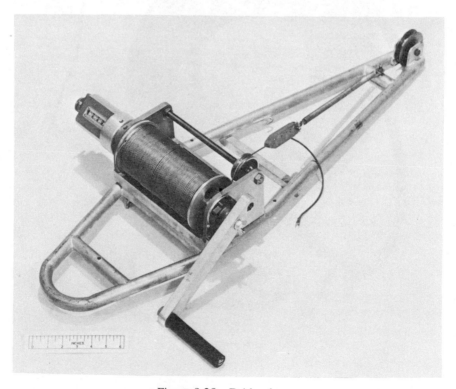

Figure 8.28   Bridge frame

The lightweight frame assembly can easily be transferred from one measuring point to another by lowering the sounding weight to the stream bed and moving the frame assembly to the new location while the weight is at rest. By using this technique, it is not necessary to completely raise the weight and meter each time a bridge truss member is encountered.

8.4.4 Reels

Cable-suspension equipment is necessary on deep streams where wading cannot be done. A sounding reel or winch is used for storing and metering the cable in and out. The sounding reel consists of a drum, crank-and-ratchet assembly, and cable metering and indicating device (Figures 8.29 and 8.30). A small reverse-lay cable with insulated inner conductor is wound on the drum of the reel with a connector and hanger bar on its outer end. Reverse-lay cable is used to prevent the twisting that is typical of single-lay cable. The current meter and sounding weight are mounted on the hanger bar and are raised, lowered, or held at the desired position with the crank-and-ratchet assembly.

Figure 8.29   Canfield reel

Figure 8.30   A-Pak reel

Figure 8.31   Heavy-duty reel

Some heavy-duty reels (Figure 8.31) that are designed for heavy sounding weights are driven by electric motors. Reels are used on cranes for bridge measurements, on cableway carriages, and on boats.

8.4.5 Cranes

Cranes are used in making discharge measurements from bridges. They provide a base on which to mount the sounding reel and a means for extending the weight and current meter over the handrail and beyond bridge members.

Figure 8.32   Three-wheel crane

A lightweight three-wheel crane (Figure 8.32) is often used with sounding weights of 25 kg or less. The front two wheels of the crane are placed close to the bridge rail and parallel to it. The crane is then tilted forward until it rests against the bridge rail, extending the weight and current meter over and beyond the rail. In the transport position the crane rests on all three wheels, the weight and meter being suspended directly above the front wheels.

Four-wheel cranes (Figure 8.33) are used with heavier weights. The bridge rail is not essential as a point of support for the four-wheel crane. A linkage system between the superstructure and the base allows the boom to be extended over and retrieved from the railing while the truck remains stationary. Counterweights are required.

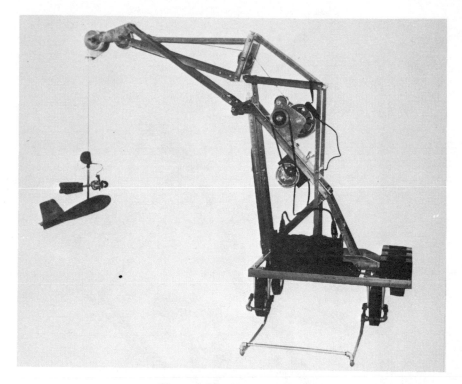

Figure 8.33   Four-wheel crane

In addition to the three- and four-wheel cranes, many special arrangements to overcome certain obstacles have been devised. Truck-mounted cranes (Figure 8.34a and 8.34b) are often used to measure from bridges over larger rivers.

Cranes have a protractor incorporated to measure the angle the sounding line makes with the vertical when the weight and meter are dragged downstream by the force of the water.

Figure 8.34(a)   Truck-mounted crane

Figure 8.34(b)   Truck-mounted crane

8.4.6 Cableways

Cableways are commonly employed for making discharge measurements. They provide a track for a carriage from which the sounding weight and current meter are suspended. There are two basic types of cableways, namely: (a) those with a carriage in which the operator travels and makes observations, and (b) those with an instrument carriage controlled from the stream bank.

In the first type, the carriage supports the operator, the sounding reel, and other necessary equipment. The carriage is propelled along the track from one position to another by means of a carriage puller. It is held in position by a brake on the carriage wheels or by wedging a strap (part of the puller) between the wheel and the cable. Several types of carriage are used. Typical examples are illustrated in Figures 8.35 and 8.36.

The carriage of the second type supports only the sounding cable with the weight and current meter attached. The lateral position of the carriage and the vertical position of the weight and meter are controlled by the operator from the stream bank.

8.4.7 Boat measurement equipment

Special equipment is necessary to suspend the weight and current meter and to hold the boat in position when depths are such that rod suspension cannot

Figure 8.35   Sit-down type cable carriage

Figure 8.36　Stand-up type cable carriage

be used. Typical boat measurement equipment is illustrated in Figure 8.37. A crosspiece reaching across the boat is clamped to the sides of the boat, and a boom attached to the centre of the crosspiece extends over the bow. The crosspiece is equipped with a guide sheave and clamp arrangement at each end to attach the boat to a tag line and makes it possible to slide the boat along the tag line from one measuring vertical to the next. A small rope can be attached to these clamps so that in an emergency a tug on the rope will release the boat from the tag line. The boom is equipped with a reel plate on one end and a sheave over which the meter cable passes on the other.

8.4.8　Ice measurement equipment

Current meter measurements on ice covered streams require that holes be cut in the ice through which the current meter is suspended. These holes are frequently cut by use of a gasoline-powered ice auger (Figure 8.38). Such an auger will drill through 1 m of ice in less than 2 min. Where it is not practical to use the powered auger or one is not available, ice chisels are used to cut the holes. These chisels are usually about $1\frac{1}{2}$ m in length and weigh approximately 5 kg.

When holes are cut through the ice, the water below is usually under pressure from the weight of the ice and rises in the hole. In order to determine

Figure 8.37   Boat measurement equipment in operation

the effective depth of the stream, ice measuring sticks are used to measure the distance from the water surface to the underside of the ice. The entire depth of water is then measured and corrected by the distance determined with the ice measuring stick.

Water depth in most ice-covered streams is usually such that a rod can be used for the sounding and meter suspension (Figure 8.39). In a few larger streams, heavier suspension equipment is required, and devices such as the collapsible reel support sled (Figure 8.40) are used.

Figure 8.38   Ice auger

8.4.9   Echo-sounders

The echo-sounder (Figure 8.41) is an electroacoustic instrument which indicates the depth of water by measuring the time differential between the transmission of a burst of acoustic energy and the reception of the reflection of this energy from the stream bed. Depth is determined from the equation

$$d = \frac{t}{2} c$$

in which d is the distance from the transducer to the stream bed, t is the travel

Figure 8.39   Measurement of ice-covered stream by rod suspension

time of the acoustic pulse, and c is the velocity of sound in water. Since the depth determined varies directly with the velocity of sound in water and the velocity of sound varies both with density and the elasticity of water, the accuracy of the depth measurement is highly dependent upon calibrating for the correct velocity of sound. Calibration is best accomplished by suspending a plate at a known distance directly below the transducer in the water in which the unit is to be used.

Echo-sounders are very useful for making rapid profiles of cross-sections. They are also useful in instances where the bed of the stream is so soft that

Figure 8.40   Collapsible reel support sled

sounding weights sink deeply into the silt or where the velocity is so high that other means are not practicable.

8.4.10   Sounding weights

Sounding weights are used to hold the sounding line as nearly vertical as practicable and to keep the current meter stable in the flowing water. They are streamlined in shape to present the least possible resistance to the flow and usually have a set of tail fins to assist in aligning them with the direction of flow. Figure 8.42 shows typical sounding weights in various sizes.

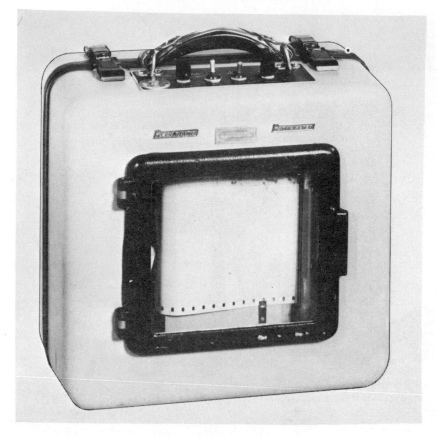

Figure 8.41   Echo-sounder

There is no set rule for determining the size of sounding weight to be used, but the following formula has been suggested:

$$\text{velocity (m/s)} \times \text{depth (m)} = \frac{\text{weight (kg)}}{5}$$

Hanger bars and weight pins (Figure 8.43) are used for attaching the sounding weight and current meter to the sounding line.

8.5   WIDTH MEASURING EQUIPMENT

The distance to any point in a cross-section is measured from an initial point on the bank. Manned cableways and bridges used regularly for making discharge measurements are commonly marked at frequent intervals with permanent markings. Distance between markings is estimated, or measured

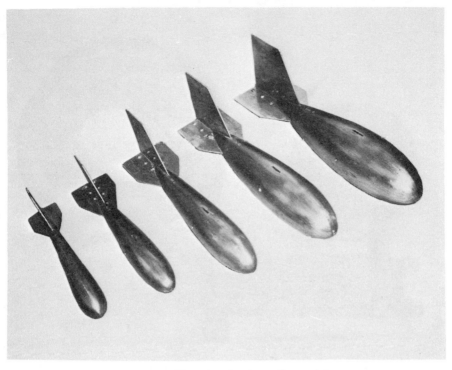

Figure 8.42   Sounding weights

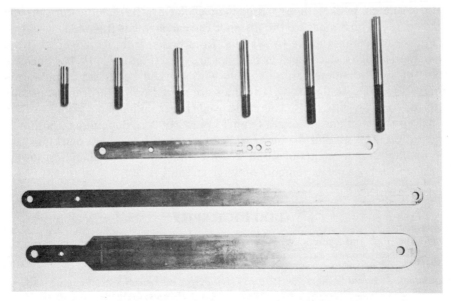

Figure 8.43   Hanger bars and weight pins

Figure 8.44   Tag lines

with a rule or pocket tape. Unmanned cableways have indicators on the carriage reel which measure the distance the carriage has travelled.

For measurements made by wading, from boats, from ice cover, or from unmarked bridges, steel tapes or tag lines are used (Figure 8.44). Tag lines are made of small-diameter aircraft cable with marked beads set at measured intervals to indicate distances. Tag lines are stored on metal frames, usually referred to as tag-line reels. The cable used ranges in diameter from less than 1 mm up to 5 mm and in length from 15 m to 1000 m, depending upon their application. The smaller diameter, shorter ones, with frequent markings, are for wading measurements of small steams; while the larger diameter, longer ones, are for boat measurements on large streams.

BIBLIOGRAPHY

Buchanan, T. J. and Somers, W. P., 1968. Stage measurement at gaging stations, US Geological Survey, Techniques of Water Resources Investigations, Book 3, chap. A7.
Buchanan, T. J. and Somers, W. P., 1969. Discharge measurements at gaging stations, US Geological Survey, Techniques of Water Resources Investigations, Book 3, chap. A8.

Chow, Ven Te, 1964. *Handbook of Applied Hydrology*, McGraw-Hill, New York.

Corbett, D. M. *et al.*, 1943. Stream-gaging procedure, US Geological Survey, Water Supply Paper 888, Washington D.C.

Frazier, A. H., 1974. *Water Current Meters*, Smithsonian Institution Press, Smithsonian Studies in History and Technology, No. 28.

Guide to Hydrometeorological Practices, 1974. World Meteorological Organization, Geneva.

ISO: 2537, 1974. Liquid flow measurement in open channels—cup-type and propeller-type current meters, International Organization for Standardization, Geneva.

ISO: 3454, 1976. Liquid flow measurement in open channels—sounding and suspension equipment, International Organization for Standardization, Geneva.

ISO: 3455, 1975. Liquid flow measurement in open channels—calibration of current meters in straight open tanks, International Organization for Standardization, Geneva.

ISO: 4373, 1978. Liquid flow measurement in open channels—water-level measuring equipment, International Organization for Standardization, Geneva.

ISO: 4375, 1978. Liquid flow measurement in open channels—cableway system, International Organization for Standardization, Geneva.

ISO: 1978. Liquid flow measurement in open channels—echo sounders, International Organization for Standardization, Geneva.

Smoot, G. F., 1974. A review of velocity-measuring devices, US Geological Survey, Open-File Report, Washington D.C., 34 pp.

Smoot, G. F. and Novak, C. E., 1968. Calibration and maintenance of vertical-axis type current meters, US Geological Survey, Techniques of Water Resources Investigations, Book 8, chap. B2.

Troskolanski, A. T., 1960. *Hydrometry*, Pergamon Press, Oxford.

CHAPTER 9

Weed Growth—a Factor of Channel Roughness

K. E. C. POWELL

9.1 INTRODUCTION

In the lowland areas of the Anglian Water Authority river channels tend to be of a standard form, that is to say embanked earth channels, grassed and treeless and of regular trapezoidal section. The values of Manning's coefficient of channel roughness commonly given to this form of embanked channel is 0.030. The practice of assessing values of n from standard works, notably Chow (1959) and Barnes (1967), was questioned when the possibility of river gauging by means of the slope-area method was considered. This prompted further questions. How in fact does n vary with changes in stage and season? Could a knowledge of n values and measurements of water slope be the means of obtaining continuous measurements of flow, particularly in channels where there is no stable relationship between stage and discharge? Can the constraint imposed by channel vegetation be quantified and expressed in terms of extreme values of Manning's n?

In order to attempt answers to these questions a measuring reach was constructed on the river Bain just upstream of the river gauging station at

Fulsby Lock, such that values of n could be assessed and related to specific hydraulic conditions. It was assumed that the values of n given to the measuring reach would vary with stage and season; that during winter months, and throughout the full range of stage, variations in n would be small; that winter values of roughness might be attributed to (1) the surface roughness of the channel, i.e. the grain size of the material which forms the wetted perimeter, (2) bank growth and such plant material that exists during winter, (3) irregularities in wetted perimeter and channel alignment, and (4) the energy expended in the movement of bed material; that during summer months, variations in n would be very large and that this departure from the winter condition is almost entirely due to the mass of aquatic weed and reeds that occupy the channel during these months.

9.2 THE MEASURING REACH

9.2.1 Measurement of water slope, mean depth, water area, and wetted perimeter

The measuring reach was set up in January 1967 when a stilling well, complete with vernier depth gauge, was placed at a site 1549 m upstream of the primary river gauging station at Fulsby Lock. This is the Haltham stilling well. In January 1968 a second stilling well and vernier depth gauge were sited 1341 m downstream of the Haltham well and 207.3 m upstream of the gauging station. This is the Fulsby stilling well. In September 1971 both stilling wells were fitted with Fischer Porter punched paper tape recorders. These instruments record water levels at 15-min intervals, and have a timing accuracy of ±4 min per month (see Chapter 8).

The reach has a trapezoidal section which expands with its downstream direction, it is without major bends, and its design flow is given as 38.8 cumecs. A take-off channel exists in the right bank, downstream of the Fulsby well but upstream of the gauging station. These features are shown in Figure 9.1.

The stilling wells consist of circular steel piles which are 0.60 m in diameter and driven into the toe of the bank. They are connected to the river by means of two 0.05 m diameter intake pipes. The recorders and vernier depth gauges are housed in steel cabinets and access to the cabinets is obtained by means of walkways. The river gauging station, which was commissioned in September 1962, consists of a round-crest weir that operates throughout a full range of flows. By virtue of a lockpit which is immediately downstream, the weir remains modular throughout its range. The station is equipped with a Fischer Porter punched paper tape recorder and a Munro IH95 autographic recorder. The weir calibration was obtained by model tests, and current meter checks indicate an accuracy of ±5 per cent. Flows in the take-off channel are

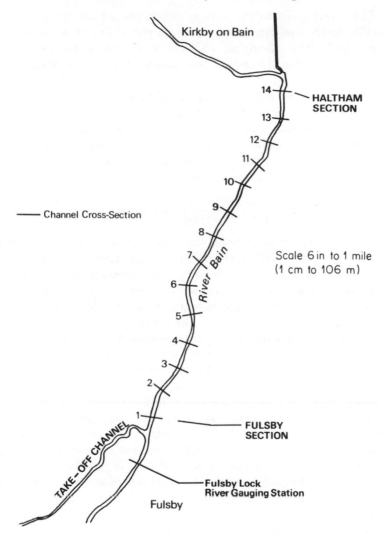

Figure 9.1   Location plan of measuring reach

measured by means of a thin-plate weir which does, in fact, form an integral part of the river gauging station.

Particular care has been taken in preserving the accuracy of measurements within the reach. The vernier depth gauges at Haltham, Fulsby and the river gauging station were levelled to within ±9 mm by means of a Watts Microptic Parallel Plate Level. The levels were checked in May 1970, September 1971 and September 1976. It was found that the Haltham well had subsided some 13 mm between 1967 and 1970, and 30 mm between 1971 and 1976; the Fulsby well, 5 mm between 1967 and 1970, and 35 mm between 1971 and

1976. The levels given to the vernier depth gauges are the means of the 1967–1970 levels, and later, the means of the 1971–1976 levels. That is, Haltham 1967–1970, 12.967 m, 1971–1976 13.396 m; Fulsby 1967–1970 12.903 m, 1971–1976 13.295 m. All levels refer to Ordnance Datum, Newlyn.

The length of the measuring reach was obtained by means of a steel wire and floats placed at intervals along the centreline of the channel; this measurement is considered accurate to within ±1 m. With each vernier depth gauge levelled to ±4.5 mm, the vernier reading of water level accurate to ±3 mm and the Fischer Porter recorders accurate to ±3 mm, the measurement of water slope is taken to be within

$$\frac{2 \times (4.5 + 3 + 3)}{(1341 \pm 1) \times 1000} = 0.000\ 001\ 5$$

Fourteen channel cross-sections were taken at intervals of 95.5 m and a longitudinal section plotted; it was noted that the mean bed slope is 0.000 34. The mean cross-section was obtained by superimposing one channel section upon another, using mean bed level as a reference. From the mean section, relationships between mean depth, water area and wetted perimeter were obtained. The mean depth is taken as the mean of the depths at the Haltham and Fulsby section. Due to the method of measuring channel sections and to the fact that the channel is scoured out annually by excavator, values of water area and wetted perimeter are considered to be no better than ±10 per cent at high flows and ±20 per cent at low flows, and values of mean depth, ±5 per cent at high flows and ±30 per cent at low flows. The mean cross-section and the Haltham and Fulsby sections are shown diagrammatically in Figure 9.2.

During the period January 1967 to January 1968, measurements of water slope were taken from levels at the stilling wells at Haltham and the gauging station. A correction was made to account for the flow lost to the take-off channel, whereby the gauging station level was taken as the head-on-weir that related to the sum of the station and the take-off channel discharges. During the period January 1967 to September 1971, measurements of water slope were taken by vernier depth gauge at approximately weekly intervals but more frequently during period of high flows. With such measurements an error is induced in that the time interval between the reading at Haltham and the reading at Fulsby is about 11–12 min. This error was excluded when the punched paper tape recorders were installed.

9.2.2 Channel vegetation

The measuring reach consists of an embanked channel which is formed of clays, gravel and alluvial spoil. There is a lining of bed muds and silt and substrata of gravels and clays. During the growing season 1 April to 31 July the modal discharge, daily mean flow, and its associated mean velocity are

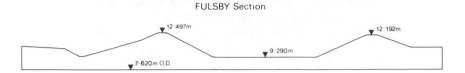

FULSBY Section

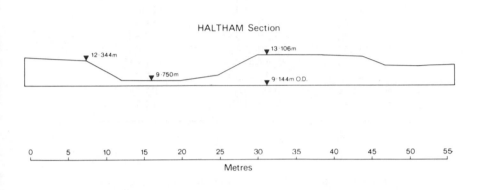

HALTHAM Section

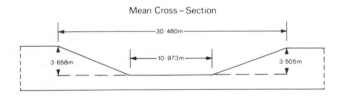

Mean Cross – Section

Figure 9.2   The mean cross-section and the Haltham and Fulsby
sections

approximately 0.25 cumec and 0.03 m s^{-1}, while at flood flows velocities may
rise to 1.0 m s^{-1}. Modal flows for the individual growing seasons 1967–1977
are given in Table 9.1. As observed by Haslam (1978), the type and proliferation of plants are directly related to the occurrence of a growing season
modal flow, while under flood conditions, plants are battered and their
underground parts are alternately scoured out and inundated by moving bed
material. The nature and growth of plant material in the measuring reach is
indicated in Figures 9.3 to 9.10 and a species list is given in Table 9.2.

The cycle of growth and decay in plant material is controlled by roding, a
process whereby during the months of August and September excavators
remove growth and muds from the bed and bottoms of banks. Figures 9.3 and
9.4 show the channel state that is likely to exist after roding and throughout

Table 9.1   Modal flows for the growing season 1 April
to 31 July for the years 1967–1977

Year	Modal class group (cumecs)	Frequency of occurrence as a percentage
1967	0.35–0.40	9.01
1968	0.50–0.55	10.65
1969	0.35–0.40	7.37
1970	0.35–0.40	11.47
1971	0.20–0.25	11.47
1972	0.30–0.35	9.83
1973	0.20–0.25	13.93
1974	0.15–0.20	21.31
1975	0.25–0.30	17.21
1976	0.00–0.05	24.59
1977	0.15–0.20	13.11

Table 9.2   Species list of plant material in the river Bain

Site	Water column	Marginal
Haltham (upstream of intake)	*Potamogeton pectinatus* (fennel-like pondweed) *Elodea canadensis* (Canadian pondweed) *Vaucheria sp.* (cott) *Enteremorpha sp.*	*Glyceria maxima* (reed sweet grass) *Juncus effusus* (soft rush) *Typha latifolia* (reedmace) *Ranunculus sceleratus* (celery-leaved crowfoot) *Rorippa nasturtium aquaticum* (watercress) *Myosotis scorpioides* (water forget-me-not)
Fulsby lock	As above	As above + *Erysimum cheiranthoides* (treacle mustard)

the winter months; the masses of weed and reed have been removed and bank grasses and nettles are reduced to dry stalks. Then, in April, the signs of new growth appear on the banks and in the reed beds (Figure 9.5 and 9.6). Figure 9.7 shows full growth, when bank grasses reach average heights of 1 m, reeds

Figure 9.3 Upstream of river gauging station (November), $Q = 0.29 \text{ m}^3 \text{ s}^{-1}$

Figure 9.4 Upstream of river gauging station (March), $Q = 3.41 \text{ m}^3 \text{ s}^{-1}$

Figure 9.5   Upstream of river gauging station (April), $Q = 0.93 \text{ m}^3 \text{ s}^{-1}$

Figure 9.6   Upstream of river gauging station (June), $Q = 2.08 \text{ m}^3 \text{ s}^{-1}$

Figure 9.7 Upstream of river gauging station (July), $Q = 0.49 \, \mathrm{m}^3 \, \mathrm{s}^{-1}$

Figure 9.8 Upstream of Haltham (January frost), $Q = 1.84 \, \mathrm{m}^3 \, \mathrm{s}^{-1}$

Figure 9.9   Upstream of Haltham (June), $Q = 0.65 \text{ m}^3 \text{ s}^{-1}$

Figure 9.10   Upstream of Haltham (July), $Q = 0.38 \text{ m}^3 \text{ s}^{-1}$

reach average heights of 0.65 m and some seven-tenths of the river surface is covered by weed. Figures 9.8, 9.9 and 9.10 show the channel upstream of the Haltham section and are representative of the winter condition, new growth in June and the full growth in July, respectively.

9.3 CALCULATION OF MANNING'S n

From the measurement of water slope it is found that at low flows the water slope (0.000 12) is approximately one-third of the bed slope (0.000 34), while at high flows the water slope and bed slope are approximately equal. Also, the cross-section of flow expands with its distance downstream. Thus the conditions for uniform flow as given by Chow (1959) are not strictly met. In order to accommodate situations in which there is a significant change in velocity head throughout the reach, the energy slope in the Manning equation is taken to be the difference between the total heads at each end of the reach divided by the length of the reach as shown below (ISO, 1973).

From Figure 9.11 and using the following notation:

A = water area, mean section
α = energy coefficient = 1.0
β = momentum coefficient
$\bar{d}$ = depth at mean section
d_F = depth at Fulsby section
d_H = depth at Haltham section
H_L = energy loss between cross-sections
l = length, Haltham to Fulsby section = 1341 m
L_F = level at Fulsby section
L_H = level at Haltham section
n – Manning's coefficient of roughness
P = wetted perimeter of mean section
Q = discharge of measuring reach

Q_F = discharge at gauging station
Q_T = discharge at take-off channel
R_h = hydraulic mean depth
S = energy slope
S_B = bed slope
S_W = water slope
$\bar{V}$ = mean velocity at mean section
V_F = mean velocity at Fulsby section
V_H = mean velocity at Haltham section
Z_F = bed level, Fulsby section = 9.290 m ODN
Z_H = bed level, Haltham section = 9.750 m ODN

$$\bar{V} = \frac{R_h^{2/3} S^{1/2}}{n}$$ (Manning's equation—see Chapter 5, equation 5.1a)

(9.1)

then

$$Q = \frac{A\left(\frac{A}{p}\right)^{2/3}\left(\frac{H_L}{l}\right)^{1/2}}{n}$$

(9.2)

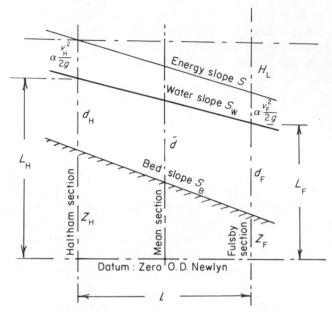

Figure 9.11   Derivation of equation for calculating values
of n

and

$$n = \frac{A\left(\frac{A}{p}\right)^{2/3}\left(\frac{H_L}{l}\right)^{1/2}}{Q}$$

(9.3)

where

$$Q = Q_F + Q_T$$

and

$$H_L = \left(\frac{\alpha_H V_H^2}{2g} + d_H + Z_H\right) - \left(\frac{\alpha_F V_F^2}{2g} + d_F + Z_F\right)$$

but which Chow modifies as follows:

$$H_L = \frac{\beta}{2g}\left(\alpha_H V_H^2 - \alpha_F V_F^2\right) + [(d_H + Z_H) - (d_F + Z_F)]$$

where

$$\beta = 0.5 \text{ if } V_H > V_F \quad \text{and} \quad \beta = 1.0 \text{ if } V_H < V_F$$

Experimental data indicate that the value of the energy coefficient α varies from about 1.03 to 1.36 for fairly straight prismatic channels and will exceed 2.0 in the vicinity of obstructions or near pronounced irregularities (Chow, 1959). However, in the months of June, July and August, and at low flows, the grouping of dense plant material can be considered as a pronounced irregularity. This condition of reduced water area and greatly increased meander length would indicate values of α in excess of 2.0. But in so far as a pronounced irregularity can also be considered as a factor of channel roughness, α is given the value of 1.0 and the loss in energy due to the irregularity is expressed in the value of n.

Now, from equation (9.1),

$$AR_h^{2/3} - \frac{nQ}{S^{1/2}}$$

where $R_h = A/P$, it is seen that $AR_h^{2/3}$ is a function of depth and that depth is proportional to n, Q and S. Thus, in this chapter, variations in the inter-relationship $\bar{d}$, Q, S, n are examined, as well as variation in the value of n.

9.4 ESTIMATION OF UNCERTAINTIES

An estimate of the total uncertainty likely to be found in the values of n as calculated for the reach is taken as 26 per cent, over the range $\bar{d} = 1.0$–3.0 m, and 52 per cent when $\bar{d}$ is less than 1.0 m but greater than 0.5 m (see Table 9.3). This represents a total uncertainty in that it assumes the rare occurrence when each variable is subject to an extreme error. No distinction is drawn between the winter and summer condition of the channel because although in summer groups of dense plant material cause a reduction in effective water area and an increase in the meander length, such constraints are seen as an extreme limit of the vegetation factor of channel roughness and not as a source of error in the variables A, P and l.

Table 9.3 Estimate of total uncertainty in n

Range		Q_F	Q_T	A	P	l	d_F	d_H	r.m.s.	Uncertainty
Error given to	$1 \leq \bar{d} \leq 3$ m	5	5	10	10	1	15	15	26.4	±26%
variables as	$0.5 \leq \bar{d} \leq 1$ m	8	8	20	20	1	30	30	52.2	±52%
per cent value										

9.5 PROCESSING OF DATA

The records from the Haltham and Fulsby stilling wells are divided into two groups: (1) the irregular readings from vernier depth gauges which extend from January 1967 to September 1971, and (2) the 15-min punchings on paper tape which, apart from data losses, are ongoing from September 1971. The reason for the loss of records is largely due to the fact that the Haltham and Fulsby recorders were the standby recorders for the primary river gauging stations; as such, they were sometimes required at other gauging stations.

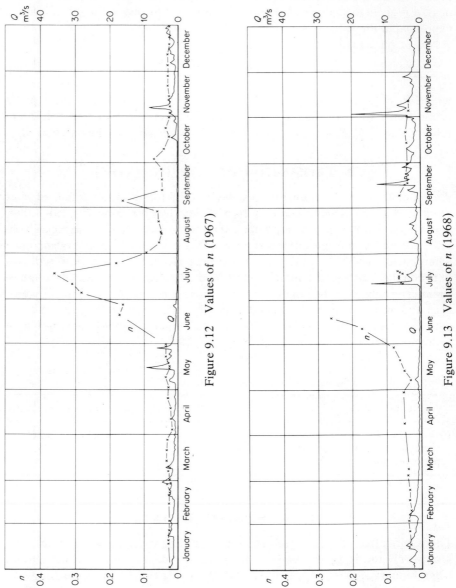

Figure 9.12　Values of *n* (1967)

Figure 9.13　Values of *n* (1968)

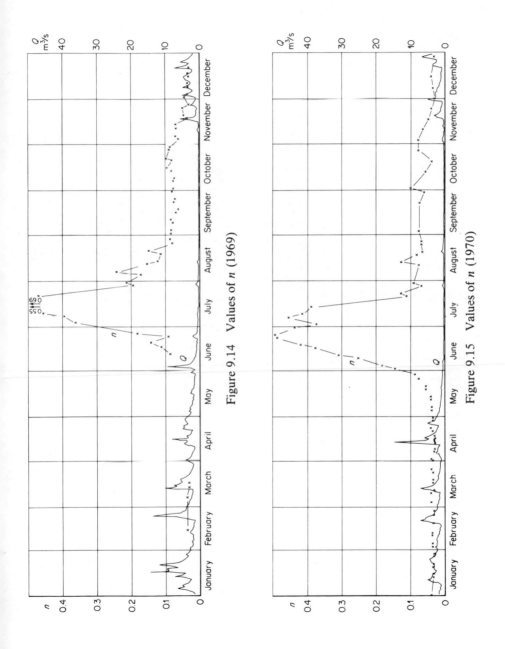

Figure 9.14 Values of *n* (1969)

Figure 9.15 Values of *n* (1970)

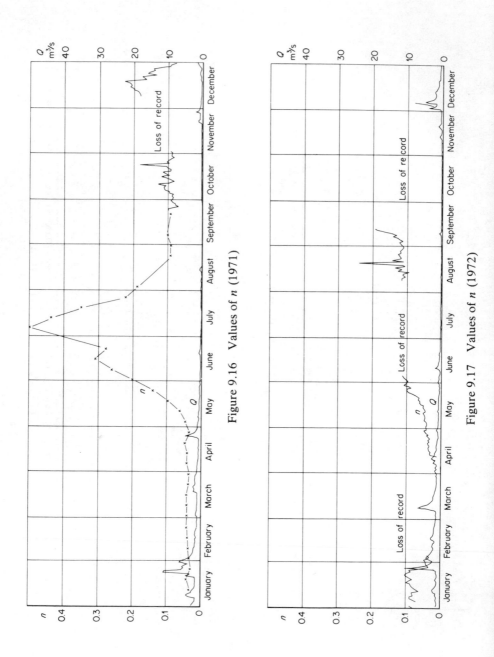

Figure 9.16 Values of *n* (1971)

Figure 9.17 Values of *n* (1972)

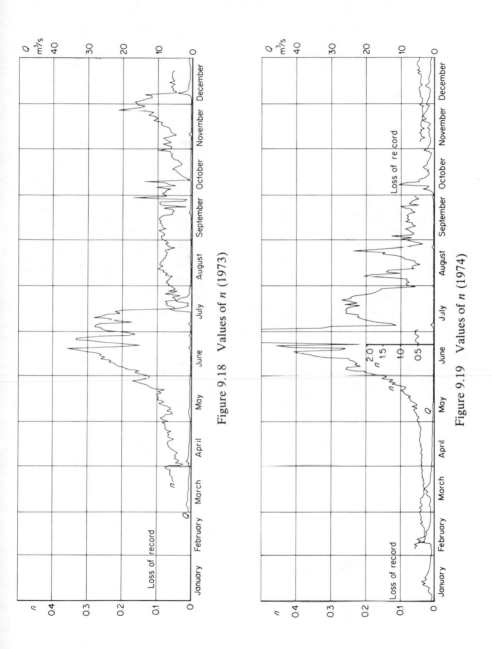

Figure 9.18 Values of *n* (1973)

Figure 9.19 Values of *n* (1974)

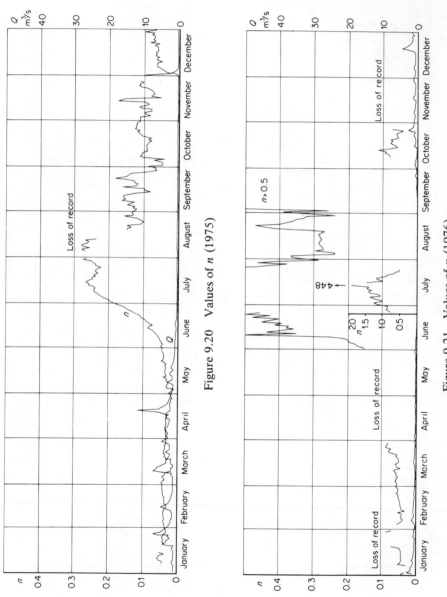

Figure 9.20   Values of *n* (1975)

Figure 9.21   Values of *n* (1976)

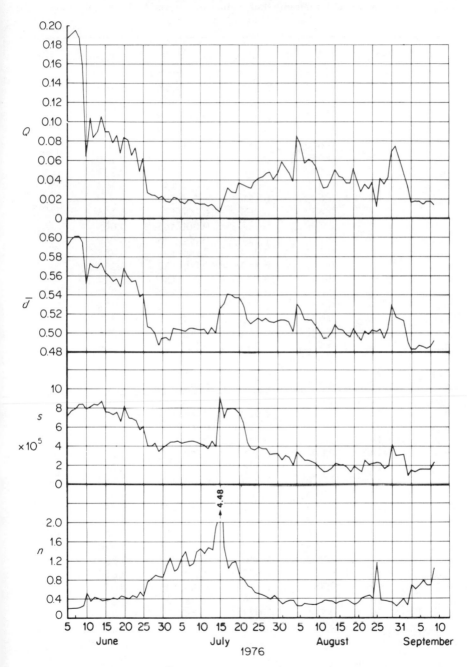

Figure 9.22   Values of $\bar{d}$, Q, S and n (9 a.m., 5 June to 10 September, 1976)

Table 9.4   Maximum and minimum daily values of n based on 15-min readings, 1–20 July, 1973

	Day	$\bar{d}$	Q	S	n	n_{diff}
n_{max}	1	0.680	0.203	0.000 185	0.367	
n_{min}	1	0.683	0.286	0.000 178	0.257	0.110
n_{max}	2	0.677	0.231	0.000 180	0.316	0.057
n_{min}	2	0.653	0.243	0.000 152	0.259	
n_{max}	3	0.645	0.183	0.000 161	0.346	0.086
n_{min}	3	0.650	0.240	0.000 151	0.260	
n_{max}	4	0.643	0.180	0.000 163	0.353	0.060
n_{min}	4	0.639	0.205	0.000 149	0.293	
n_{max}	5	0.642	0.169	0.000 161	0.370	0.095
n_{min}	5	0.637	0.208	0.000 142	0.275	
n_{max}	6	0.790	0.225	0.000 365	0.605	0.516
n_{min}	6	0.816	1.114	0.000 176	0.089	
n_{max}	7	0.571	0.059	0.000 131	0.785	0.562
n_{min}	7	0.642	0.259	0.000 137	0.223	
n_{max}	8	0.641	0.196	0.000 156	0.314	0.069
n_{min}	8	0.633	0.227	0.000 133	0.245	
n_{max}	9	0.635	0.178	0.000 154	0.338	0.081
n_{min}	9	0.630	0.215	0.000 134	0.257	
n_{max}	10	0.622	0.160	0.000 142	0.350	0.094
n_{min}	10	0.622	0.203	0.000 124	0.256	
n_{max}	11	0.622	0.170	0.000 138	0.323	0.085
n_{min}	11	0.619	0.211	0.000 117	0.238	
n_{max}	12	0.614	0.163	0.000 130	0.320	0.082
n_{min}	12	0.616	0.207	0.000 114	0.238	
n_{max}	13	0.611	0.171	0.000 131	0.303	0.086
n_{min}	13	0.604	0.210	0.000 104	0.217	
n_{max}	14	0.602	0.156	0.000 122	0.314	0.102
n_{min}	14	0.601	0.208	0.000 100	0.212	
n_{max}	15	0.672	0.328	0.000 172	0.214	0.166
n_{min}	15	1.226	5.096	0.000 245	0.048	
n_{max}	16	1.426	6.600	0.000 328	0.057	0.016
n_{min}	16	1.393	6.851	0.000 204	0.041	
n_{max}	17	0.829	1.465	0.000 160	0.067	0.022
n_{min}	17	0.936	2.612	0.000 148	0.045	
n_{max}	18	0.725	0.810	0.000 131	0.086	0.022
n_{min}	18	0.802	1.340	0.000 137	0.064	
n_{max}	19	0.754	0.972	0.000 135	0.078	0.005
n_{min}	19	0.781	1.153	0.000 145	0.073	
n_{max}	20	0.741	0.916	0.000 131	0.080	0.009
n_{min}	20	0.761	1.074	0.000 130	0.071	

Readings from the first group are plotted on to Figures 9.12 to 9.15 in terms of Q and n. With the second group, it was not possible to process all available data; instead, items were selected as follows:

(1) The 0900-hour readings each day, 24 September, 1971 to 31 December, 1976. These are plotted onto Figures 9.16 to 9.21 in terms of Q and n, and on to Figure 9.22 in terms of $\bar{d}$, Q, S and n for June–September 1976. The 0900-hour readings were chosen because it is thought they represent the mean of the daily variations in n.
(2) 15-min readings each day, 1–20 July, 1973. From these $\bar{d}$, Q, S, n and n_{diff}, for the maximum and minimum values of n, were calculated and are given in Table 9.4. The month of July 1973 was chosen because its records contained periods of low flows with pronounced variations and a summer flood peak.
(3) 15-min readings each day, 1–28 February, 1977. From these $\bar{d}$, Q, S, n and n_{diff}, for maximum and minimum values of n, were calculated and are given in Table 9.5. The month of February 1977 was chosen because February is a time when the factor of vegetation has least influence on the value of n, and also the record contains a flood peak that almost equals the channel designated design flow.

Table 9.5   Maximum and minimum daily values of n based on 15-min readings 1–28 February, 1977

	Day	$\bar{d}$	Q	S	n	n_{diff}
$n_{\max}$	1	0.922	2.959	0.000 115	0.340	0.202
$n_{\min}$	1	0.894	3.347	0.000 027	0.138	
$n_{\max}$	2	1.014	3.860	0.000 138	0.034	0.005
$n_{\min}$	2	1.305	7.054	0.000 138	0.029	
$n_{\max}$	3	1.361	7.500	0.000 184	0.034	0.008
$n_{\min}$	3	1.340	7.733	0.000 119	0.026	
$n_{\max}$	4	1.211	6.083	0.000 141	0.030	0.004
$n_{\min}$	4	1.211	6.279	0.000 116	0.026	
$n_{\max}$	5	1.304	7.066	0.000 143	0.030	0.003
$n_{\min}$	5	1.259	6.722	0.000 121	0.027	
$n_{\max}$	6	1.215	5.957	0.000 152	0.032	0.005
$n_{\min}$	6	1.214	6.163	0.000 121	0.027	
$n_{\max}$	7	1.200	5.867	0.000 148	0.031	0.004
$n_{\min}$	7	1.149	5.525	0.000 117	0.027	
$n_{\max}$	8	1.003	3.836	0.000 126	0.032	0.004
$n_{\min}$	8	1.079	4.794	0.000 120	0.028	
$n_{\max}$	9	1.136	5.053	0.000 165	0.035	0.004
$n_{\min}$	9	0.995	3.794	0.000 118	0.031	
$n_{\max}$	10	1.923	15.500	0.000 256	0.038	0.013
$n_{\min}$	10	2.616	31.494	0.000 137	0.025	

[*table continued overleaf*

Table 9.5　*(cont.)*

	Day	$\bar{d}$	Q	S	n	n_{diff}
$n_{\max}$	11					
$n_{\min}$	11					
n_{mav}	12					
$n_{\min}$	12		[Haltham recorder at fault]			
$n_{\max}$	13					
$n_{\min}$	13					
$n_{\max}$	14					
$n_{\min}$	14					
$n_{\max}$	15	1.046	4.382	0.000 126	0.030	0.001
$n_{\min}$	15	1.056	4.561	0.000 120	0.029	
$n_{\max}$	16	1.002	3.855	0.000 124	0.031	0.002
$n_{\min}$	16	1.040	4.380	0.000 115	0.029	
$n_{\max}$	17	1.003	3.811	0.000 130	0.033	0.003
$n_{\min}$	17	1.010	4.025	0.000 124	0.030	
$n_{\max}$	18	1.176	5.485	0.000 162	0.034	0.008
$n_{\min}$	18	1.425	8.790	0.000 121	0.026	
$n_{\max}$	19	1.427	8.733	0.000 135	0.028	0.005
$n_{\min}$	19	1.363	8.291	0.000 104	0.023	
$n_{\max}$	20	1.293	7.204	0.000 133	0.028	0.004
$n_{\min}$	20	1.204	6.404	0.000 099	0.024	
$n_{\max}$	21	1.348	7.731	0.000 143	0.029	0.004
$n_{\min}$	21	1.358	8.103	0.000 113	0.025	
$n_{\max}$	22	1.313	7.345	0.000 141	0.029	0.003
$n_{\min}$	22	1.308	7.434	0.000 123	0.026	
$n_{\max}$	23	1.438	8.835	0.000 151	0.029	0.003
$n_{\min}$	23	1.522	10.136	0.000 131	0.026	
$n_{\max}$	24	1.440	8.880	0.000 143	0.028	0.002
$n_{\min}$	24	1.336	7.737	0.000 124	0.026	
$n_{\max}$	25	1.285	7.062	0.000 145	0.029	0.002
$n_{\min}$	25	1.269	6.926	0.000 123	0.027	
$n_{\max}$	26	1.226	6.239	0.000 137	0.029	0.002
$n_{\min}$	26	1.156	5.683	0.000 116	0.027	
$n_{\max}$	27	1.008	3.970	0.000 125	0.031	0.004
$n_{\min}$	27	1.071	4.796	0.000 115	0.027	
$n_{\max}$	28	0.998	3.805	0.000 130	0.032	0.002
$n_{\min}$	28	0.991	3.757	0.000 112	0.030	

9.6　DISCUSSION

It can be said that of the factors that determine n in channels such as the Bain, surface roughness, channel irregularities, channel alignment and obstructions have a more or less constant effect for a given stage and throughout the seasons, whereas by comparison the channel vegetation factor is the cause of a considerable variation in n because the vegetative material shows a seasonal variation in biomass. This variable biomass offers resistance to river flow in the following ways:

(a) with changes in growth and decay of the vegetative biomass;
(b) with changes in depth of the water column, whereby the plant material extends throughout the water column in shallow conditions but in only a part of the water column when greater depths occur;
(c) with the responses of the plant material to sunlight, whereby the plants become rigid due to excess oxygen production from photosynthesis during the day and markedly flacid at night, creating a diurnal variation in the values of n and, as a consequence, variations in $\bar{d}$, Q and S (Wetzel, 1975).

Once it was realized that the channel vegetation factor is the single cause of the large variations in n, a pattern emerged for the values of n collected over the 10-year period. As expected with a vegetation factor it varied from year to year due to the influence of parameters such as the hours of direct sunshine, the water temperature, any self-shading effects, the occurrence of certain modal flows that promote growth, flood flows that scour and uproot plant material, the availability of nutrients, and the ambient herbivore population.

From Figures 9.12 to 9.21 it appears that February/March is the period when the factor of channel vegetation has least influence. Channel roding operations of the previous autumn have removed most of the plant material and such growth that occurred after roding has become dormant. Under these conditions, the 0900-hour values of n vary with flow over the range 0.03 to 0.06, while the 15-min values for February 1977 show a maximum range of 0.023 to 0.038 and a mean diurnal variation of 0.004, typically 0.027 to 0.031. Then in April, with a new growth, there is a steady increase in plant material and a corresponding increase in the range of n. At some time in July or early August, when plants go to seed and their vegetative parts reduce, there is a marked fall in the range of n. In most years this fall predates the start of roding operations and is the beginning of the cycle: roding—second growth—in some years second roding—third growth.

From the plots it is seen that the extreme values of n occur in July at low flows, the 0900-hour 1976 series shows extreme values of n of 1.91 and 4.48 (Figure 9.22); the July 1973 15-min readings shows a maximum diurnal variation of 0.562, over the range 0.223 to 0.785 (Table 9.4). The validity of these extreme values is questioned and there is uncertainty regarding the limit to the condition whereby the Manning coefficient is relevant. The calculations $n = 1.91$ and $n = 4.48$ derive from flows of 0.10 and 0.007 cumecs, respectively, and mean depths of 0.50 and 0.525 m, respectively. Thus, apart from the question of error, and although a mean depth of 0.50 m is sufficient to provide an unbroken water slope, Haltham to Fulsby sections, there are occasions when groups of emergent plant material will create a series of moving controls that interrupt the water slope and the calculation of n becomes invalid (see Figure 9.10).

How then to reject data that is invalid? The argument is advanced, albeit tentatively, that in this case high n values represent an extreme condition of channel growth whereby water area is greatly reduced, meander length greatly increased, and energy is expended overcoming the buoyancy and resilience of plant material. Although the limit to the Manning application may not be known, the relevance of trends and a continuity in the interrelationship $\bar{d}$, Q, S, n should be sought.

The year 1976 was notable for a historic drought and a record number of hours of sunshine, when flow in the Bain was reduced to 0.007 cumec. Plant material proliferated and it is certain that the water slope between Haltham and Fulsby was interrupted.

From Figure 9.22 it is shown that n increases from 0.20 on 5 June to 1.91 on 14 July, when an erratic value of 4.48 on the 15 July is followed by a recession from which n has a value of 0.286 on 31 August. A visual inspection of the plots indicates that, apart from the erratic value of 15 July and with due allowance for times of travel, an interrelationship $\bar{d}$, Q and S appears to be maintained and that values of n up to 1.9 are meaningful. The plots can be divided into distinct movements: (1) $\bar{d}$, Q and S, a fall from 8 to 30 June followed by a steady variation until 15 July; n, inversely, a rise from 8 June to the 15 July; (2) $\bar{d}$, Q and S, a sharp rise from 15 July followed by a fall; n, inversely a fall followed by a rise; (3) a response shown by all variables during the last week of August.

With reference to Figure 9.22, the 1976 plot of n, 5 June to 10 September, the marked saw-tooth form suggests an oscillatory component to the interrelationship $\bar{d}$, Q, S, n. That is:

(1) Increased resistance offered by vegetation; increase in channel storage ($+d\bar{d}$, $-dQ$, $+dS$, $+dn$).
(2) *High $\bar{d}$* overwhelms vegetation; decrease in channel storage ($-d\bar{d}$, $+dQ$, $-dS$, $-dn$).
(3) Increased resistance offered by vegetation; increase in channel storage ($+d\bar{d}$, $-dQ$, $+dS$, $+dn$).

From the July 1973 15-min readings, hourly plots of Q, S and n were made for selected 3-day intervals. Table 9.4, 1, 2 and 3 July, shows the diurnal variations that may occur under steady flow conditions. These are attributed to a combination of photosynthesis and evaporation, a twofold response to sunshine; the process of photosynthesis producing a diurnal turgidity in plants and a consequent variation in n; secondly, the loss of water from the channel due to evaporation from the water surface and transpiration from plants, particularly the reeds.

Table 9.4, 6, 7 and 8 July, shows the effects of small departures from a steady flow condition. An increase in flow from 0.225 to 1.114 cumecs on 6 July causes n to fall from 0.605 to 0.089. A similar effect is noticed with the 0900-hour values of 1973 (Figure 9.18). Here, the very small peaks of the 20 and 28 June and 6 July caused notable decreases in n, whereas a small reduction in flow such as occurred at 0900 hours on 7 July produced a very large increase in n. Here, flow decreased from a steady 0.3 cumec to an average of 0.06 cumec and then increased to 0.25 cumec; during this period of lower flow n rose from 0.215 to a peak of 0.785 and then, as the flow increased to 0.25 cumec, n fell to 0.25.

Table 9.4, 15–20 July, shows the effect of a summer flood peak of 7.4 cumecs; n fell from a value of 0.214 to a value of 0.048 and remained at this lower value thereafter. This instance appears to be a combination of effects: (1) the results of higher velocities when a quantity of plant material is uprooted and carried away, and (2) the annual going to seed of plants and the reduction in their vegetative parts. It is noted that the extreme rise of S and fall of n occur with the greatest rate of change of Q (actually between 0600 and 0700 hours). The peak Q, however, is not reached until about 7 h later.

9.7 METHODS OF MEASURING FLOW AND THE EFFECTS OF CHANNEL VEGETATION

In the case of the slope-area method, it is assumed that n remains constant for a specified period of time and, possibly, for a specified range of stage. In the case of the velocity-area method, it is assumed that $\bar{d}$ and S are directly related and that n remains constant for a specified period of time. With channels such as the Bain it is shown that n may vary considerably throughout a 24-hour period and over a range of flows. In the case of gauging weirs and flumes, crest and invert levels that assure modular operation under winter conditions may be insufficient to maintain modular operation when channel vegetation produces high values of n and $\bar{d}$. There is a risk that such structures will become drowned in their low and medium ranges during summer months.

9.8 CHANNEL DESIGN

The flood peak of 31.4 cumecs on 10 February, 1977 (Table 9.5) indicates that, at a flow of 38.8 cumecs, the designated channel design flow, and under winter conditions, n is likely to be within the range 0.025 to 0.030, whereas the summer peak of 7.4 cumecs, on 16 July, 1973, indicates a value of n of 0.05. However, as 7.4 cumecs is only 19 per cent of the design flow, and as the higher velocities of the design flow can be expected to uproot more plant material than occurs at 7.4 cumecs, n is likely to be within 0.030 to 0.035. But because energy is expended transporting the uprooted vegetation and bearing in mind a possible error of ±26 per cent, a value of 0.040 might be appropriate.

9.9 CONCLUSIONS

With respect to channels such as the Bain, where plant material extends over the greater part of the water area, it is concluded that:

(1) Annual roding is essential to the proper maintenance of the watercourse.
(2) The application of slope-area and velocity-area methods are inappropriate.
(3) The channel should be designed against a coefficient of roughness of 0.035, which at design flows is considered accurate to within ±26 per cent.
(4) Under summer conditions and at low flows the value of n may reach 2.0 and that such a value is considered accurate to within ±52 per cent.
(5) The value of n is related to the biomass and can indicate states of growth.

ACKNOWLEDGEMENTS

The author wishes to thank the Anglian Water Authority for permission to publish this paper and to express his indebtedness to his colleages in the Lincolnshire River Division.

REFERENCES

Barnes, Harry H., 1967. Roughness characteristics of natural channels, US Geological Water Supply Paper 1849, US Department of the Interior.
Chow, Ven Te, 1959. *Open-Channel Hydraulics*, McGraw-Hill, New York.
Haslam, S. M., 1978. *River Plants*, Cambridge University Press.
ISO: 1070, 1973. Liquid flow measurement in open channels—slope-area method, International Organization for Standardization, Geneva.
Wetzel, R. G., 1975. *Liminology*, W. B. Saunders, Philadelphia.

CHAPTER 10

Accuracy

R. W. HERSCHY

10.1 INTRODUCTION

The accuracy or, more correctly, the error of a measurement of discharge may be defined as the difference between the measured flow and the true value. The true value of the flow is unknown and can only be ascertained (within close limits) by weighing or by volumetric measurement. An estimate of the true value has therefore to be made by calculating the uncertainty in the measurement, the uncertainty being defined as the range in which the true value is expected to lie expressed at the 95 per cent confidence level (Figure 10.1). Although the error in a result is therefore, by definition, unknown the uncertainty may be estimated if the distribution of the measured values about the true mean is known.

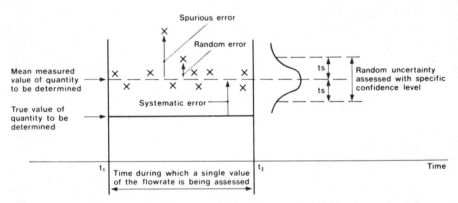

Figure 10.1 Basic statistical terms. (This figure is an extract from draft International Standard ISO/DIS: 5168. When published as an International Standard, copies may be purchased from the ISO Central Secretariat, Geneva, or any ISO member body (in the case of the United Kingdom, the British Standards Institution))

The uncertainty in the measurement of an independent variable is normally estimated by taking n observations and calculating the standard deviation, where n is preferably at least 20. Using this procedure to calculate the uncertainty in a gauging, however, would require n consecutive measurements of discharge at constant stage which is clearly impracticable. An estimate of the true value has therefore to be made by examining all the various sources of errors in the measurement.

In applying the theory of statistics to river gauging data it is assumed that the observations are independent random variables from a statistically uniform distribution. This ideal condition is seldom met in hydrometry. For example, the measurement of velocity by current meter cannot be independent since the velocity itself is fluctuating with time due mainly to pulsations in flow. The measurement of stage at any particular moment in time is not strictly independent since the flow passing the reference gauge is continuous.

River flow is by nature, therefore, non-random; each hourly, daily monthly and annual discharge being dependent on the previous hourly, daily, monthly and annual discharge. The cause of this is due mainly to the lag in the rainfall runoff relation, although in most catchments this may be an oversimplification. However, it is generally accepted by statisticians that the departure of river gauging data from the theoretical concept of errors is not serious provided care and attention is exercised in sampling techniques.

It should be stressed that the statistical analysis of river flow data is only applicable if the field data has been obtained by acceptable hydrometric principles and practices, i.e. to the relevant national or international standards (for the relevant international standards available see the Bibliography). In this connection, a large responsibility is imposed on the hydrometric observer who has to ensure that his current meters, water level recorders, reference gauges and other equipment are all in good order; that his measurements are as precise as he is able to make them and that they are meticulously recorded. If the raw data is questionable for any reason it is his duty to record this suspicion in the records. No amount of computer processing or statistical analysis will correct a wrong measurement. In addition he should avoid getting himself into the position of producing a very accurate measurement of the wrong water level, or other parameter, as may well happen under certain circumstances in the field. Statistical analysis is an aid to improving the presentation of the hydrometric data for the user's benefit but the final quality of the data depends on the observer.

10.2 STATISTICAL TERMS AND DEFINITIONS AS APPLIED TO RIVER GAUGING

10.2.1 Error

The error in a result is the difference between the measured and true value of the quantity measured (Figure 10.1).

10.2.2 Uncertainty

Uncertainty is the range within which the true value of a measured quantity can be expected to lie expressed at the 95 per cent confidence limits (see Figure 10.1).

10.2.3 Laplace–Gauss normal law (the error law)

The distribution conforming to this law, known as the normal distribution, is bell shaped and symmetrical about a maximum central ordinate. If a sample of data fits this curve then by statistical inference an estimate may be made of the dispersion characteristics of the population.

10.2.4　Standard deviation (*s*)

This is a measure of the scatter of the observations about the arithmetic mean of the sample. It is sometimes referred to as the root mean square (r.m.s.) of the deviations:

$$s = \left(\frac{\Sigma(x)^2}{N-1}\right)^{1/2}$$

where x is the deviation of the item X from the arithmetic mean $\bar{X}$, i.e. $x = X - \bar{X}$; and N the number of observations.

In a normal distribution the dispersion of the items about the average is measured in standard deviations, and 68 per cent of the observations lie within one standard deviation of the mean (average), 95 per cent lie within two standard deviations of the mean and almost all (actually 99.73 per cent) lie within three standard deviations of the mean (Table 10.1). The International Standards Organization (ISO) have recommended that the probability level used should be the 95 per cent level (two standard deviations). All uncertainties therefore in river gauging are generally estimated at two standard deviations.

Table　10.1　Relation between range and confidence level

Range	Confidence level
$\bar{Q} \pm 0.675s$	0.50
$\bar{Q} \pm 1s$	0.68
$\bar{Q} \pm 2s$	0.95
$\bar{Q} \pm 3s$	0.99

10.2.5　Standard error of estimate (*S_e*)

This is a measure of the variation or scatter of the observations about the line of average relation. It is numerically similar to standard deviation except that the curve of relation (regression) replaces the arithmetic mean:

$$S_e = \left(\frac{\Sigma(d)^2}{N-2}\right)^{1/2}$$

where d is the deviation of the actual value from the computed value taken from the curve of relation.

In the case of a stage–discharge curve of relation, for example, the 'd' value is the difference between a current meter gauging and the corresponding discharge taken from the stage–discharge relation at the same stage. Dispersion can be measured in the same way as for standard deviation and lines drawn at a distance of two standard errors on each side of the line of relation and parallel to it will include 95 per cent of the observations.

10.2.6 Standard error of the mean relation (S_{mr})

The standard error of the mean is an estimate of the probable accuracy of the computed mean relation. It is therefore the range within which the true mean would be expected to lie:

$$S_{mr} = \frac{S_e}{\sqrt{N}}$$

In the case of a stage–discharge curve, the S_{mr} indicates the probable accuracy of discharges taken from the relation for corresponding stages. The S_{mr} uncertainty band is therefore much narrower than the S_e band which refers to individual gaugings. Since the divisor in the right-hand side of the equation is $\sqrt{N}$ it follows that the more gaugings and check gaugings taken to define the curve of relation the more accurate the relation becomes, the proviso being that the check gaugings should fall within the S_e limits without bias.

In modern textbooks on statistics S_{mr} is sometimes referred to as the 95 per cent confidence interval, the value varying along the regression line being close to the regression at the centre and wider at the extremes. S_{mr} is computed by both methods in subsection 10.5.4.

In the case of correlation of gauging stations the standard error of the mean is not strictly applicable since the data is derived from two or more populations having different characteristics. Moreover the data is not synchronous and no regard is paid to changes in river regime, catchment characteristics, etc.

10.2.7 Variance

Variance is the square of the standard deviation. It is a property of variance that if the variances of the individual observations are known then the variance of the mean is equal to the sum of the individual variances:

$$s_T^2 = s_1^2 + s_2^2 + s_3^2 \ldots$$

where s_1, s_2, etc., are standard deviations of the mean. Also, the variance of the sum or difference of two independent random variables is equal to the sum of the variance:

$$s^2 = \frac{s_1^2}{N_1} + \frac{s_2^2}{N_2}$$

If the samples s_1 and s_2 are assumed to be derived from the same population then the standard error of the difference of the mean is

$$S_e = s\left(\frac{1}{N_1} + \frac{1}{N_2}\right)^{1/2}$$

$$= s\left(\frac{N_1 + N_2}{N_1 N_2}\right)^{1/2}$$

10.2.8   Coefficient of variation (CV)

This is the standard error of estimate expressed as a percentage of the mean:

$$CV = \frac{S_e}{\bar{X}} \times 100$$

or

$$CV = \frac{s}{\bar{X}} \times 100$$

(if standard deviation is used).

The advantage of using the coefficient of variation to estimate dispersion is that it is independent of units and can therefore be useful for comparison purposes. It is still common practice, however, to refer to 'percentage standard deviation' (or error).

10.2.9   Mean deviation

Mean deviation is a measure of scatter, making use of all the observations by taking the difference between each observation and the average, adding these differences without regard to their sign and dividing this total by the number of observations N.

10.2.10   Degrees of freedom

This is a concept in statistical analysis imposing constraints on the system. In computing standard deviation one degree of freedom is lost and N is replaced by $N-1$ (see subsection 10.2.4). For a straight-line relation two degrees of freedom are lost and in computing standard error N is replaced by $N-2$.

If two straight lines are required to locate a relation four degrees of freedom are lost, and if a transition curve is required to fit the two lines a further two degrees of freedom are lost and N is replaced by $(N-6)$. In a stage–discharge curve, however, where two or more lines may be used to determine the relation, each relation is treated separately and N is replaced in each by $(N-2)$.

10.2.11 Student's *t* distribution

The *t*-statistic is designed to compensate for using the estimated rather than the population standard deviation (or standard error). In practice it is only possible to obtain an estimate of the standard deviation since an infinite number of measurements would be required to determine it precisely and the confidence limits must therefore be based on this estimate. The *t* distribution for small samples is then used to relate the required confidence level to the range. In Table 10.2 it can be seen, for example, that at the 95 per cent level a value of $t = 2$ may be taken for a sample size of 15 or more in computing

Table 10.2   Values of Student's *t*

Degrees of freedom	Confidence level			
	0.500	0.900	0.950	0.990
1	1.000	6.314	12.706	63.657
2	0.816	2.920	4.303	9.925
3	0.765	2.353	3.182	5.841
4	0.741	2.132	2.776	4.604
5	0.727	2.015	2.517	4.032
6	0.718	1.943	2.447	3.707
7	0.711	1.895	2.365	3.499
10	0.700	1.812	2.228	3.169
15	0.691	1.753	2.131	2.947
20	0.687	1.725	2.086	2.845
30	0.683	1.697	2.042	2.750
60	0.679	1.671	2.000	2.660
∞	0.674	1.645	1.960	2.576

standard deviation or standard error. The relation between range of uncertainty and confidence level may be found as follows:

(1) Where N is a number of measurements, $N - 1$ is taken as the number of degrees of freedom, and where N is the number of observations on a line of relation the number of degrees of freedom taken should be as recommended in subsection 10.2.10.
(2) Obtain the value of t as a function of the confidence level and the degrees of freedom from Table 10.2.
(3) Calculate the standard deviation (s) or standard error (S_e) of the distribution of the measurements of the quantity Q say.
(4) The range of values within which a given observation would be expected to lie is $\bar{Q} + ts$ or $\bar{Q} \pm tS_e$.

(5) The range of values within which the true mean would be expected to lie
is

$$\bar{Q} \pm \frac{ts}{\sqrt{N}} \quad \text{or} \quad \bar{Q} \pm \frac{tS_e}{\sqrt{N}}$$

at the confidence level chosen. In river gauging, as stated in Subsection
10.2.2, the confidence level chosen is 0.95 and t is taken as 2.

The t-statistic can be conveniently applied to stage–discharge relations in
assessing check observations on the curve. The test indicates whether or not
the check gaugings might be considered to have derived from the same
homogeneous population as the gaugings making up the stage–discharge
relation. By definition, the Student's t distribution is the ratio of the deviation
of the mean $\bar{X}$ of a particular sample of N observations from an expected
value E say, to its standard deviation (or standard error) of the mean, that is:

$$t = \frac{\bar{X} - E}{s/\sqrt{N}}$$

In the case of two means the expression becomes

$$t = \frac{\bar{X}_1 - \bar{X}_2}{s/\sqrt{N}}$$

where $\bar{X}_1$ and $\bar{X}_2$ are the means of the two samples.

10.3 NATURE OF ERRORS

Errors of observation are usually grouped as random (or stochastic), systematic and spurious.

10.3.1 Random errors

Random errors are sometimes referred to as experimental errors and the
observations deviate from the mean in accordance with the laws of chance
such that the distribution usually approaches a normal distribution. They are
the most important errors to be considered in river gauging (Figure 10.1).

10.3.2 Systematic errors

Systematic errors are those which cannot be reduced by increasing the
number of observations if the instruments and equipment remain unchanged.
In river gauging, systematic errors may be present in the water level recorder,
in the reference gauge or datum and in the current meter. These errors may
be generally small but in some cases their effect may cause a systematic error

in the stage–discharge relation. It is also possible that the crest of a weir may be levelled-in incorrectly to the station datum, so producing a systematic error in head measurement which might have serious effect on low values of head (Figure 10.1).

10.3.3 Spurious errors

These are human errors or instrument malfunction and cannot be statistically analysed. The observation must therefore be discarded (Figure 10.1).

It is sometimes difficult to distinguish between random and systematic errors and some errors may be a combination of the two. For instance, where a group rating is used for current meters each of the meters forming the group may have a plus or minus systematic error which is randomized to obtain the uncertainty in the group rating. A construction error in the length of a weir crest may be randomized by measuring the crest say 10 times after construction to obtain the mean and standard deviation, each measuurement being independent of the other.

10.4 THEORY OF ERRORS

If a quantity Q is a function of several measured quantities $x, y, z \ldots$, the error in Q due to errors $\delta x, \delta y, \delta z \ldots$ in $x, y, z \ldots$, respectively, is given by

$$\delta Q = \frac{\partial Q}{\partial x}\,\delta x + \frac{\partial Q}{\partial y}\,\delta y + \frac{\partial Q}{\partial z}\,\delta z + \cdots \tag{10.1}$$

The first term in equation (10.1), $(\partial Q/\partial x)\delta x$, is the error in Q due to an error δx in x only (i.e. corresponding to $\delta y, \delta z \ldots$, all being zero). Similarly the second term $(\partial Q/\partial y)\,\delta y$ is the error in Q due to an error δy in y only. Squaring gives

$$\delta Q^2 = \left(\frac{\partial Q}{\partial x}\,\delta x\right)^2 + 2\,\frac{\partial Q}{\partial x}\frac{\partial Q}{\partial y}\,\delta x\,\delta y + \left(\frac{\partial Q}{\partial y}\,\delta y\right)^2 + \cdots \tag{10.2}$$

Now the terms $(\partial Q/\partial x)(\partial Q/\partial y)\,\delta x\,\delta y$, etc., are covariance terms and, since they contain quantities which are as equally likely to be positive or negative, their algebraic sum may be conveniently taken as being either zero or else negligible as compared with the squared terms. Very little work to confirm this assertion has been carried out on field observations but in some limited studies performed by the ISO Working Group on errors in flow measurement it was concluded that the covariance terms could be neglected.

Equation (10.2) then becomes

$$\delta Q^2 = \left(\frac{\partial Q}{\partial x}\,\delta x\right)^2 + \left(\frac{\partial Q}{\partial y}\,\delta y\right)^2 + \left(\frac{\partial Q}{\partial z}\,\delta z\right)^2 + \cdots \tag{10.3}$$

i.e. the error in Q, δQ is the sum of the squares of the errors due to an error in each variable. Now

$$\frac{\partial Q}{\partial x} = yz; \qquad \frac{\partial Q}{\partial y} = xz; \qquad \frac{\partial Q}{\partial z} = xy$$

and

$$\delta Q = [(yz\delta x)^2 + (xz\delta y)^2 + (xy\delta z)^2 + \cdots]^{1/2} \tag{10.4}$$

and dividing by $Q = xyz$

$$\frac{\delta Q}{Q} = \left[\left(\frac{yz}{xyz}\delta x\right)^2 + \left(\frac{xz}{xyz}\delta y\right)^2 + \left(\frac{xy}{xyz}\delta z\right)^2 + \cdots\right]^{1/2} \tag{10.5}$$

and

$$\frac{\delta Q}{Q} = \left[\left(\frac{\delta x}{x}\right)^2 + \left(\frac{\delta y}{y}\right)^2 + \left(\frac{\delta z}{z}\right)^2 + \cdots\right]^{1/2} \tag{10.6}$$

where $(\delta x)/x$, $(\delta y)/y$ and $(\delta z)/z$ are fractional values of the errors (standard deviations) in x, y and z, and if they are each multiplied by 100 they become percentage standard deviations. Let X_Q be the percentage standard deviation of Q and

$$X_x = \text{percentage standard deviation of } x$$

$$X_y = \text{percentage standard deviation of } y$$

$$X_z = \text{percentage standard deviation of } z$$

then

$$X_Q = \pm(X_x^2 + X_y^2 + X_z^2 \ldots)^{1/2} \tag{10.7}$$

which is generally referred to as the root-sum-squares equation for the estimation of uncertainties. It is this equation that is employed to estimate the uncertainty in a current meter gauging.

Similarly, if (i)

$$Q = \frac{x}{y}$$

then

$$X_Q = \pm(X_x^2 + X_y^2)^{1/2} \tag{10.8}$$

(ii)

$$Q = x \pm y$$

$$\delta Q^2 = \left(\frac{\partial Q}{\partial x}\delta x\right)^2 \pm \left(\frac{\partial Q}{\partial y}\delta y\right)^2$$

$$\delta Q = (\delta x^2 \pm \delta y^2)^{1/2}$$

$$X_Q = \frac{100\delta Q}{Q} = \frac{100}{Q}\left[\frac{x^2\,\delta x^2}{x^2} \pm \frac{y^2\,\delta y^2}{y^2}\right]^{1/2}$$

therefore

$$X_Q = \pm\frac{(x^2 X_x^2 \pm y^2 X_y^2)^{1/2}}{(x \pm y)} \tag{10.9}$$

(iii)

$$Q = Cbh^\beta \text{ (weir equation)}$$

then

$$X_Q = \pm(X_c^2 + X_b^2 + \beta^2 X_h^2)^{1/2} \tag{10.10}$$

which is the error equation for weirs, where C is the coefficient of discharge; b the length of crest; β the exponent of h, usually $3/2$ for a weir and $5/2$ for a V-notch; and h is the head over the weir.

Uncertainties in hydrometry are generally expressed as percentages. This practice has been recommended by the International Standards Organization (ISO 5168) and experience in the field has proved this approach to be convenient both in the statistical analysis of the data and in the use to which data is put.

10.5 THE ERROR EQUATION

10.5.1 Velocity-area method

The general form of the working equation for computing discharge in the cross-section is

$$Q - \sum_{i=1}^{m} (b_i d_i \bar{v}_i) \tag{10.11}$$

where Q is the total discharge in the cross-section, and b_i, d_i and $\bar{v}_i$ are the width, depth, and mean velocity of the water in the ith of the m verticals or segments into which the cross-section is divided.

Equation (10.11) assumes that a sufficient number of verticals have been taken in the cross-section but if this is not the case then equation (10.11) should be multiplied by a factor F so that

$$Q = F \sum_{i=1}^{m} (b_i d_i \bar{v}_i) \tag{10.12}$$

where F may be greater or less than unity. Equation (10.12), therefore, requires to be optimized until sufficient verticals are employed so as to make F unity.

Let the following contributing random uncertainties, in making a single determination of discharge by current meter, be expressed as percentage standard deviations at the 95 per cent confidence levels:

$$X_b = \text{uncertainty in width measurement}$$

$$X_d = \text{uncertainty in depth measurement}$$

$$X_{\bar{v}} = \text{uncertainty in the mean velocity in the vertical}$$

$$X_Q = \text{overall uncertainty in discharge}$$

$$X_m = \text{uncertainty due to the limited number of verticals}$$

and

$$\text{let } m = \text{number of verticals}$$

Now

$$X_Q = \frac{\text{sum of the percentage errors in the segment discharges}}{\text{sum of the segment discharges}}$$

(10.13)

Then

$$X_Q = \pm \left[\frac{\sum_{i=1}^{m} (b_i d_i \bar{v}_i)(Xb_i^2 + Xd_i^2 + X\bar{v}_i^2)^{1/2}}{\left(\sum_{i=1}^{m} b_i d_i \bar{v}_i \right)} \right]$$

(10.14)

or

$$X_Q = \pm \left[\frac{\sum_{i=1}^{m} [(b_i d_i \bar{v}_i)^2 (Xb_i^2 + Xd_i^2 + X\bar{v}_i^2)]}{\left(\sum_{i=1}^{m} b_i d_i \bar{v}_i \right)^2} \right]^{1/2}$$

(10.15)

The random uncertainty, X_m, in using a limited number of verticals, has to be allowed for in equation (10.15). In addition, the uncertainty in the mean velocity $(X_{\bar{v}})$ depends on the exposure time necessary to minimize the uncertainty due to pulsations in flow (X_e), the number of points taken in the vertical (X_p) and the uncertainty in the current meter rating (X_c). $X_{\bar{v}}^2$ should therefore be replaced by $(X_e^2 + X_p^2 + X_c^2)$ and the final equation now becomes

$$X_Q = \left[X_m^2 + \frac{\sum [(b_i d_i \bar{v}_i)^2 (X^2 b_i + X^2 d_i + X^2 e_i + X^2 p_i + X^2 c_i)]}{\left(\sum_{i=1}^{m} b_i d_i \bar{v}_i \right)^2} \right]^{1/2}$$

(10.16)

Now if the segment discharges $(b_i d_i \bar{v}_i)$ are nearly equal and the random uncertainties X_{b_i} are nearly equal and of value X'_b, and similarly for X_{d_i}, X_{e_i}, X_{p_i} and X_{c_i}, then

$$X'_Q = \pm \left[X'^2_m + \frac{1}{m}(X'^2_b + X'^2_d + X'^2_e + X'^2_p + X'^2_c) \right]^{1/2} \qquad (10.17)$$

which is the simplified error equation. Note that the X' refers to a random uncertainty as distinct from a systematic uncertainty X''—see equation (10.21).

Equation (10.17) is in a very much simplified form compared with equation (10.16) and its development from the latter can be easily demonstrated by a simple example. In Table 10.3 the segment discharges and the uncertainties are shown as being equal, respectively, in 14 verticals. Let

$$X = (X'^2_b + X'^2_d + X'^2_e + X'^2_p + X'^2_c) \qquad (10.18)$$

where X is the average value per vertical for the contributing random uncertainties which in this case, from Table 10.3, is 174.

From equation (10.16),

$$\frac{\Sigma(bd\bar{v})^2(X)}{(\Sigma bdv)^2} = \frac{126(X)}{42^2}$$

$$= \frac{1}{14}(X)$$

Table 10.3 Simulated gauging; equal segment discharges and equal uncertainties

1	2	3	4		5
Vertical	$bd\bar{v}$	$(bd\bar{v})^2$	$(X'^2_b + X'^2_d + X'^2_e + X'^2_p + X'^2_c)$ percentages		3×4
1	3	9	$0.5^2 + 1^2 + 12^2 + 5^2 + 2^2$	174	1 566
2	3	9	$0.5^2 + 1^2 + 12^2 + 5^2 + 2^2$	174	1 566
3	3	9	$0.5^2 + 1^2 + 12^2 + 5^2 + 2^2$	174	1 566
4	3	9	$0.5^2 + 1^2 + 12^2 + 5^2 + 2^2$	174	1 566
5	3	9	$0.5^2 + 1^2 + 12^2 + 5^2 + 2^2$	174	1 566
6	3	9	$0.5^2 + 1^2 + 12^2 + 5^2 + 2^2$	174	1 566
7	3	9	$0.5^2 + 1^2 + 12^2 + 5^2 + 2^2$	174	1 566
8	3	9	$0.5^2 + 1^2 + 12^2 + 5^2 + 2^2$	174	1 566
9	3	9	$0.5^2 + 1^2 + 12^2 + 5^2 + 2^2$	174	1 566
10	3	9	$0.5^2 + 1^2 + 12^2 + 5^2 + 2^2$	174	1 566
11	3	9	$0.5^2 + 1^2 + 12^2 + 5^2 + 2^2$	174	1 566
12	3	9	$0.5^2 + 1^2 + 12^2 + 5^2 + 2^2$	174	1 566
13	3	9	$0.5^2 + 1^2 + 12^2 + 5^2 + 2^2$	174	1 566
14	3	9	$0.5^2 + 1^2 + 12^2 + 5^2 + 2^2$	174	1 566
Σ	42	126			21 924

Therefore when the segment discharges and the uncertainties in each panel are equal, respectively, then

$$\frac{[\Sigma(b_i d_i \bar{v}_i)^2 (X_{b_i}^2 + X_{d_i}^2 + X_{e_i}^2 + X_{p_i}^2 + X_{c_i}^2)]^{1/2}}{(\Sigma b_i d_i \bar{v}_i)^2}$$

$$= \frac{1}{m}(X_b'^2 + X_d'^2 + X_e'^2 + X_p'^2 + X_c'^2)^{1/2} \tag{10.19}$$

Although the segment discharges or the contributory uncertainties in a gauging are seldom, if ever, equal, it has been found in practice that equation (10.17) gives results which are not significantly different from those given by equation (10.16) (Herschy, 1975a). Table 10.4 shows results of such comparison tests and it can be seen that the most significant factor in the results is the dependency on the number of verticals used in the gauging. Where the verticals number 20 or more the X_Q value using either equation is less than 7 per cent. The discrepancies therefore between the two equations become rather academic and it is clear that for routine gauging X_Q can be conveniently calculated from equation (10.17). For special studies, however, which require more precise estimation of the uncertainty X_Q, equation (10.16) should, of course, be used.

It will be evident that equations (10.16) and (10.17) can be employed to obtain any required value of X_Q by giving special consideration to the

Table 10.4   Results of comparison tests between the error equation (10.16) and the simplified error equation (10.17)

Gauging No.	Number of verticals	Maximum percentage difference in discharge in adjacent verticals	Average velocity (m s^{-1})	X_Q, random uncertainty in a single determination of discharge—percentage uncertainty at 95 per cent confidence level eqn (10.16)	eqn (10.17)
1	14	22	—	5.99	5.98
2a	20	23	0.116	6.80	6.97
2b	10	35	0.116	11.14	11.75
3a	20	21	0.301	5.33	5.61
3b	10	32	0.301	9.39	9.69
4a	21	9	0.344	6.26	6.10
4b	10	33	0.344	10.32	9.52
5a	29	16	0.160	4.54	4.66
5b	14	33	0.160	7.59	7.29
6a	58	12	0.568	3.20	2.92
6b	18	41	0.568	7.47	7.25

individual uncertainties in the equations. Individual uncertainties may usually be reduced by increasing the number of verticals, increasing the number of points in the verticals or, in the case of measuring structures, by reducing the uncertainty in the head measurement—equation (10.10).

Systematic uncertainties may be combined by the root-sum-square method as in the case of random uncertainties unless the uncertanty is known or has been measured precisely, in which case it becomes a systematic error and should be added algebraically to the result. The uncertainty due to this source is then taken as zero. An example of this is a zero error E_z'' in a gauge datum. If E_z'' is not known, the uncertainties are taken as being random and the following equation is used:

$$X_h' = \pm \frac{100}{h}(E_g'^2 + E_z'^2)^{1/2} \qquad (10.20)$$

where X_h' is the percentage random uncertainty in head or stage measurement (95 per cent level); E_g' the random uncertainty in recorder reading (mm); and E_z' is the random uncertainty in gauge zero (mm).

If, however, E_z'' is known then

$$X_h' = \frac{100}{h}(E_g')$$

and the zero error E_z'' is added algebraically to each reference gauge reading. If both the recorder and the gauge zero have known systematic errors then E_g'' and E_z'' should be added algebraically to each reference gauge reading.

In the velocity-area method the error equation for the systematic uncertainty X_Q'' is

$$X_Q'' = \pm(X_b''^2 + X_d''^2 + X_c''^2)^{1/2} \qquad (10.21)$$

where X_b'' is the percentage systematic uncertainty in the instrument measuring width; X_d'' the percentage systematic uncertainty in the instrument measuring depth; and X_c'' is the percentage systematic uncertainty in the current meter rating tank.

In weirs and flumes any systematic uncertainties are generally negligible as it is the practice to check carefully the dimensions after installation and amend any which do not conform to the design dimensions, thus randomizing any uncertainties in length of crest, width of throat, height of wing walls or divide walls and gauge zero. Generally great care is exercised in levelling-in the crest level to the gauge zero. If the crest is level then these two values should be equal. Any departure, if it is measurable, is a systematic error; if it is not, the departure is estimated as a systematic uncertainty. If, in addition, the sign is unknown then the uncertainty has to be randomized (E_z' in equation 10.20). This latter value is generally taken as ±3 mm. If a punched tape recorder is used to record head or stage, E_g' is also taken as ±3 mm (Herschy, 1975a).

The overall random and systematic uncertainties in discharge are then combined by the root-sum-square method:

$$X_Q = \pm(X_Q'^2 + X_Q''^2)^{1/2} \qquad (10.22)$$

where X_Q is the overall uncertainty in a single determination of discharge.

Generally, systematic uncertainties are negligible compared with random uncertainties but it is important that they are investigated and if possible rectified. Recommended values for both random and systematic contributing uncertainties are summarized in Section 10.6.

The final presentation of the result of a single determination of discharge is then made by one of the following methods (ISO 5168):

(1) Discharge $= Q \pm X_Q$; random uncertainty $= \pm X_Q'$.
(2) Discharge $= Q$; random uncertainty $= \pm X_Q'$; systematic uncertainty $= \pm X_Q''$.

10.5.2 Weirs and flumes

The error equation for the estimation of the uncertainty in a single determination of discharge for a weir, equation (10.10), was established in section 10.4 as follows: if

$$Q = Cbh^\beta$$

then

$$X_Q' = \pm(X_c'^2 + X_b'^2 + \beta^2 X_h'^2)^{1/2} \qquad (10.10)$$

where C = coefficient of discharge, b = length of crest, h = gauged head, β = exponent of h, usually $3/2$ for a weir and $5/2$ for a V-notch, X_Q' = percentage random uncertainty in a single determination of discharge, X_c' = percentage random uncertainty in the value of the coefficient of discharge, X_b' = percentage random uncertainty in the measurement of the length of crest, X_h' = percentage random uncertainty in the measurement of gauged head.

All values of uncertainties are percentage standard deviations at the 95 per cent level.

It is usually convenient to include any systematic uncertainty components in the terms X_c', X_b' and X_h' in equation (10.10). If the coefficient C, for example, has been established in the laboratory this is usually performed by a graphical relation of Q versus h. The uncertainty of the relation may therefore be taken as random, any systematic bias being due to systematic errors in the instrumentation. These should be insignificant in calibrations carried out in national laboratories (Francis and Miller, 1968). Also, it is usual to investigate, on site, any suspicion of a systematic uncertainty in the gauge zero or in the recorder and correct these. If, however, an allowance is made in equation (10.20) for any zero error, E_z, this should ensure that any systematic bias in the head measurement is included in X'_h.

An allowance of 0.1 per cent in X'_b should more than compensate for any systematic error in the steel tape or other means for measuring the length of crest.

If, however, there is any suspicion of significant systematic errors in X_c, X_b or X_h, equation (10.10) should be used to compute both the random and the systematic uncertainties in discharge and X'_Q and X''_Q combined by the root-sum-squares to obtain the overall uncertainty in discharge X_Q as before—equation (10.22).

Thin-plate weirs

The value of β in equation (10.10) is taken as 1.5 for rectangular thin-plate weirs and 2.5 for V-notches, and the uncertainty in the coefficient X_c is taken as ± 1.0 per cent (ISO 1438).

Triangular profile (Crump) weirs

The random uncertainty X_c is given by the following equation:

$$X'_c = \pm(10C_v - 9) \text{ per cent} \tag{10.23}$$

Values for C_v for the purpose of this equation are given in Table 10.5, where C is taken as 0.633, A is the cross-sectional area of flow, b is the length of crest, and h is the head. However, for normal field installations X_c can generally be taken as ± 2 per cent (Herschy *et al.*, 1976).

Table 10.5 Values of coefficient of approach velocity C_v for long-base weirs (A = area of cross-section of flow)

$\dfrac{Cbh}{A}$	C_v
0.1	1.003
0.2	1.010
0.3	1.020
0.4	1.039
0.5	1.057
0.6	1.098
0.7	1.146
0.8	1.217

Round-nosed horizontal crest weir

The random uncertainty in the value of the coefficient is

$$X'_c = \pm 2(21 - 20C) \text{ per cent} \tag{10.24}$$

and

$$C = \left(1 - \frac{0.01L}{b}\right)\left(1 - \frac{0.005L}{h}\right)^{3/2} \tag{10.25}$$

where L is the length of the horizontal section of the crest in the direction of flow. The appropriate value of C is then inserted in equation (10.24).

Rectangular profile weirs

The random uncertainty in the value of the coefficient is

$$X'_c = \pm(10F - 8) \text{ per cent} \tag{10.26}$$

where F is the coefficient correction factor. Values of F for given values of h/L and $h/(h+p)$ are obtained from Table 10.6.

Table 10.6 Values of coefficient correction factor F for values of h/L and $h/(h+p)$ for rectangular profile weirs (L = length of crest in direction of flow)

$\dfrac{h}{L}$	$\dfrac{h}{h+p}$			
	0.600	0.500	0.400	0.350
0.35	1.059	1.032	1.011	1.001
0.40	1.062	1.035	1.014	1.002
0.45	1.066	1.040	1.018	1.007
0.50	1.074	1.047	1.025	1.014
0.60	1.094	1.068	1.044	1.034
0.70	1.120	1.092	1.070	1.058
0.80	1.144	1.115	1.093	1.080
0.95	1.152	1.123	1.101	1.089

For a flume (rectangular, trapezoidal, U-shaped) the error equation may be expressed as follows:

$$X'_Q = (X_c'^2 + A^2 X_b'^2 + B^2 X_h'^2 + C^2 X_s'^2)^{1/2} \tag{10.27}$$

where X'_Q is the random uncertainty in a single determination of discharge; X'_c the random uncertainty in the value of the coefficient; X'_b the random uncertainty in the width of throat; X'_s the random uncertainty in the side slope (m) of the throat (m horizontal to 1 vertical), and A, B and C are numerical coefficients depending on the flume geometry, e.g. for a rectangular flume $A = 1$, $B = 1.5$ and $C = 0$ and for a U-shaped flume $C = 0$. Values of A, B and C can be obtained from Table 10.7 for trapezoidal and U-shaped flumes, where b and D are the width of throat for trapezoidal and U-shaped flumes, respectively.

Table 10.7 Values of numerical coefficients for trapezoidal and U-shaped flumes

	Trapezoidal flumes			U-shaped flumes		
$\dfrac{mh}{b}$	A	B	C	$\dfrac{h}{D}$	A	B
0.01	0.99	1.51	0.01			
0.03	0.97	1.53	0.03			
0.10	0.94	1.57	0.07	0.10	0.53	1.97
0.20	0.88	1.62	0.12	0.20	0.55	1.94
0.50	0.74	1.77	0.27	0.50	0.65	1.85
1.00	0.58	1.93	0.43	1.00	0.81	1.69
2.00	0.40	2.09	0.59	2.00	0.91	1.59
5.00	0.21	2.30	0.80	5.00	0.97	1.53
10.00	0.12	2.39	0.89	10.00	0.98	1.51
20.00	0.07	2.44	0.94	20.00		
50.00	0.03	2.48	0.98	50.00		
100.00	0.01	2.49	0.99	100.00		

The equation for the random uncertainty in the value of the coefficient for all three flumes is

$$X'_c = \pm[1+20(C_v - C)] \text{ per cent} \qquad (10.28)$$

The values of C_v and C for the purpose of estimating the uncertainty X'_c may be found from Tables 10.8 and 10.9, respectively. In Table 10.8, B is the width of the approach channel.

It is clear from equation (10.20) that the percentage uncertainty in the head measurement, X_h, increases as h decreases and at low values of h the value of

Table 10.8 Values of velocity coefficient C_v for estimating the uncertainty X_c in flumes

	$\dfrac{h}{h+P}$				
$\dfrac{b}{B}$	1.0	0.8	0.6	0.4	0.2
0.10	1.002	1.001	1.001	1.000	1.000
0.20	1.009	1.006	1.003	1.001	1.000
0.30	1.021	1.013	1.007	1.003	1.001
0.40	1.039	1.024	1.013	1.006	1.001
0.50	1.064	1.039	1.021	1.009	1.002
0.60	1.098	1.058	1.031	1.013	1.003
0.70	1.147	1.083	1.043	1.018	1.004
0.80		1.115	1.058	1.024	1.006
0.90			1.076	1.031	1.007
1.00			1.098	1.039	1.009

Table 10.9 Values of discharge coefficient C for estimating the uncertainty X_c in flumes (L = length of flume throat)

$\dfrac{L}{b}$	0.70	$\dfrac{h}{L}$ 0.50	0.30	0.10
0.2	0.992	0.990	0.984	0.954
1.0	0.988	0.985	0.979	0.950
2.0	0.982	0.979	0.973	0.944
3.0	0.976	0.973	0.968	0.938
4.0	0.970	0.968	0.962	0.932
5.0	0.965	0.962	0.956	0.927

the term βX_h in equation (10.10) is manifestly the most significant component uncertainty. This is demonstrated in Table 10.10 which gives values of βX_h for corresponding values of E_g, the values of E_z, X_c and X_b being taken as ±3 mm, 2 per cent and 0.1 per cent, respectively. Table 10.10 shows that while at high heads the uncertainty in discharge approaches the uncertainty in the coefficient X_c, at low heads the uncertainty in the head measurement is critical. For example, at $h = 0.900$ m an uncertainty in the head measurement of 10 mm produces an uncertainty in discharge of only 2.7 per cent, but the same uncertainty in the head at 0.150 m produces an uncertainty in discharge of over 10 per cent.

Table 10.10 Comparison of uncertainties in head measurement with uncertainties in weir discharge

E_g ± (mm)	E_z ± (mm)	$h = 0.150$ m βX_h (%)	X_Q (%)	$h = 0.450$ m βX_h (%)	X_Q (%)	$h = 0.900$ m βX_h (%)	X_Q (%)
3	3	4.2	4.7	1.4	2.4	0.7	2.1
4	3	5.0	5.4	1.7	2.6	0.8	2.2
5	3	5.8	6.2	1.9	2.8	1.0	2.2
6	3	6.7	7.0	2.2	3.0	1.1	2.3
7	3	7.6	7.9	2.5	3.2	1.3	2.4
8	3	8.5	8.8	2.9	3.5	1.4	2.5
9	3	9.5	9.7	3.2	3.7	1.6	2.6
10	3	10.4	10.6	3.5	4.0	1.7	2.7

During the 1976 drought in England many Crump weirs were operating under heads where the constant coefficient no longer applied. Heads as low as 10 mm or less were recorded. If the values of E_g and E_z were no better than

±3 mm the uncertainty in the head measurement from equation (10.20) would be

$$X_h = \pm \frac{100}{10} (3^2 + 3^2)^{1/2}$$

$$= \pm 42 \text{ per cent}$$

The modified coefficient for this value of head is approximately (Herschy *et al.*, 1977)

$$C = 0.633 \left(1 - \frac{0.0003}{h}\right)^{3/2} = 0.604 \tag{10.29}$$

If the coefficient is not adjusted to this figure the additional error in the coefficient is

$$\frac{0.633-0.604}{0.633} \times 100 = 4.6 \text{ per cent}$$

and the uncertainty in C now becomes

$$X_c = \pm (2^2 + 4.6^2)^{1/2}$$

$$= \pm 5 \text{ per cent}$$

The uncertainty in discharge from equation (10.10) is

$$X_Q = \pm (5^2 + 0 + 1.5^2 \times 42^2)^{1/2}$$

(note: X_b has been taken as being negligible), then

$$X_Q = \pm 63 \text{ per cent}$$

Clearly in a situation such as this the head requires to be measured with greater accuracy and every endeavour made to reduce E_g and E_z to ±1 mm by employing more sophisticated techniques. It is evident that the uncertainty in the coefficient is almost negligible compared with the uncertainty in head, and little advantage is gained by adjusting the coefficient. Although the uncertainty in discharge will still be high (of the order of ±20 per cent) by reducing E_g and E_z to ±1 mm, it is possible at least to attach reasonably reliable uncertainty limits to the discharge under such extreme conditions of flow.

The user should, wherever possible, estimate the contributing uncertainties for each particular gauging station but recommended values are given in Section 10.6. These values are the result of several investigations which are listed in the Bibliography.

To confirm the reliability of the error equations (10.10), (10.16) or (10.17) in the field would require either a volumetric method or the use of an alternative device to measure discharge. This is clearly impracticable in rivers.

However, in an investigation in Yorkshire (Herschy, 1975b; Herschy *et al.*, 1978), where up to 40 current meters were continuously operated over periods of 24 hours, it was possible to make a close estimate of the true discharge. The error equation, equation (10.17), could therefore be compared against known values of the actual errors in discharges. The results showed that about 95 per cent of the values as given by equuation (10.17) were well within the actual errors recorded.

10.5.3 Examples

Velocity-area method

The following details are given of a gauging taken on the River Thames at Sutton Courtenay; calculate the uncertainty in the discharge X_Q, using (i) equation (10.16) and (ii) equation (10.17):

discharge $Q = 24.012 \ \text{m}^3\,\text{s}^{-1}$

verticals = 20 No.

average velocity = $0.304 \ \text{m s}^{-1}$

method used = 5-point method

$X'_m = 5$ per cent (from Table 10.15)

$X'_b = 0.1$ per cent (from Table 10.23)

$X'_d = 1.0$ per cent (from Table 10.23)

$X'_e = 5$ per cent (from Table 10.16—time of exposure 3 min)

$X'_p = 5$ per cent (from Table 10.17)

$X'_c = 1$ per cent (from Table 10.18—individual rating)

$X''_b = 0.5$ per cent (from Table 10.24)

$X''_d = 0.5$ per cent (from Table 10.24)

$X''_c = 1.0$ per cent (from Table 10.24)

Taking (ii) first and using equation (10.17),

$$X'_Q = \pm\left[X'^2_m + \frac{1}{m}(X'^2_b + X'^2_d + X'^2_e + X'^2_p + X'^2_c) \right]^{1/2}$$

$$= \pm\left[5^2 + \frac{1}{20}(0.1^2 + 1^2 + 5^2 + 5^2 + 1^2) \right]^{1/2}$$

$$= \pm\left[25 + \frac{1}{20}(52) \right]^{1/2}$$

$$X'_Q = \pm 5.2 \text{ per cent}$$

Table 10.11 Calculation of the uncertainty in a discharge measurement, X'_Q, from the error equation (10.16)

1	2	3	4	5	6	7	8	9
Verti-cal	Dis-tance (m)	b (m)	d (m)	$\bar{v}$ (m s^{-1})	$bd\bar{v}$ (m^3 s^{-1})	$(bd\bar{v})^2$	$X'^2_b + X'^2_d + X'^2_e + $ $X'^2_p + X'^2_c$	7×8
RB	0	0.5	1.44	0.160	0.013	0.013	$0+1+144+11+4 = 160$	2.08
1	1	1.5	1.63	0.186	0.455	0.207	$0+1+144+11+4 = 160$	33.12
2	3	2	1.73	0.274	0.948	0.899	$0+1+ 49+11+1 = 62$	55.74
3	5	2	1.86	0.304	1.131	1.279	$0+1+ 36+11+1 = 49$	62.67
4	7	2	2.17	0.363	1.575	2.481	$0+1+ 25+ 4+1 = 31$	76.91
5	9	2	2.23	0.364	1.623	2.634	$0+1+ 25+ 4+1 = 31$	81.65
6	11	2	2.13	0.383	1.632	2.663	$0+1+ 25+ 4+1 = 31$	82.55
7	13	2	2.23	0.377	1.681	2.826	$0+1+ 25+ 4+1 = 31$	87.60
8	15	2	2.38	0.353	1.680	2.822	$0+1+ 25+ 4+1 = 31$	87.48
9	17	2	2.41	0.347	1.673	2.799	$0+1+ 25+ 4+1 = 31$	86.77
10	19	2	2.58	0.320	1.651	2.726	$0+1+ 25+ 4+1 = 31$	84.50
11	21	1.5	2.41	0.327	1.182	1.397	$0+1+ 25+ 4+1 = 31$	43.31
12	22	1	2.40	0.339	0.814	0.663	$0+1+ 25+ 4+1 = 31$	20.55
13	23	1.5	2.40	0.327	1.177	1.385	$0+1+ 25+ 4+1 = 31$	42.94
14	25	2	2.37	0.324	1.536	2.359	$0+1+ 25+ 4+1 = 31$	73.13
15	27	2	2.35	0.317	1.490	2.220	$0+1+ 25+ 4+1 = 31$	68.82
16	29	2	2.26	0.274	1.238	1.533	$0+1+ 49+ 4+1 = 55$	84.32
17	31	2	1.90	0.265	1.007	1.014	$0+1+ 64+ 4+1 = 70$	70.98
18	33	1.75	1.66	0.256	0.744	0.555	$0+1+ 64+ 4+1 = 70$	38.85
19	34.5	1.5	1.24	0.273	0.508	0.258	$0+1+ 49+11+1 = 62$	16.00
20	36	1.5	1.24	0.105	0.193	0.037	$0+1+400+11+25$	16.17
							$= 437$	
LB	37.4	0.7	1.20	0.089	0.075	0.006	$0+1+ 400+11+25$	2.62
							$= 437$	

$$\Sigma\ 24.128 \qquad X'_Q = \left(25 + \frac{1219}{24.128^2}\right)^{1/2} \qquad \Sigma\ 1219$$

$$= \mp 5.2 \text{ per cent}$$

(i) Using equation (10.16),

$$X'_Q = \left[X'^2_m + \frac{\Sigma(b_i d_i \bar{v}_i)^2 (X'^2_{bi} + X'^2_{di} + X'^2_{ei} + X'^2_{pi} + X'^2_{ci})}{(\Sigma b_i d_i \bar{v}_i)^2} \right]^{1/2}$$

The right-hand terms of the equation are tabulated in Table 10.11 and it can be seen that $X'_Q = \pm 5.2$ per cent as before. The uncertainties X'_d, X'_e and X'_p in (i) were evaluated on site. The remainder of the uncertainties were taken from the tables in section 10.6 and X'_b was taken as being negligible.

Now, from equation (10.21) the systematic uncertainty X''_Q, in the discharge is

$$X''_Q = \pm (X''^2_b + X''^2_d + X''^2_c)^{1/2}$$

$$= \pm (0.5^2 + 0.5^2 + 1^2)^{1/2}$$

$$X''_Q = \pm 1.0 \text{ per cent}$$

Then

$$X_Q = (5.2^2 + 1^2)^{1/2}$$

$$= \pm 5.3 \text{ per cent}$$

The result may be presented as follows (Subsection 10.5.1):

(1) Discharge = 24.012 m^3 s^{-1} ± 5.3 per cent
    random uncertainty = ±5.2 per cent
or (2) Discharge = 24.012 m^3 s^{-1}
    random uncertainty = ±5.2 per cent
    systematic uncertainty = ±1.0 per cent

Weirs and flumes

(1) Details of a Crump weir discharge are given as follows; calculate the uncertainty in the discharge, X_Q:

$$\text{discharge} = 48.610 \text{ m}^3 \text{ s}^{-1}$$

$$\text{gauged head, } h = 1.219 \text{ m}$$

$$\text{weir height, } P = 0.61 \text{ m}$$

$$\text{crest length, } b = 15.24 \text{ m}$$

$$X_b' = 0.1 \text{ per cent}$$

$$E_g' = \pm 3 \text{ mm}$$

$$E_z' = \pm 3 \text{ mm}$$

$$\beta = 1.5$$

From equation (10.20),

$$X_h' = \pm \frac{100}{h} (E_g^2 + E_z^2)^{1/2}$$

$$X_h' = \pm \frac{100}{1219} (3^2 + 3^2)^{1/2}$$

$$= \pm 0.35 \text{ per cent}$$

From equation (10.23),

$$X_c' = \pm (10C_v - 9) \text{ per cent}$$

To obtain C_v from Table 10.3,

$$\frac{Cbh}{A} = \frac{0.633 \times 15.24 \times 1.219}{(1.219 + 0.61)15.24}$$

$$= 0.42$$

From Table 10.5,

$$C_v = 1.043$$

and

$$X'_c = \pm(10 \times 1.043 - 9) \qquad \text{(from equation 10.23)}$$

$$= \pm 1.43 \text{ per cent}$$

It is realistic to take X'_c as ± 2 per cent which allows for any under-estimation by equation (10.23) and also allows for any systematic uncertainty in the coefficient (Subsection 10.5.2).

Now equation (10.10) for the uncertainty in discharge, X'_Q, is

$$X'_Q = \pm(X'^2_c + X'^2_b + \beta^2 X'^2_h)^{1/2} \text{ per cent}$$

$$= \pm(2^2 + 0.1^2 + 1.5^2 \times 0.35^2)^{1/2} \text{ per cent}$$

$$= \pm 2.2 \text{ per cent}$$

An allowance of ± 3 mm has been made for any systematic uncertainty in the gauge zero, E_z, and an allowance of 0.1 per cent has been made to include both random and systematic uncertainties in measuring the length of crest. The result may therefore be presented as follows:

$$\text{discharge} = 48.610 \text{ m}^3 \text{ s}^{-1} \pm 2.2 \text{ per cent}$$

(including random and systematic uncertainties).

(2) Details of a rectangular flume discharge are given as follows; calculate the uncertanty in the discharge, X_Q:

$$\text{discharge} = 1.712 \text{ m}^3 \text{ s}^{-1}$$

$$\text{gauged head, } h = 1.0 \text{ m}$$

$$\text{throat width, } b = 1.0 \text{ m}$$

$$\text{approach channel width, } B = 2.0 \text{ m}$$

$$\text{hump height, } p = 0.25 \text{ m}$$

$$\text{throat length, } L = 3.0 \text{ m}$$

$$X'_b = 0.2 \text{ per cent (2 mm in 1.0 m)}$$

$$E'_g = \pm 3 \text{ mm}$$

$$E'_z = \pm 3 \text{ mm}$$

$$\beta = 1.5$$

From equation (10.20),

$$X'_h = \pm \frac{100}{1000} (3^2 + 3^2)^{1/2} \text{ per cent}$$

$$= \pm 0.4 \text{ per cent}$$

Now

$$\frac{b}{B} = 0.50$$

$$\frac{h}{h+p} = 0.80$$

and C_v from Table 10.8 = 1.039.

$$h/L = 0.33$$

$$L/b = 3.0$$

and C from Table 10.9 = 0.968. Now, from equation (10.28),

$$X'_c = \pm[1 + 20(C_v - C)] \text{ per cent}$$

$$= \pm[1 + 20(1.039 - 0.968)] \text{ per cent}$$

$$= \pm 2.4 \text{ per cent}$$

say ± 3 per cent.

Again this makes allowance for an under-estimation in equation (10.28) and any systematic uncertainty in C.

Now equation (10.27) gives

$$X'_Q = (X'^2_c + A^2 X'^2_b + B^2 X'^2_h + C^2 X'^2_s)^{1/2}$$

where, in the case of a rectangular flume, $A = 1$, $B = 1.5$, and $C = 0$. Then, $X'_Q = \pm(3^2 + 0.2^2 + 1.5^2 \times 0.4^2)^{1/2}$ per cent $= \pm 3$ per cent. (Note: The remarks regarding systematic uncertainties in the Crump weir example also apply in this case.) The result may then be presented as follows:

$$\text{discharge} = 1.712 \text{ m}^3 \text{ s}^{-1} \pm 3 \text{ per cent}$$

(including random and systematic uncertainties).

(3) Details of a triangular thin-plate weir (V-notch) discharge are given as follows; calculate the uncertainty in the discharge X_Q:

$$\text{discharge} = 0.007 . \text{m}^3 \text{ s}^{-1}$$

$$\text{gauged head, } h = 0.121 \text{ m}$$

$$\alpha \text{ (notch angle)} = 90 \text{ deg}$$

$$\text{height of notch, } P = 0.30 \text{ m}$$

$$E'_g = \pm 3 \text{ mm}$$

$$E'_z = \pm 3 \text{ mm}$$

$$X'_c = \pm 1 \text{ per cent (see Subsection 10.5.2)}$$

depth of notch = 0.440 m

top width of notch = 0.440 m

uncertainty in depth of notch = ±1.0 mm

uncertainty in top width of notch = ±0.5 mm

$$\beta = 2.5$$

From equation (10.20),

$$X'_h = \pm \frac{^-100}{121} (3^2 + 3^2)^{1/2} \text{ per cent}$$

then

$$X'_h = \pm 3.5 \text{ per cent}$$

now

$$\tan \frac{\alpha}{2} = \frac{\frac{1}{2} \text{ top width of notch}}{\text{height of notch}} = \frac{b_t}{h_t} \text{ say}$$

and X'_b in equation (10.10) may be found from

$$X' \tan \alpha/2 = \pm (X'^2_{b_t} + X'^2_{h_t})^{1/2} \text{ per cent}$$

$$= \pm 100 \left[\left(\frac{1.0}{220} \right)^2 + \left(\frac{0.5}{440} \right)^2 \right]^{1/2} \text{ per cent}$$

then

$$X' \tan \frac{\alpha}{2} = \pm 0.5 \text{ per cent}$$

and

$$X'_Q = \pm (1.0^2 + 0.5^2 + 2.5^2 \times 3.5^2)^{1/2}$$

Uncertainty in discharge $X'_Q = \pm 9$ per cent.

Note 1: The remarks regarding systematic uncertainties in the previous examples also apply in this case.

Note 2: It will be observed that the large uncertainty in X'_Q of 9 per cent in this example is due to the significant uncertainty in the measurement of the low head, where values of E'_g and E'_z have been taken as ±3 mm as before. The result may be presented as follows:

discharge $= 0.007 \text{ m}^3 \text{ s}^{-1} \pm 9$ per cent (including random and systematic uncertainties)

By measuring the head with a point gauge and by assessing and correcting the zero error, significant improvement could be obtained. For example, if E'_g and E'_z could be measured to ±1 mm then X'_h would be reduced to 1.2 per cent and X'_Q to ±3.2 per cent.

Note 3: In the above examples on measuring structures it will be observed that as h increases the uncertainty in discharge will tend towards the uncertainty in the coefficient X_c.

10.5.4 Uncertainty in the stage–discharge relation

The equation for a single determination of discharge by current meter, equation (10.17), does not include a term for the uncertainty in the measurement of stage, as does equation (10.10) for measuring structures.

A current meter gauging may be required for several purposes; for special studies for example, or for *ad hoc* measurement. In these situations the actual water level may not be required. However, if a continuous record of discharge is required, as is usually the case, a stage–discharge relation is necessary and the uncertainty in the stage measurement becomes as important as for measuring structures. The equation for the stage–discharge relation may be expressed in the general form

$$Q = Kh^\beta \tag{10.30}$$

where Q is the discharge; K is a coefficient for the relation; h is the stage; and β is an exponent, which is in the same form as the weir equation $Q = Ch^\beta$ (where K corresponds to C) and is obtained in a similar manner, i.e. by equating Q against h. In this equation the discharge Q is given as the discharge per unit length of crest ($b = 1$).

Letting X_{rd} be the uncertainty in the recorded discharge, the error equation is then

$$X_{rd} = (X'^2_k + \beta^2 X'^2_h)^{1/2} \tag{10.31}$$

In practice, X_k is the standard error of the mean relation, S_{mr}, at the 95 per cent confidence level (Subsection 10.2.6) and h is replaced by $(h + a)$, where a is the gauge reading (stage) at zero flow. Then equation (10.31) becomes

$$X_{rd} = [S^2_{mr} + \beta^2 X'^2_{(h+a)}]^{1/2} \tag{10.32}$$

The uncertainty in each recorded discharge may then be ascertained by employing equation (10.32).

The uncertainty S_{mr} in the stage–discharge relation may be found by first calculating the standard error of estimate, S_e, from one of the following equations:

$$S_e(\log_e Q) = t\left[\left[\frac{N-1}{N-2}\right][S^2_{\log_e Q} - \beta^2 S^2_{\log_e h}]\right]^{1/2} \tag{10.33}$$

$$S_e(\log_e Q) = t\left[\frac{\Sigma(\log_e Q_i - \log_e Q_c)^2}{N-2}\right]^{1/2} \tag{10.34}$$

$$S_e(Q) = t\left[\frac{\Sigma\left[\frac{Q_i - Q_c}{Q_c} \times 100\right]^2}{N-2}\right]^{1/2} \tag{10.35}$$

where

N = number of current meter (discharge) observations

$S_{\log_e Q}$ = standard deviation of the natural logarithms of the discharge observations

$$S^2_{\log_e Q} = \frac{\Sigma[\log_e Q_i - \overline{\log_e Q_i}]^2}{N-1}$$

Q_i = current meter observation

$\overline{\log_e Q_i}$ = average value of $\log_e Q_i$

$S_{\log_e h}$ = standard deviation of the stage values $(h + a)$

$$S^2_{\log_e h} = \frac{\Sigma[\log_e (h+a) - \overline{\log_e (h+a)}]^2}{N-1}$$

$(h + a)$ = stage values corresponding to values of Q_i

$\overline{\log_e (h + a)}$ = average value of $\log_e (h + a)$

β = exponent of the stage–discharge equation

Q_c = discharge taken from rating curve corresponding to Q_i and $(h + a)$ (note $Q_c = C(h + a)^\beta$)

t = Student's t correction (see Subsection 10.2.11) at the 95 per cent confidence level for N gaugings. May be taken as 2 for 20 or more gaugings (Table 10.2)

It will be noted that natural logarithms are used in equations (10.33) and (10.34). The calculations may, however, be carried out using logarithms to the base 10 and multiplying the answer by 2.3 and then by 100 to give the standard error as a percentage.

The standard error so calculated is an estimate of the uncertainty of the spread or dispersion of the gaugings about the stage–discharge relation. In fact, 19 out of 20 of the gaugings should be within the S_e band. Further, the statistical assumption is that 19 out of every 20 future gaugings should also be within this band if there are no hydraulic or physical changes in the river section or the control.

Parallel straight lines are drawn on either side of the stage–discharge relation and distance S_e from it, as shown in the example in Chapter 1 (Figure 1.13).

The decision as to which of the three equations to use depends on the circumstances and the availability of calculators. Equation (10.33) is the most statistically acceptable equation but for the best results requires a computer program, since six places of decimals are desirable. Equation (10.35) is the least statistically acceptable but has worked well in practice and it is convenient for rapid calculations by pocket calculator. All three equations give the same result, as can be seen by the example presented in Subsection 10.5.6.

The standard error of the mean, expressed as a percentage, may then be calculated from the following equation:

$$S_{mr} = tS_e(\log_e Q) \left[\frac{1}{N} + \frac{[\log_e (h+a) - \overline{\log_e (h+a)}]^2}{\Sigma[\log_e (h+a) - \overline{\log_e (h+a)}]^2} \right]^{1/2} \times 100 \qquad (10.36)$$

S_{mr} is calculated for each observation of $(h+a)$ and the limits will therefore be curved on each side of the stage–discharge relation and shall be minimum at the value of $\log (h+a) = \overline{\log (h+a)}$ when equation (10.36) reduces to

$$S_{mr} = \frac{tS_e}{\sqrt{N}} \qquad (10.37)$$

If the stage–discharge relation comprises of one or more break points, S_e and S_{mr} are calculated for each range and $(N-2)$ degrees of freedom allowed for each range. It should be noted that in textbooks on statistics S_{mr} is often referred to as the 95 per cent confidence limits (in this case of Q).

For the best results, at least 20 observations should be available in each range before a statistically acceptable estimate can be made of S_e and S_{mr}.

10.5.5 Uncertainty in the daily mean discharge

The value of discharge most commonly required for design and planning purposes is the daily mean discharge, which may be calculated by taking the average of the number of observations of discharge during the 24-hour period.

The uncertainty in the daily mean discharge for a velocity area station may be calculated from the following equation:

$$X_{dm} = \pm \frac{\Sigma[(S_{mr}^2 + \beta^2 X_{(h+a)}^2)^{1/2} Q_h]}{\Sigma Q_h} \tag{10.38}$$

where X_{dm} is the uncertainty in the daily mean discharge (95 per cent confidence level) and Q_h is the discharge corresponding to $(h+a)$.

The corresponding equation for a measuring structure is similar and may be expressed as follows:

$$X_{dm} = \pm \frac{\Sigma[(X_c^2 + \beta^2 X_{(h+a)}^2)^{1/2} Q_h]}{\Sigma Q_h} \tag{10.39}$$

where X_c is the uncertainty in the coefficient of discharge. Note that X_b, uncertainty in the length of crest (width of throat), has been neglected and note also that in this case the value of a is zero.

The uncertainty in the monthly mean and annual discharge may be estimated from the following equations:

$$X_{mm} = \frac{\Sigma(X_{dm} Q_{dm})}{\Sigma Q_{dm}} \tag{10.40}$$

and

$$X_{aa} = \frac{\Sigma(X_{mm} Q_{mm})}{\Sigma Q_{mm}} \tag{10.41}$$

where X_{mm} is the uncertainty in the monthly mean discharge (95 per cent level); X_{aa} the uncertainty in the annual discharge; Q_{dm} the daily mean discharges; and Q_{mm} is the monthly mean discharges.

The percentage uncertainty in stage (or head) in the above equations may be found from equation (10.20).

10.5.6 Example calculation for S_e and S_{mr}

Using the stage–discharge curve in Chapter 1 (Figure 1.13) the required tabulation for S_e and S_{mr} is given in Table 10.12, where the logarithms are to the base 10.

S_e from equation (10.33) gives

$$S_e = t \left[\frac{31}{30} \left[\frac{12.336\,4}{31} - \frac{1.530\,1^2 \times 5.266\,5}{31} \right] \right]^{1/2}$$

$$= t\,(0.000\,2)^{1/2}$$

$$= 2 \times 0.014\,14 \times 2.3 \times 100$$

$$= \pm 6.2 \text{ per cent}$$

Table 10.12 Tabulated values required to calculate S_e and S_{mr}

Obs. No.	$h+a$ ($a = -0.115$)	$\log(h+a)$	$[\log(h+a) -\log(h+a)]^2$	Q_i (gauging) (m³ s⁻¹)	$\log Q_i$	$[\log Q_i -\log Q_i]^2$	Q_c (from rating equation) (m³ s⁻¹)	$\log Q_c$	$[\log Q_i -\log Q_c]^2$	$\left[\dfrac{Q_i-Q_c}{Q_c}\times100\right]^2$ (per cent)	$2S_{mr}$ ± per cent
1	0.157	−0.804 1	0.351 2	2.403	0.391 46	0.776 789	2.323	0.366 0	0.000 65	36.00	2.0
2	0.158	−0.801 3	0.347 9	2.325	0.366 42	0.821 561	2.345	0.370 1	0.000 01	0.81	2.0
3	0.188	−0.725 8	0.264 6	2.923	0.465 83	0.651 233	3.060	0.485 7	0.000 40	20.25	1.8
4	0.192	−0.716 7	0.255 3	3.242	0.510 81	0.580 659	3.160	0.499 7	0.000 12	6.76	1.8
5	0.219	−0.659 6	0.200 8	3.841	0.584 44	0.473 592	3.865	0.587 1	0.000 01	0.36	1.7
6	0.259	−0.586 7	0.140 8	4.995	0.698 54	0.329 798	4.996	0.698 6	0.000 00	0.00	1.5
7	0.278	−0.556 0	0.118 7	5.410	0.733 20	0.291 189	5.568	0.746 7	0.000 16	7.84	1.5
8	0.279	−0.554 4	0.117 6	5.422	0.734 16	0.290 155	5.598	0.748 0	0.000 19	10.24	1.5
9	0.287	−0.542 1	0.109 3	5.883	0.769 60	0.253 230	5.846	0.766 9	0.000 01	0.36	1.4
10	0.295	−0.530 2	0.101 6	6.154	0.789 16	0.233 297	6.097	0.785 1	0.000 02	0.81	1.4
11	0.348	−0.458 4	0.061 0	7.376	0.867 82	0.164 025	7.851	0.894 9	0.000 74	36.00	1.3
12	0.405	−0.392 5	0.032 8	9.832	0.992 64	0.078 501	9.902	0.995 7	0.000 01	0.49	1.2
13	0.433	−0.363 5	0.023 1	11.321	1.053 88	0.047 935	10.968	1.040 1	0.000 19	10.24	1.2
14	0.461	−0.336 3	0.015 6	12.372	1.092 44	0.032 537	12.072	1.081 8	0.000 11	6.25	1.2
15	0.465	−0.332 5	0.014 7	11.825	1.072 80	0.040 008	12.233	1.087 5	0.000 22	10.89	1.2
16	0.501	−0.300 2	0.007 9	13.826	1.140 70	0.017 456	13.711	1.137 1	0.000 01	0.64	1.2
17	0.511	−0.291 6	0.006 4	14.102	1.149 28	0.015 262	14.132	1.150 2	0.000 00	0.04	1.1
18	0.606	−0.217 5	0.000 0	19.020	1.279 21	0.000 040	18.345	1.263 5	0.000 24	13.69	1.1

Table 10.12—contd.

Obs. No.	$h + a$ $(a = -0.115)$	$\log(h + a)$	$[\log(h+a) - \log(\overline{h+a})]^2$	Q_i (gauging) $(m^3\ s^{-1})$	$\log Q_i$	$[\log Q_i - \log \overline{Q_i}]^2$	Q_c (from rating equation) $(m^3\ s^{-1})$	$\log Q_c$	$[\log Q_i - \log Q_c]^2$	$\left[\dfrac{Q_i - Q_c}{Q_c}\times100\right]^2$ (per cent)	$2S_{mr}$ ± per cent
19	0.624	-0.204 8	0.000 0	19.790	1.296 44	0.000 558	19.185	1.283 0	0.000 18	10.24	1.1
20	0.632	-0.199 3	0.000 1	20.280	1.307 10	0.001 175	19.563	1.291 4	0.000 24	13.69	1.1
21	0.681	-0.166 9	0.002 0	21.204	1.326 41	0.002 872	21.931	1.341 1	0.000 21	10.89	1.1
22	0.731	-0.136 1	0.005 7	23.996	1.380 13	0.011 515	24.442	1.388 1	0.000 06	3.24	1.1
23	0.926	-0.033 4	0.031 7	36.242	1.559 21	0.082 019	35.098	1.545 3	0.000 19	10.89	1.2
24	1.225	0.088 1	0.089 7	54.591	1.737 12	0.215 574	53.855	1.731 2	0.000 03	1.96	1.4
25	1.411	0.149 5	0.130 3	67.327	1.828 19	0.308 436	66.859	1.825 2	0.000 01	0.49	1.5
26	1.646	0.216 4	0.183 1	79.050	1.897 90	0.390 725	84.631	1.927 5	0.000 88	43.56	1.6
27	1.895	0.277 6	0.239 2	110.783	2.044 47	0.595 444	104.989	2.021 1	0.000 54	30.25	1.8
28	2.517	0.400 9	0.375 1	162.814	2.211 69	0.881 477	162.095	2.209 8	0.000 00	0.16	2.0
29	3.150	0.498 3	0.503 7	227.600	2.357 17	1.175 814	228.478	2.358 8	0.000 00	0.16	2.2
30	3.165	0.500 4	0.506 7	228.800	2.359 46	1.180 786	230.145	2.362 0	0.000 01	0.36	2.3
31	3.191	0.503 9	0.511 7	228.500	2.358 89	1.179 548	233.044	2.367 4	0.000 07	3.61	2.3
32	3.225	0.508 5	0.518 3	236.600	2.374 01	1.212 619	236.854	2.374 5	0.000 00	0.01	2.3
	$\overline{\log(h+a)}$ -0.211 45		Σ5.266 5	$\underset{\log Q_i}{\Sigma}$	40.730 38 1.272 82	Σ12.336 459			Σ0.005 48	Σ 291.18	

S_e from equation (10.34) gives

$$S_e = t\left(\frac{0.005\ 48}{30}\right)^{1/2}$$

$$= 2 \times 0.135 \times 2.3 \times 100$$

$$= \pm 6.2 \text{ per cent}$$

S_e from equation (10.35) gives

$$S_e = t\left(\frac{291.18}{30}\right)^{1/2}$$

$$= 2 \times 3.115$$

$$= \pm 6.2 \text{ per cent}$$

The calculation for S_{mr} for each of the 32 observations proceeds from equation (10.36) as follows:

for observation No. 1,

$$S_{mr} = t \times S_e\left(\frac{1}{32} + \frac{0.351\ 2}{5.266\ 5}\right)^{1/2}$$

$$= 6.2 \times 0.313$$

$$= \pm 1.94 \text{ per cent}$$

for observation No. 18,

$$S_{mr} = t \times S_e\left(\frac{1}{32} + 0\right)^{1/2}$$

$$= 6.2 \times 0.177$$

$$= \pm 1.1 \text{ per cent}$$

for observation No. 32,

$$S_{mr} = t \times S_e\left(\frac{1}{32} + \frac{0.518\ 3}{5.266\ 5}\right)^{1/2}$$

$$= 6.2 \times 0.360$$

$$= \pm 2.23 \text{ per cent}$$

Note 1: The S_{mr} values in Table 10.12 have been rounded.

Note 2: The equation $S_{mr} = \dfrac{S_e}{\sqrt{N}}$ gives $S_{mr} = \dfrac{6.2}{32} = 1.1$ per cent, which is the value at $\log(h + a) = -0.211\ 45$

10.5.7 Example calculation for the uncertainty in the daily mean discharge X_{dm}

The calculation proceeds as follows:

(1) Calculate $X_{(h+a)}$ for each of the N values of discharge, used to calculate the daily mean, from equation (10.20).
(2) Calculate X_{dm} from equation (10.38) or (10.39) using the appropriate value for S_{mr}

A typical calculation for hourly values of discharge is given in Table 10.13.

10.5.8 Systematic error in the stage–discharge curve

A systematic error in the stage–discharge curve has not been included in the above example because (1) the curve is usually established using at least three different current meters, the inference being that any systematic error in the meters is randomized, and (2) an allowance of ± 3 mm has been made in equation (10.32) for a possible systematic error in the stage measurement. If these precautions are not taken, however, then the possible presence of a systematic error in the stage–discharge curve requires to be investigated. In effect the curve may shift along both the stage axis and the discharge axis, the resultant shift being maximum when stage and discharge shifts have opposite signs. The direction of shift, however, will be unknown and plus and minus values require to be assigned to both and the maximum allowed for of ± 3 mm for stage and ± 1 per cent for discharge (Tables 10.22 and 10.24). The resultant of these two factors produce a bandwidth within which the curve will be expected to lie due to the systematic uncertainties alone. The analysis for S_e and S_{mr} is unaffected and is carried out independently. The result, however, should then be presented as (see Figure 10.1):

$$\text{discharge} = Q \pm S_{mr} \pm E_s, \quad \text{where } E_s \text{ is the estimated systematic uncertainty}$$

An investigation of the reference gauge zero, however, would determine the sign and value of the stage shift and if possible the error could be corrected on site or allowed for in plotting $(h+a)$ values. The systematic error in the current meter rating tank, however, can only be found from a direct comparison of a selection of rating tanks by rating the same current meters in each (Herschy, 1975a).

The complexities involved in estimating systematic uncertainties in the stage–discharge curve would seem to indicate the desirability of following the procedure in (1) and (2) above.

Note that, by tradition, the dependable variable Q in the stage–discharge equation is plotted on the abscissa with stage on the ordinate. This procedure has no effect on the calculations or on the analysis of the uncertainties.

Table 10.13 Typical computation for the uncertainty in the daily mean discharge using hourly values of discharge

Time	h (m)	$(h-0.115)$ (m)	Q_h (m³ s⁻¹)	$X'_{(h+a)}$ (per cent)	∂S_{mr} (per cent)	$[2S_{mr}^2 + \beta^2 X_{(h+a)}^2]^{1/2} Q_h$
0900	1.225	1.110	46.314	0.4	1.3	64.84
1000	1.565	1.450	69.707	0.3	1.5	115.53
1100	1.971	1.856	101.699	0.2	1.8	183.06
1200	2.293	2.178	129.906	0.2	1.9	246.82
1300	2.520	2.405	151.186	0.2	2.0	302.37
1400	2.670	2.565	165.850	0.2	2.0	331.70
1500	2.789	2.674	177.814	0.2	2.0	355.63
1600	2.872	2.767	186.328	0.2	2.1	391.29
1700	2.929	2.814	192.255	0.2	2.1	403.74
1800	2.981	2.876	197.717	0.1	2.1	415.21
1900	3.034	2.929	203.339	0.1	2.2	447.34
2000	3.067	2.952	206.867	0.1	2.2	455.11
2100	3.082	2.967	208.478	0.1	2.2	458.65
2200	3.065	2.950	206.653	0.1	2.2	454.64
2300	3.026	2.911	202.487	0.1	2.2	445.47
2400	2.975	2.860	197.084	0.1	2.1	413.88
0100	2.915	2.800	190.793	0.2	2.1	400.66
0200	2.845	2.730	183.543	0.2	2.1	385.44
0300	2.747	2.632	173.558	0.2	2.0	347.11
0400	2.628	2.513	161.697	0.2	2.0	323.39
0500	2.495	2.380	148.788	0.2	2.0	297.57
0600	2.365	2.250	136.534	0.2	1.9	259.41
0700	2.257	2.142	126.635	0.2	1.9	240.61
0800	2.164	2.049	118.320	0.2	1.8	212.98
			Σ 3883.56			Σ 7952.45
		Daily mean	161.815			

$$X_{dm} = \frac{\Sigma[(2S_{mr}^2 + \beta^2 X_{(h+a)}^2)^{1/2} Q_h]}{\Sigma Q_h}$$

$$= \frac{795\,2.45}{388\,3.56}$$

$$= \pm 2 \text{ per cent}$$

Then daily mean discharge $= 161.815$ m³ s⁻¹ ± 2 per cent

It should be noted that, in an ideal situation, the current meter observations would in fact fall on the stage–discharge curve, the stage–discharge relation being permanent and S_e and S_{mr} both zero. The scatter about the curve experienced in practice, however, is due principally to the uncertainty in the current meter observations, the uncertainty in stage measurement, the instability of the station control, changing conditions in the channel due to scour or accretion and to seasonal changes in the river regime. In view of these factors it is possible in some cases for S_e to be larger than the uncertainty in the current meter measurements (X'_O). The current meter measurements may indeed be of high accuracy for the conditions prevailing at the time of measurement but they may be influenced by one or more of the above factors when plotted on the stage–discharge curve.

Practice in the United States Geological Survey is to ensure that current meter observations plot within ±5 per cent of the stage-discharge curve. In statistical terms this implies an S_e value of about ±5 per cent at the 99 per cent confidence level (three standard deviations) which, in view of the above remarks, is a stringent requirement in many cases.

10.5.9 Student's *t* test for check gaugings

So long as the check gaugings plot within the S_e uncertainty band without significant bias, the established stage–discharge relation may be considered valid. This means that, in each range, out of 10 check gaugings about five should plot on either side of the stage–discharge curve (see check gaugings 33–36 plotted in Figure 1.13 (Chapter 1). Check gaugings should relate to a homogeneous period of time, i.e. a water year, or part of a water year, between the changes in control. In most cases, however, check gaugings will amount to perhaps only a few per water year and in such circumstances a majority of the points may plot to one side of the curve without necessarily indicating bias, since there is no reason to suppose that additional data, if they had been available, might not have restored the balance.

A suitable statistical test to indicate whether or not the two sets of data may be considered to derive from the same statistical population is the Student's *t* test. The test is based on the property of sample means being distributed about the population mean with standard deviation $s\sqrt{N}$ (see Subsection 10.2.11). If a sample is taken at random, therefore, and it is found that it lies further from the population mean than $1.96s/\sqrt{N}$ (for the 95 per cent confidence level), then it can be concluded that the sample is not likely to belong to the same population.

In the case under consideration, it is required to examine whether the two sets of data can be regarded as being drawn from the same population, i.e. the difference in their means should not be significantly greater than zero. The above equation, therefore, becomes (see Subsection 10.2.11)

$$t = \frac{\bar{X}_1 - \bar{X}_2}{s/\sqrt{N}} \tag{10.42}$$

where $\bar{X}_1$ and $\bar{X}_2$ are the means of the two samples, respectively.

It can be seen from equation (10.42) that the larger the discrepancy between the means the greater the value of *t*. Also, *t* has its own probability

Table 10.14   Student's t test of check gaugings for bias

Observation ref. No.	Stage $(h-0.115)$ (m)	Q_i gauging (cumecs)	Q_c Discharge from rating equation (cumecs)	$\dfrac{Q_i-Q_c}{Q_c}\times100=d_1$ percentage deviation	$d_1-\bar{d}_1$	$(d_1-\bar{d}_1)^2$
33	0.267	4.996	5.234	−4.5	−3.625	13.141
34	0.722	25.609	23.984	+6.8	+7.675	58.906
35	1.583	77.940	79.725	−2.2	−1.325	1.756
36	3.222	227.900	236.510	−3.6	−2.725	7.426

$$\Sigma d_1 = -3.5 \qquad\qquad\qquad\qquad \Sigma = 81.229$$

$$\bar{d}_1 = -0.875$$

$$\Sigma d^2 = 291.18 \text{ (from Table 10.12)}$$

$$s = \left[\frac{\Sigma(d^2)+\Sigma(d_1-\bar{d}_1)^2}{N+N_1-2}\right]^{1/2}$$

$$= \left[\frac{291.18+81.229}{32+4-2}\right]^{1/2} = 3.31$$

Then S = standard error of the difference of the means

$$= s\left[\frac{N+N_1}{N\times N_1}\right]^{1/2} = 3.31\times0.53 = 1.754$$

$$t = \frac{\bar{d}}{S} = \frac{0.875}{1.754} = 0.50$$

Now for $(N+N_1-2)$ degrees of freedom (i.e. 34) t at the 95 per cent probability level = 2.0 (see Table 10.2); therefore no action required.

Note: If $(t=\bar{d}/S)$ is greater than t at the 95 per cent probability level then further action may be required (e.g. additional check gaugings and $\bar{d}/S$ re-calculated) to ensure that there has been no shift in the stage–discharge relation.

distribution, i.e. any specified values of t being exceeded with a calculated probability. Table 10.2 shows the value t may reach for given probability levels. The probability level can be tentatively set at 5 per cent (0.05), i.e. there is a 95 per cent (0.95) chance of being right, equivalent to a 5 per cent chance of being wrong. The odds therefore against values of t as large as or larger than these occurrences by chance are 19 : 1.

The computed value of t is therefore compared with the 95 per cent value in Table 10.2 using the appropriate number of degrees of freedom, in this case $N_1 + N_2 - 2$ (see Subsection 10.2.10). If t, as computed, is less than the tabulated value in Table 10.2 it may be accepted that both sets of data belong to the same population. If it is not, then additional check gaugings should be taken and the calculation repeated to ensure that there has been no shift in the stage–discharge relation. If there is still no improvement then recalibration may be indicated unless the reason for the bias can be ascertained. Table 10.14 shows a typical calculation for t for the check gaugings shown in Figure 1.13.

10.6 VALUES OF UNCERTAINTIES

Suggested values for the uncertainty components in the error equations (10.10), (10.16), (10.17), (10.20) and (10.21) are given in Tables 10.15 to 10.24. They are based on investigations by Carter and Anderson (1963), Dementev (1962), Smoot and Carter (1968), Grindley (1970–72), Herschy (1969–76) and ISO Data 2 (1978), and others. It is always advisable, however, for the user either to confirm these values for his own particular station or to establish his own values. The values of the uncertainties in the tables are percentage standard deviations at the 95 per cent confidence limits (two standard deviations), except where stated otherwise.

10.6.1 Random uncertainties

The random uncertainties listed in Tables 10.15 to 10.23 are as follows:

(1) Random uncertainty X'_m, due to the limited number of verticals used (Table 10.15).
(2) Random uncertainty in velocity due to pulsations, X'_e (Table 10.16).

Table 10.15   Uncertainty X'_m

Number of verticals	Percentage uncertainty X'_m (95% confidence level)
5	20
10	10
15	7
20	5
25	5
30	3
35	3
40	3
45	3

Table 10.16   Uncertainty X'_e

Velocity $(m\,s^{-1})$	0.2, 0.4 or 0.6D				0.8 or 0.9D			
	Exposure time (min)							
	0.5	1	2	3	0.5	1	2	3
	Percentage uncertainty X'_e (95% level)							
0.050	50	40	30	22	80	60	50	40
0.075	33	26	19	16	50	40	28	23
0.100	27	22	16	13	33	27	20	17
0.125	22	19	14	11	27	22	16	14
0.150	19	16	12	9	22	20	14	12
0.175	17	14	10	8	19	16	12	10
0.200	15	12	9	7	17	14	10	8
0.225	13	10	8	6	15	12	9	7
0.250	12	9	7	6	13	10	7	6
0.275	11	8	7	5	11	8	7	6
0.300	10	7	6	5	10	7	6	5
0.400	8	6	6	5	8	6	6	5
0.500	8	6	6	4	8	6	6	4
0.5–1.0	7	6	6	4	7	6	6	4
over 1.0	7	6	5	4	7	6	5	4

(3)   Random uncertainty in the average velocity in the vertical due to the limited number of points taken, X'_p (Table 10.17).

Table 10.17   Uncertainty X'_p

Method of measurement	Percentage uncertainty X'_p (95% level)
Velocity distribution	1
5 point	5
2 point	7
1 point	15

(4a)   Random uncertainty in the current meter rating X'_c for an individual rating—cup-type and propeller meters (Table 10.18).

(4b)   Random uncertainty in the current meter rating X'_c for a standard rating—cup-type only (Table 10.19).

Table 10.18 Uncertainty X'_c (individual rating, cup-type and propeller current meters)

Velocity (m s^{-1})	Percentage uncertainty (95% level)
0.031	20
0.100	5
0.152	2.5
0.229 and over	1.0

Table 10.19 Uncertainty X'_c (standard rating, cup-type current meters only)

Velocity (m s^{-1})	Percentage uncertainty X'_c (95% level) new meters (Price)	new meters (Watts)	used meters (Watts)
0.076	6	9	20
0.152	3	5	13
0.228	2	3	10
0.305	2	3	8
0.458	1	3	7
0.610	1	3	6
1.524	1	3	5
2.438	1	3	5

(4c) Random uncertainty in the current meter rating X'_c for a standard rating—propeller current meters only (Table 10.20).

Table 10.20 Uncertainty X'_c (standard rating, propeller current meters only)

Velocity (m s^{-1})	Percentage uncertainty (95% level)
0.038	20
0.076	10
0.152	6
0.229	4
0.305	3
0.456 and over	2

(5a) Random uncertainty in measuring stage or head, E'_g (Table 10.21).

Table 10.21 Uncertainty E'_g in stage or head measurement

Method	Uncertainty
By float-operated punched tape recorder	$= \pm 3$ mm
By float-operated autographic recorder	$= \pm 10$ mm
By point gauge, electrical tape gauge, tape gauge, etc.	$= \pm 1$ mm
By reference (vertical or inclined) gauge	$= \pm 3$ mm

(5b) Random uncertainty in the uncorrected zero in stage or head, E'_z (Table 10.22).

Table 10.22 Zero error E'_z

Method	Uncertainty
Set by surveyor's level	$= \pm 3$ mm
Set by surveyor's precision level	$= \pm 1$ mm

(6) Random uncertainties in measuring length or breadth, X'_b, or depth, X'_d (Table 10.23).

Table 10.23 Uncertainties X'_b (length) and X'_d (depth)

Term	Percentage uncertainty (95% confidence level)
X'_b	± 0.1 to ± 0.5 depending on the actual length
X'_d	± 1 to ± 3 depending on the actual depth

10.6.2 Systematic instrument uncertainties

Systematic instrument uncertainties in measuring length or breadth X''_b, depth X''_d, and in the current meter rating tank X''_c are given in Table 10.24.

Table 10.24 Systematic uncertainties in X_b'' (length), X_d'' (depth) and X_c'' (rating tank)

Term	Percentage uncertainty (95% confidence level)
X_b''	±0.5
X_d''	±0.5
X_c''	±1.0

10.7 CONCLUSIONS

Gaugings carried out to the relevant ISO Standard and employing the techniques for the calculation of uncertainties outlined in Sections 10.1 to 10.6 may be expected to have the following upper limits of uncertainties at the 95 per cent confidence limits:

(1) Single determination of discharge (X_Q) ±7 per cent
(2) Standard error of estimate of the stage–discharge relation $(2S_e)$
                                                                                 ±10 per cent
(3) Standard error of the mean of the stage–discharge relation $(2S_{mr})$
                                                                                 ±5 per cent
(4) Recorded discharge (15-min value, etc.) (X_{rd}) ±5 per cent
(5) Daily mean discharge (X_{dm}) monthly mean discharge or annual discharge
                                                                                 ±5 per cent

These values may have to be modified for extremely low values of discharge or for other difficult conditions.

The determination of stream flow accuracy by employing statistical methods affords a satisfactory means of presentation of the data for the benefit of the user. The overall quality of a measurement, however, depends largely on the vigilance and efforts of the field observers. This has proved to be the case in the past and it will also be the case in the future.

ACKNOWLEDGEMENT

This material is published with the permission of the Director, Water Data Unit, Reading, UK.

REFERENCES AND BIBLIOGRAPHY

Botma, H. and Struijk, A. J., 1970. Errors in measurement of flow by velocity area methods, Proc. International Symposium on Hydrometry, Koblenz, Unesco/WMO/IAHS, Pub. No. 99, pp. 86–98.

Carter, R. W. and Anderson, I. E., 1963. Accuracy of current meter measurements, *Proc. Soc. Civ. Eng.*, **89** (HY4).

Dementev, V. V., 1962. Investigation of pulsations of velocities of flow of mountain streams and of its effect on the accuracy of discharge measurements; translated from *Soviet Hydrology* by D. B. Krimgold and published by the *American Geophysical Union*, No. 6, 558–623.

Francis, J. R. D. and Miller, J. B., 1968. The accuracy of calibration of model gauging structures, *Proc. ICE* Paper 7053 (Feb.), vol. 39—Discussion (July 1968), vol. 40.

Grindley, J., 1970–72. The calibration and behaviour of current meters:
HRS Report INT 80, 1970. Method of calibration.
HRS Report INT 87, 1971. Effect of oblique flow.
HRS Report INT 93, 1971. Effect of suspension.
HRS Report INT 95, 1971. Accuracy.
HRS Report INT 96, 1971. Drag.
HRS Report INT 99, 1972. Tests in flowing water.

Herschy, R. W., 1969. The magnitude of errors at flow measurement stations, Water Resources Board, TN No. 11, revised 1971, Reading, UK.

Herschy, R. W., 1970. The magnitude of errors at flow measurement stations, Proc. International Symposium on Hydrometry, Koblenz, Unesco/WMO/IAHS, Pub. No. 99, pp. 109–126.

Herschy, R. W., 1975a. The accuracy of existing and news methods of river gauging, *Ph. D thesis*, University of Reading, UK.

Herschy, R. W., 1975b. The effect of pulsations in flow on the measurement of velocity, International Seminar on Modern Developments in Hydrometry (World Meteorologidal Organization), Padova.

Herschy, R. W., 1976. An evaluation of the Braystoke current meter, TM No. 7, Department of the Environment, Water Data Unit, Reading, UK.

Herschy, R. W., 1978. The accuracy of current meter measurements, *Proc. ICE*, Part 2, **65,** TN 187.

Herschy, R. W., White, W. R. and Whitehead, E., 1977. The design of Crump weirs, TM No. 8, Department of the Environment, Water Data Unit, Reading, UK.

Herschy, R. W., Hindley, D. R., Johnson, D. and Tattersall, K. H., 1978. The effect of pulsations on the accuracy of river flow measurements, TM No. 10, Department of the Environment, Water Data Unit, Reading, UK.

ISO 748, 1973. Liquid flow measurement in open channels by velocity area methods, International Organization for Standardization, Geneva.

ISO: 1088, 1973. Collection of data for determination of errors in measurement by velocity area methods, International Organization for Standardization, Geneva.

ISO: 1438, 1975. Thin plate weirs and flumes. International Organization for Standardization, Geneva.

ISO: 3455, 1976. Liquid flow measurement in open channels—calibration of current meters in straight open tanks, International Organization for Standardization, Geneva.

ISO: 3846, 1977. Liquid flow measurement in open channels by weirs and flumes—rectangular broad crested weirs, International Organization for Standardization, Geneva.

ISO: 3847, 1977. Liquid flow measurement in open channels by weirs and flumes—end depth method, International Organization for Standardization, Geneva.

ISO: 4359, 1977. Liquid flow measurement in open channels using flumes, International Organization for Standardization, Geneva.

ISO: 4360, 1977. Liquid flow measurement in open channels by weirs and flumes—triangular profile weirs, International Organization for Standardization, Geneva.

ISO: 4361, 1977. Liquid flow measurement in open channels by weirs and flumes—round nosed broad crested weirs, International Organization for Standardization, Geneva.

ISO: 4373, 1977. Water level measuring devices, International Organization for Standardization, Geneva.

ISO: 4377, 1977. Flat V weirs, International Organization for Standardization, Geneva.

ISO: 5168, 1978. Calculation of the uncertainty of a measurement of flow-rate, International Organization for Standardization, Geneva.

ISO: Data 2, 1978. Investigation on the total error in measurement of flow by velocity area methods, International Organization for Standardization, Geneva.

Smoot, G. F. and Carter, R. W., 1968. Are individual current meter ratings necessary?, *Proc. ASCE*, **94,** Paper 5848.

Spencer, E. A., Smith, K. V. H., Crewe, P. R., Kinghorn, F. C. and Struijk, A. J. Correspondence with and discussion at ISO and BSI meetings.

CHAPTER 11

The Acquisition and Processing of River Flow Data

TERRY MARSH

11.1 INTRODUCTION

A century ago Lord Kelvin postulated that scientific knowledge began at the point when man could measure or quantify. The accumultaion of data has proceeded at an ever quickening pace throughout the twentieth century and the information business is now recognized as one of the world's major growth industries. Today's scientists and engineers must wrestle with a challenge as fundamental as that faced by their forebears, as intuition gave precedence to quantification. An indiscriminate approach to the acquisition of data can no longer be countenanced when the need is for a concision able to provide the information necessary for the effective surveillance and

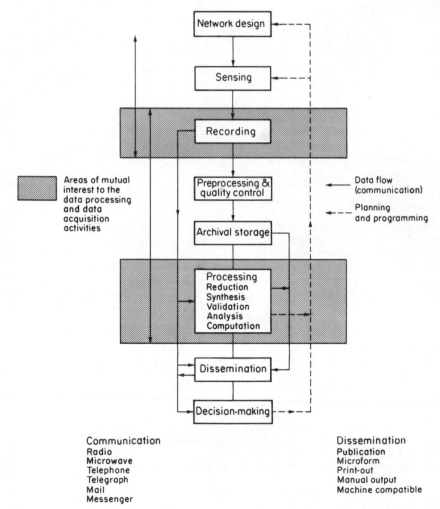

Communication
Radio
Microwave
Telephone
Telegraph
Mail
Messenger

Dissemination
Publication
Microform
Print-out
Manual output
Machine compatible

Figure 11.1   Flow chart for a hydrological information system (after Whetstone and Grigoriev, 1972)

management of our environment. Hydrologists and hydrometricians have long been aware of this problem; central to their discipline is the art of collecting sufficient data to realistically represent the spatial and temporal variations in the elements comprising the hydrological cycle.

While the need for river flow data continues to grow at an impressive rate, increasing thought is being devoted to a reappraisal of existing monitoring networks to ascertain whether the distribution of measurement stations is truly cost-beneficial. Economic constraints are demanding that the maximum value is derived from the data that are collected and that a more prudent and more selective approach is adopted toward river flow data acquisition. Modern technology has equipped man with a powerful battery of recording instruments and data processing hardware with which to monitor hydrometric variables. Unfortunately, man has less tangible assistance when deciding what flow data should be collected, where it should be collected, and how this data should be reduced and presented. Nevertheless, the need to be selective is vital; flow data required at the local, regional, national and international level need to be identified and catered for. There is a heavy responsibility on the designers of data acquisition and processing systems to ensure that the information they provide will make a positive contribution to the efficient management of a river network and thereby justify the expense of the monitoring exercise.

The processing of river flow data spans the considerable interface between the sensing on the one hand and the final decision-making on the other. However, the information necessary to frame the required decisions and the available monitoring equipment both have important implications for the overall system (Figure 11.1) and the flow processing component should not be considered in isolation. In judging the effectiveness of any such system a critical examination should be made of its ability to allow for the differing demands of data users and, in particular, to ensure that suitably filtered information is available at the right time and at appropriate accuracy level.

11.2 THE NEED FOR FLOW DATA

The advanced state of industrial and urban development in the UK, allied to the growing public concern for the maintenance of environmental standards, dictates the need for a dense flow-monitoring network. Even in less-developed countries where the requirements of agriculture and water supply are paramount, there is an inevitable expansion in the types of flow information required, as new uses and needs are recognized. However, it is still possible to distinguish two basic categories of streamflow data:

(1) Data required for the efficient day-to-day management and regulation of a river system.
(2) Data required for design, planning and modelling purposes.

11.2.1 Operational data

This first category includes considerable volumes of data which are of short-term relevance only and are not, in their raw state, destined for archival storage. The overriding operational requirement is for rapid access to the appropriate data, commonly to recognize alarm or emergency situations; the growth of telemetry systems can be traced to the immediacy of the information requirements. In operational work the value of a phenomenon at the present moment and its variation over the preceding few hours are, generally, the data necessary for decision-making. For instance, if the possibility of flooding appears imminent on a particular river a rapid assessment is needed of (a) the water level, (b) the rate of rise, and (c) a measure of the antecedent rainfall.

The main characteristic of these data if they are to be useful is that they must be measured and communicated to the decision-maker in the shortest possible time. There is little point in making elaborate provision for permanently recording the data, especially if this delays data communication.

11.2.2 Planning data

The planning needs for data generally require long, quality controlled, records of river flow to permit assessments of variability to be made with confidence. A secondary distinction can be drawn between those purposes requiring information about the total quantity of water flowing past a point (usually expressed as the daily, monthly or annual mean flow) and those purposes best served by information concerning extremes. Extreme values have been much used historically in frequency/magnitude studies of flood or drought return periods, but it is now being realized that the mean, maximum and minimum of a flow series are, often, inadequate to provide answers to many of the questions being posed.

Designers and planners of water-related facilities are commonly turning to the statistical characteristics of a streamflow series rather than examining flows at specific times. A knowledge of the statistical properties of a flow time series permits the prediction of future streamflows in terms of the probability of occurrence over a period of years. In addition, there has been considerable progress in the simulation of streamflows utilizing the statistical characteristics associated with a recorded flow sequence. Such data extension requires a lengthy, and continuous, flow record to establish the appropriate characteristics. The traditional scene in hydrometric data processing of the assembly of continuous records, principally daily mean flows, is now being complemented by techniques designed to extend the range of information which can be derived from a limited time series. However, the ultimate success of such techniques will strongly reflect the precision with which current flow data is collected, processed and archived.

11.3 DEVELOPMENT OF THE RIVER GAUGING NETWORK IN THE UNITED KINGDOM

11.3.1 Historical background

The quantitative study of river flow appears to have started with Galileo in the early sixteenth century, but it was another 300 years before systematic gauging programmes were instigated in Europe and the USA. In Britain, although flow records were maintained by some water undertakings and private individuals, for many years previously the routine collection and publication of flow data on a national scale did not begin until the establishment of the Inland Water Survey in 1935. The number of gauging station records published in the early volumes of the *Surface Water Year Book* was small; only 27 were included in the first volume (1935–36) and the quality of a number of these was conjectural, especially towards the extremities of the flow range. Approximately 100 further records were added during the next 20 years and a considerable expansion occurred following the formation of the River Boards in 1948.

Economic stringency resulted in a downturn in gauging station construction until the Water Resources Act of 1963 provided a welcome impetus to extend the monitoring network. The Act called upon the river authorities to plan hydrometric schemes for data collection and empowered central government to contribute towards capital costs of installations. Between 1965 and 1974 the number of primary flow measurement stations in operation increased dramatically and many established stations were improved. Figure 11.2 illustrates the growth in the UK gauging station network.

11.3.2 The existing network

The UK now has a gauging station density of 1 per $250\,km^2$ and all major rivers, plus tributaries with mean flows in excess of 5 cumecs, are gauged at least once. To what degree does such a network meet current needs? Is the current level of flow monitoring cost-beneficial?

Definitive answers to these fundamental questions cannot yet be given but progress should result from the developing dialogue between the hydrometricians, who can assess the costs of obtaining data, and the planners and managers who are in a position to estimate the benefit of flow information. In a simple situation it is straightforward to assess the very tangible benefit associated with a flow record. Clearly where the accurate monitoring of river flow, over a period, justifies a significant reduction in the margin of safety allowed when setting a residual flow for a particular river, the value of that information is very high. Given the increasing complexity of water use and re-use, however, the creation of meaningful cost–benefit relationships will require ever more sophisticated analysis. In particular, the worth of flow data

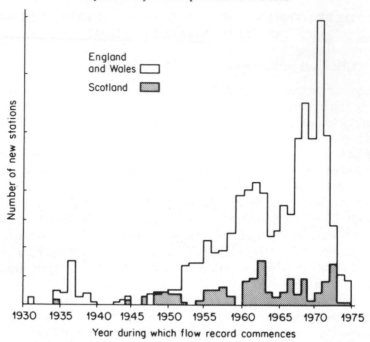

Figure 11.2   Annual increase in gauging stations providing
continuous flow records

in the research field will remain difficult to evaluate and the system designer
will need to tread a careful path between providing an abundance of redundant data and adopting the strategically inadequate approach of assembling
particular types of flow data only when the need is recognized.

11.3.3   Information transfer mechanisms

Flow data are not only valuable in themselves but serve also to calibrate
information transfer mechanisms which facilitate the estimation of flow
parameters for ungauged sites; models of this type may eventually reduce the
need for high density monitoring. Significant advances have been made in the
development of practical information transfer techniques over the past
decade and, where a suitable base of recorded flows exist, they are able to
provide regionalized information in the absence of local gauged data. The
estimation of runoff from a knowledge of rainfall and the known proportion
which contributes to river flow at a neighbouring gauging station, demonstrates the use of a basic information transfer model. For natural catchments
the development of such models is commonly undertaken using regression
methods. Their aim is either to relate certain flow parameters to catchment
characteristics such as area, climate or geology or relate a short flow record to
a long one.

Existing techniques also provide adequate models for interpolating flows between gauging sites on a river to permit real-time flow forecasting. Nevertheless the development of models to meet the broadening spectrum of flow data needs presents formidable problems. The earnest network analyst can but rue the paradoxical situation whereby the evolution of an optimal network of flow measurement stations may well require a temporary increase in network density to satisfactorily calibrate the complementary information transfer mechanisms.

11.4 FLOW DATA ACQUISITION SYSTEMS

11.4.1 Processed data

Processed data is data manipulated or reduced in a manner to suit it for storage or further analysis. In considering acquisition and processing procedures, two major functions are demanded of any effective system:

(1) The information derived from the raw data should faithfully represent the original field measurements.
(2) The correct level of information, together with appropriate accuracy limitations, should be available to all data users.

11.4.2 'Garbage in—garbage out'

It is true of hydrometric data processing as of any information processing system that the quality and consistency of the input data will, to a great degree, determine the value of the derived information. Minimizing the amount of spurious raw data requires that the individual components of the acquisition system, from river level sensing to final archiving, should be critically examined and an assessment made of the magnitude of the individual sources of inaccurate or invalid data. Experience in the UK indicates that there are two prerequisites to the creation of a successful river flow processing system:

(1) The role of the hydrometric field staff is critical; systems should be structured to ensure that their expertise and enthusiasm are used effectively and not so designed as to leave them remote from the end product of their work.
(2) Positive steps should be taken to standardize data collection procedures and the raw data should be subject to quality control to eliminate inconsistencies. The value of a hydrometric archive can be greatly reduced if rigorous procedures are not introduced to infill missing data and validate the various archived parameters.

11.4.3   Sensing and recording equipment

Although essentially the entire flow measurement station is an instrument and should be treated accordingly, it is usual to consider the sensing and recording instrumentation as a functional unit. The rarity of direct flow recording devices leaves the measurement of river stage as the major contributor to errors in the computation of discharge. Monitoring of river levels by the routine reading of staff gauges is still practised but the dictates of accuracy, convenience and the attraction of automatic data processing has resulted in the widespread introduction of analogue recorders and the more recent spread of digital recording devices. Punched tape recorders not only offer an attractive data processing option, they have proved themselves reliable and robust field instruments; the British Antarctic Survey have maintained two such recorders to assess glacial melt-water discharges during recent Southern summers. In their traditional operating mode both chart and digital recorders are harnessed to a float system in a stilling well (Figure 11.3). Stilling well construction is, however, expensive and not justifiable for temporary reconnaissance studies, which are increasingly used in river engineering. Thus, for particular applications, pressure bulb devices have been interfaced with suitable recording equipment or recourse has been made to more sophisticated resistance, capacitance or acoustic methods of direct river level sensing (see Chapter 8).

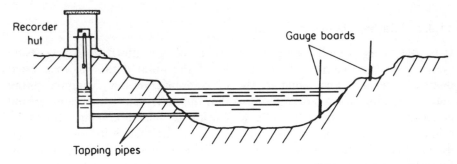

Figure 11.3   Typical level monitoring installation at a flow-measurement station

A new generation of recording instrumentation has been heralded by the use, often in experimental catchments, of sensors linked to data loggers in order to monitor a variety of meteorological and hydrometric variables. The Institute of Hydrology has developed a river stage sensor based upon a potentiometer system which can discriminate to 1 mm over almost any range of level changes. A digital signal is produced and recorded on a magnetic tape casette. Prototype systems are also being tested where sensing equipment is interfaced with microprocessors on site to perform some primary processing and control the selection of flow information for field storage or transmission to remote centres. Microprocessors allow a significant part of the processing 'intelligence' to be located in the field and flow conditions can be monitored either by interrogation or by routine reporting.

11.4.4 Complementary recording systems

The use for which flow data are required is an important factor determining the type of instrumentation suited to a particular network. For operational purposes the accuracy of measurement is at less of a premium than the early availability of the information. Planning data should be subject to precise measuring and processing procedures. This distinction is reflected in the common British practice of employing both analogue and digital recorders at important flow measurement stations. A 'quick-look' facility is provided by the chart record, enabling a rapid appraisal to be made of the hydrograph shape, and characteristic features can be easily recognized. The digital recorders punch water levels on to paper tape at predetermined time intervals. Accuracy is improved because the sensitivity of a punched tape recorder is usually finer than that of an analogue instrument. Where vigilant maintenance schedules are followed and provision is made to quality control the raw data, flow records of an impressive standard can be assembled for archival storage. Conventionally, punched tapes are removed at monthly intervals but chart recorders require more frequent site visits. Where the risk of flooding is a serious hazard, data transmission facilities may be added to give immediate notice of alarm conditions.

11.4.5 Recording media

In general, river levels are rarely recorded in a form suitable for direct input to a computer for immediate conversion to discharge; a certain amount of preprocessing or translation is required to achieve full computer compatibility. Raw river levels are usually registered on one of the following recording media: (a) manuscript; (b) analogue charts (see Figure 11.4); (c) 5-channel paper tape (see Figure 11.4); (d) 16-channel paper tape (see Figure 11.4); and (e) magnetic tape.

Where river levels, or computed discharges, are recorded on manuscript, all that is necessary is for the data to be punched on to tape or punched card, or keyed directly into the computer with suitable additional information to identify the data.

11.4.6 Analogue traces

Chart traces require the abstraction of river levels either manually or with the assistance of a trace follower which permits selected levels to be recorded, typically on paper tape or magnetic tape cassette (Figure 11.5). Provision is

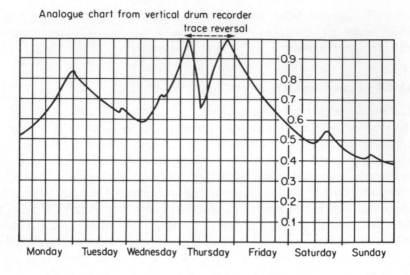

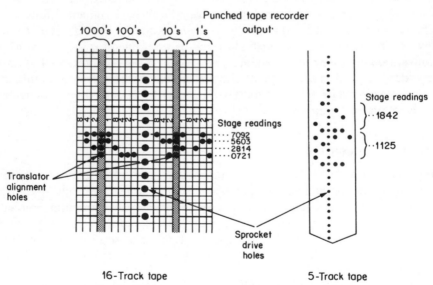

Figure 11.4   River level recording media—examples of the output from
analogue and digital stage recorders

commonly made to display the recorded levels to assist in error recognition
and editing. Although this method can effectively harness local expertise and
ensure good-quality data in the hands of a skilled operator, it can be laborious
and expensive. Recent advances in automatic pattern-recognition techniques
now allow the direct optical scanning of analogue traces which are initially
microfilmed to reduce handling problems. An essential feature of such a

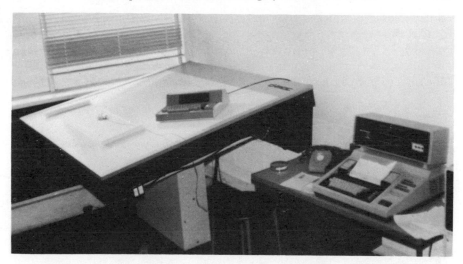

Figure 11.5   Consul and digitizing table of the D-MAC trace follower

system is the ability to distinguish the hydrograph trace from the background chart grid.

With large volumes of data positively recorded on standard charts this technique can be exceedingly economic, and considerable progress has already been made in transcribing information from recording raingauge charts. Provision is made for manual intervention to amend spurious raw data and the viability of this approach depends on the degree to which this intervention can be minimized.

11.4.7   Translation of 16-channel tapes

The abstraction of a series of river levels is, of course, unnecessary with digital recorders where stage values are punched out at predetermined intervals on to paper tape. Widespread acceptance of punched tape recorders followed the adoption by the United States Geological Survey of the instrument developed for them by Fischer and Porter Limited (see Chapter 8). Their instrument was designed to record on 16-channel tape to preclude possible obsolescence which could result from the recording media being directly compatible with a single computing system. Thus, on most processing installations, the 16-channel tape requires to be translated into a form suitable for further processing. No reader system, however, will be versatile enough to successfully handle all misaligned, torn or badly punched tapes. As a general rule the readers, having a slower more gentle tape transport, are more tolerant of misaligned tapes. They allow the tape time to settle under the sensors and for the distortions due to stress to reduce. Mechanical sensors and sprocket drives tend to cause wear on the holes of the tape and thereby impose severe limits on the number of times a tape can be put through a reader before it becomes torn or produces an unacceptably high rate of misreads. A translator is, typically, of modular construction comprising a tape reader, paper tape punch, and a console to add header information and provide an editing capability (Figure 11.6).

Figure 11.6   16-Channel to 5-channel paper tape translator

With the increased use of mini-computers, the variety of available translation methods has expanded and it is possible to interface 16-channel readers with small local computers to translate to computer compatible paper tape or directly into buffer storage for later transmission to the mainframe computer. By ensuring that river level data are correct and complete at this stage, many of the ensuing processing problems can be avoided. Increasing use is being made of visual display units to assist in the recognition of spurious hydrograph behaviour, since software alone is unlikely to detect, and correctly amend, all types of eccentric trace.

Both 5-channel paper tape and magnetic tape cassettes are nominally computer compatible, permitting the direct input of field-recorded tapes. Modern high speed readers, however, have a limited tolerance requiring that field-punched paper tapes maintain a high standard. In addition, side-stepping the translation process removes the necessity for examining the river levels prior to their conversion to discharge. Computer compatibility is only an advantage if the resulting flow information is of the required accuracy; this will seldom be achieved without strict quality controls during preprocessing.

11.4.8 Comparison of results from analogue and digital recording systems

Considerable research has been undertaken to assess the comparative accuracies of river level recorded on punched tape and chart recording instruments. Less attention has been directed towards the relative merits of the processed flow data derived from the alternative systems, but it is this information which is generally required for later analysis.

In any mechanical level recording system, the three major sources of error are inertia, friction and backlash in the moving parts, as shown in Chapter 1. It is widely accepted that the digital recorder has a greater inherent precision, since the width of pen trace, chart instability and the ensuing digitizing can but increase the overall error associated with a chart system. However, the recognition and rectification of spurious stage data is more easily performed with a graphical trace. A vigilant team of experienced hydrometric personnel can compensate for a lower basic recording accuracy by the early correction of invalid sequences of river levels which may prove too subtle to be detected in the digital system.

A meaningful comparison of the results from the two systems, outlines schematically in Figure 11.7, is hampered by the greater maintenance effort normally expended on the prime recording system. In order to overcome this problem, two gauging stations were selected where maintenance procedures were comparable for the chart and punched tape recorders. The South West Water Authority have developed a very thorough system for digitizing hourly stage values from a standard re-usable chart. These river levels are then displayed and any necessary editing of the raw data is completed prior to the output of an 8-track paper tape of stage values. For the comparison exercise, punched tapes from Ott digital recorders were forwarded to the Water Data Unit for processing. Identical state–discharge relationships were used by both systems to convert from river level to flow.

Because of its fundamental importance in water resource studies the basic parameter chosen for comparison was the daily mean flow. Figure 11.8 illustrates the very encouraging agreement between the results from the two systems. Corresponding monthly total flows were compared throughout 1975 and an analysis of the departures reveals a mean departure of 1.1 per cent with a standard deviaton of 0.95 per cent. Selected storm events were also examined and Table 11.1 indicates good agreement between the corresponding runoff volumes. These discrepancies are considerably smaller than many previously reported and they testify to the diligence of the field staff involved and the rigorous nature of the digitizing procedures. Where less sophisticated chart-following facilities are available and less experienced operators are

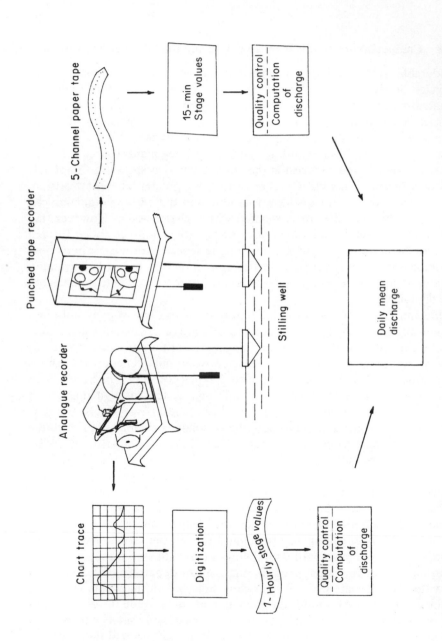

Figure 11.7   Schematic outline of the processing procedures associated with digital and analogue recorders

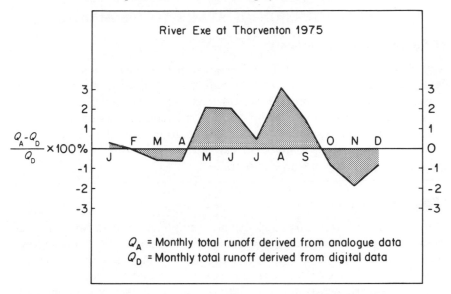

Figure 11.8 Comparison of monthly total runoffs derived from digital and analogue stage records

Table 11.1 Comparison of individual storm runoff volumes derived from corresponding digital and analogue stage records; the hydrograph peaks were synchronized for the purposes of this comparison

| River | Gauging station | Date | Runoff | | % PT > AN |
			Punch tape (mm h^{-1})	Anglogue (mm h^{-1})	
Exe	045005	18. 1.75	15 735.5	15 791.5	−0.4
Exe	045005	20. 1.75	22 672.5	21 944.0	+3.3
Exe	045005	22. 1.75	10 903.5	10 433.5	+4.5
Exe	045005	25. 1.75	5 307.0	5 108.0	+3.9
Otter	045001	27.11.74	14 616.5	14 684.3	−0.5

employed, the precision of the analogue results is likely to be markedly reduced.

Exact synchronization of the digital and analogue stages in the above tests sometimes proved difficult, reflecting the imperfect agreement between the respective clocks; differential instrument lag is likely to have been insignificant. These problems can assume importance for flood routing or conceptual modelling studies, when an assessment should be made of the possible departures of the recorded data, from their real-time occurrence.

11.5 CONVERSION OF STAGE TO DISCHARGE

11.5.1 Sources of uncertainty

Any data processing system should endeavour to reduce and synthesize raw data with the minimum of error introduced through the adopted computational procedures. Absolute accuracy is not practicable, but, as water resources are managed to ever finer limits, it becomes essential to recognize and, where possible, quantify the sources of inaccuracy in the data acquisition and processing system. Only then can the user interpret the information presented to him in a meaningful manner.

11.5.2 Sampling intervals

The selection of the most appropriate sampling interval for river stage recording has received considerable attention since the introduction of punched tape recorders. Obviously, the shorter the sampling interval the more faithfully the variation in river level is recorded. However, the collection and processing costs, together with the inconvenience of handling daunting volumes of raw data, focused attention on the degree of information loss associated with longer sampling intervals.

A study at the Institute of Hydrology in 1971 demonstrated that the accuracy of the computed runoff totals would not be significantly degraded if intervals of several hours were used to supersede the, arbitrarily selected, 15-min interval which is conventionally used throughout the UK. Using as a reference the daily mean flow computed from 96 15-min stage values the departures from this standard were derived for various sampling intervals ranging up to 8 h. Analysis of the standard deviations of the departures indicated that all flow measurement stations could be sampled at 30-min intervals with an uncertainty of only 0.25 per cent of the daily mean discharge. For the vast majority of stations a sampling interval of 1 h or over would give the same uncertainty. If a satisfactory sampling interval is required tailored to an individual catchment, then a three-parameter predictive equation (embracing area, rain and infiltration components) can be used to estimate the desired frequency for any given uncertainty.

Attractive as such an approach may appear, there are strong reasons for the continued use of a fixed-interval approach, given existing instrumentation and experience of field staff. Although the errors resulting from increasing the sampling interval can be considered small it is difficult to be certain that cumulative effects will be negligible. The effect of lengthening the sampling interval is to reduce the apparent discharge occurring during flood peaks and augment the discharge during recession (Figure 11.9). Such errors may not seriously disturb the assessment of mean discharge but could become significant in data generation exercises, time series analysis, or rainfall-runoff modelling.

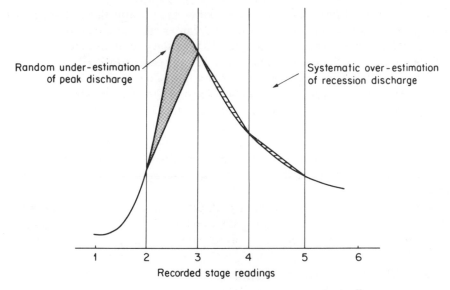

Random under-estimation of peak discharge

Systematic over-estimation of recession discharge

Recorded stage readings

Figure 11.9   Errors introduced by discrete sampling intervals

It is really against the background of a broadening spectrum of data requirements that any general abandoning of the 15-min interval could be seen as premature. The existing standardization is beneficial to field staff, allows good interpolation of stage values during restricted periods of recorder malfunction, and permits error detection by the monitoring of discontinuities to a greater degree than with longer intervals. In a modern computing environment, processing time and disc storage facilities are unlikely to be at such a premium as to dictate a lengthening of the sampling interval. In addition, a significant proportion of the data processing costs are related to tape handling and transport and these costs will not respond greatly to changes in the sampling frequency.

11.5.3   Data compression

Although there is much to be said for a standard sampling interval, at least within the UK, this does not imply that all of the recorded stage values need to be retained for later processing. Implicit in all continuous records that are finally stored by a digital computer is an assumption concerning the variation in the record between stored points. In river flow processing, the basic requirement is to store sufficient data to accurately represent any given hydrograph; clearly during periods of slow recession, many stage values could be filtered out, prior to conversion to flow, without prejudicing the accuracy of the computed discharge. In Britain, where the daily and seasonal flow variations are relatively subdued, the need for such filtering is not marked.

Less temperate regions, however, experience high proportions of their total runoff during infrequent storm events and can endure extended periods with no discharge whatsoever (see Chapter 13). In these circumstances recourse can be made to techniques which compress the recorded river levels but still allow effective linear interpolation between the values that are retained.

The level of inaccuracy which can be tolerated with the introduction of data compression procedures is a matter for local consideration. However, Ibbett (1975) suggests that if the error caused by compressing a stage record varies from zero at a stored value to $\pm r$ between two stored values, then the compression range (r) should be chosen so as to represent less than 5 per cent of the combined errors introduced by rating inadequacies and recorder precision.

An alternative, and more fundamental, approach to the problem of redundant data is to utilize recording systems which are actuated by significant changes in river level rather than at fixed time increments.

11.5.4   The stage–discharge relationship

Since most flow measurement stations do not measure flow directly, it is necessary to provide for the conversion of river level to river discharge. This requires the establishment of a relationship between stage and discharge; the derivation of such a calibration and its mathematical formulation are important factors governing the precision with which the requisite flow can be computed. The derivation of a typical stage–discharge relation is demonstrated in Chapter 1.

11.5.5   Segmented calibrations

It is often not possible to produce a satisfactory single range stage–discharge relation, especially when the flow characteristics are governed by a complex control. The overall range of river stages then requires to be split into a number of straight segments having angular discontinuities at their junctions. In these circumstances, the individual rating equations simply represent the best-fit lines through a restricted range of gaugings; the regression constants have no direct physical meaning, and experience indicates that natural river sections may exhibit exponents ranging from 0.8 to values above 8, which have been recorded for the River Ganges. Hydraulic theory would indicate values nearer to 1.5–2.5 for rectangular sections but considerable variation will be encountered, notably where a calibration is derived for a set of gaugings covering a very limited range of river stage.

11.5.6   Computer storage of calibration details

It should be the prime aim of the hydrometrician to ensure that the method chosen to transform river level to river flow faithfully represents the field situation. Of secondary importance is the facility with which calibrations can be stored in computer files and the costs involved in converting stage to discharge. Individual stage–discharge relations normally take one of the following forms: (a) look-up tables; (b) single, or segmented, rating equations; and (c) storage of structural dimensions to allow direct conversion to discharge using theoretical or laboratory evolved equations.

The relative merits of each approach are dependent on the types of gauging station which predominate in any region and on the computing facilities available to the agency involved. For manual conversion the look-up table, or rating, is most convenient but, increasingly, computers are performing the necessary conversion and on very small installations it may not be economic, or even feasible, to hold large rating tables in computer storage for occasional use. Recourse can be made to stage–discharge equations which require negligible computer storage and are easily retained to form a catalogue of the calibrations employed throughout a gauging station's history.

The use of weir or flume dimensions, in a rating relationship, often implies the adoption of an iterative procedure to compute the velocity-head increments to be added to measured stage. This successive approximation technique is elegant but it can be wasteful of computer time, especially when inordinately precise tolerances are set to required final agreement between successive interations. Table 11.2 illustrates the extravagance of the iterative approach and demonstrates that where computer storage is not a problem a look-up table is the better investment. For this comparison two river level tapes were selected for computed processing using three types of rating relations. The first tape followed a gentle recession, whereas the second monitored several storm runoff events (see diagrams with Table 11.2).

11.5.7 Sources of error

In order to minimize the archiving of incorrect or inappropriate flows, staff involved in hydrometric data processing should familiarize themselves with the methods used to establish stage–discharge relations, and also be aware of the limitations of individual ratings. Any extrapolation of a calibration above the gauged range can produce faulty discharges, especially where bank-full flow is exceeded. Similarly, downward extrapolation of a low flow range is unwarranted in the absence of gauging evidence and any such exercise should allow for the flagging of computed discharges as estimates only (see Chapter 1).

Processed flows should always be scanned by experienced personnel to ascertain that the discharges are reasonable in relation to rainfall and the flows at neighbouring stations. Attention should also be directed to ensuring that the stage–discharge relation has not been subject to recent disturbances such as to justify the revision of certain flows or of the calibration itself.

Table 11.2 Comparison of processing costs related to three methods of converting stage to discharge

Rating relation type	Tape I		Tape II	
	Mill time (s)	Total cost	Mill time (s)	Total cost
Power law equation	7	0.78	9	0.96
Look-up table	7	0.79	7	0.79
Iterative procedure	20	1.95	40	2.58

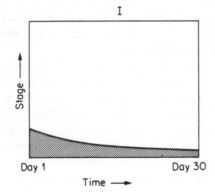

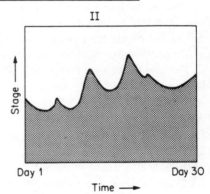

Factors which can introduce uncertainty into the processed flows include:

(a) weed growth in the measuring section or on weir crests;
(b) erosion or deposition in the channel cross-section or the build-up of accretion behind a structure;
(c) flows exceeding the modular limit for a structure;
(d) the datum correction of the record being incorrectly adjusted;
(e) malfunction of the recording instrument;
(f) hysteresis effects during periods of high flow.

Recognition, and correction of errors due to these causes will greatly lessen the burden on any later validation procedures which are applied to a flow series.

11.6 SELECTION OF FLOW DATA FOR ARCHIVAL STORAGE

11.6.1 General

It is a basic tenet of information processing that the user's requirements dictate what types of raw data should be collected and stored. Formalizing these requirements is often not easy and the necessity to provide for any anticipated demands on a data base entails that data should be assembled to offer maximum flexibility to the data user. In the case of hydrological data processing, a system which allows the storage of all recorded river level (or velocity) data, together with the appropriate mechanisms for conversion to flow, offers the possibility of responding to all requests for flow data constrained only by the small information loss arising from the stage sampling frequency. However, data can accumulate to the point where its very volume frustrates the purpose for which it was assembled, unless due consideration is given to the ease and economics of storage and retrieval.

Modern disc-storage facilities enable vast quantities of data to be efficiently stored. Indeed, all of the 15-minute stage data so far collected in the UK could easily be accommodated on 5 disc packs. Rapid access to this data is of great value for modelling work, but for the identification of long-term trends it would clearly not be appropriate to re-compute annual runoff values, say, on the basis of more than 34 000 individual stage values every time such a request was received. An archive comprising a prudent mixture of raw and derived data is best suited to meet user needs.

11.6.2 Naturalization of gauged flows

The creation of an archive based on gauged flows alone will often prove inadequate to meet the demands of further analysis; this is particularly true in the province of water resource planning. Facilities should therefore exist to compute, or estimate, natural flows on a suitable time base. It is common for water to be abstracted and discharged at numerous points in a river network, and water supply and sewerage systems do not always treat catchment divides with the respect that a hydrologist might consider proper. Water supplied from one river may find its way, via tap and treatment works, to an adjacent river basin.

Often, considerable difficulties attend the monitoring of these artificial flow components but it remains important to assess, where practicable, the degree to which a gauged flow is distorted from its natural value by net variations above the gauging station. It is unnessessary to allow for 'out and back' industrial abstractions with relatively trivial consumptive losses, but quantitative estimates should be made of significant consumptive losses such as evaporative cooling.

Storage-derived abstractions should also be accounted for, as should any import or export of water to and from the catchment. In industrial catchments, the multiplicity of abstractions and discharges may well make close monitoring impossible; given these circumstances a catalogue of the types of artificial influence on the discharge regimes can be assembled to assist in the interpretation of gauged flow results. Water supplies derived from groundwater are difficult to relate to streamflows unless the boreholes are in river gravels or otherwise in immediate hydraulic continuity with the river. It follows that, with these latter exceptions, groundwater supplies can generally be regarded as additional to natural flows.

11.7   QUALITY CONTROL OF FLOW DATA

A river flow archive depends for its success upon the provision of ready access to continuous flow records of known accuracy. The requirement for data in a consistent form is rarely matched by uniform procedures for hydrological data acquisition; network densities, instrumentation, data handling and processing techniques tend to be diverse and subject to revision. It is important therefore to develop quality control procedures to ensure that the archived flow data is of a standard to suit its potential users. Of course, only a certain proportion of the available money and expertise can reasonably be devoted to quality assurance and it should be appreciated that the production of a perfect data set is impracticable.

Within a flow data acquisition system, error sources will range from subtle misrecording of stage associated with partial intake-pipe blockage to the grotesque distortion of computed flows due to the simple expedient of mislabelling a punched paper tape. All significant errors require to be recognized and rectified. Environmental data can be most effectively validated in one, or a combination, of two modes:

(1) Temporally; fluctuations in a time series can be examined to ascertain whether they could reasonably be expected in a natural situation.
(2) Spatially; data from adjacent stations can be examined to check whether they behave sympathetically, within an appropriate tolerance range.

The continuous form of the river stage (or flow) hydrograph lends itself to time series examinations designed to detect spurious behaviour. Natural variations are, however, large and the tuning of quality control mechanisms demands effort and expertise to achieve a control procedure which will filter out only the significant errors. Detection of trivial errors serves only to jeopardize the speedy processing and archiving of data.

A trend test designed at the Institute of Hydrology to 'clean' flow data for use in hydrological research, provides a versatile technique for examining a series of river levels. The test takes advantage of the normal hydrograph characteristics to determine whether a particular sequence of stage values are acceptable for further processing. Each sequence of four stage values, 1–4, 2–5, etc., are tested successively and any doubtful combinations are rejected. All possible combinations of four successive values are illustraed in Figure 11.10, together with the corresponding action to be taken. This technique is particularly powerful when used during the preprocessing of river levels recorded with a short time interval between each reading. It is less effective

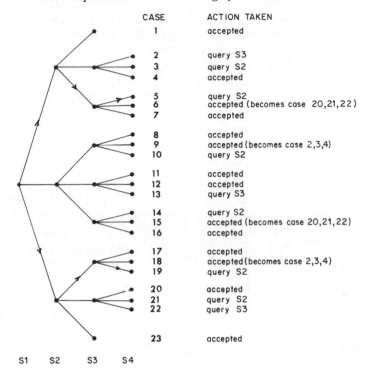

CASE	ACTION TAKEN
1	accepted
2	query S3
3	query S2
4	accepted
5	query S2
6	accepted (becomes case 20,21,22)
7	accepted
8	accepted
9	accepted (becomes case 2,3,4)
10	query S2
11	accepted
12	accepted
13	query S3
14	query S2
15	accepted (becomes case 20,21,22)
16	accepted
17	accepted
18	accepted (becomes case 2,3,4)
19	query S2
20	accepted
21	query S2
22	query S3
23	accepted

S1   S2   S3   S4

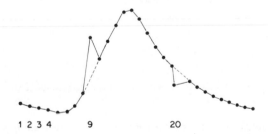

1 2 3 4   9   20

Figure 11.10   Trend test for river flow quality control. The lower diagram shows an example of a typical set of stage values indicating possible errors; the individual stage readings 9 and 20 show improbable departures from the trend which should be queried by the quality-control routine as indicated in the upper diagram; arrows indicate queried sequences

when applied to the more discrete daily mean flows which constitute the bulk of many flow archives. However, more rudimentary techniques can often be used to isolate highly erroneous data for further examination. Gross errors possess one attractive feature; they are easy to recognize. Hence, by considering the probable causes of faulty data, simple validation procedures can be designed to pinpoint the areas where corrective action is required.

Software checks to ensure that each batch of monthly data contains the correct number of days and that the monthly mean flow is, indeed, the average of the flows on the constituent days, are necessary controls, but the adoption of incorrect units or the consistent misplacement of the decimal point are more difficult to detect in isolation. Errors of the latter type are inevitably large and the validation window shown in Figure 11.11 illustrates how the restricted natural variation in annual runoff can be used to direct attention to the suspect results.

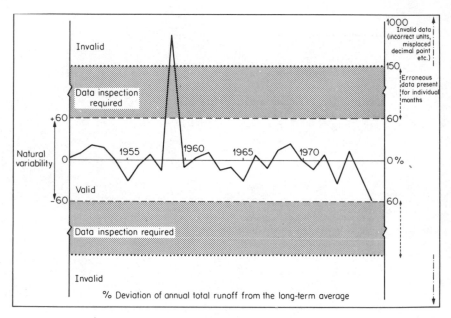

Figure 11.11   The use of a validation window to identify suspect runoff data

Despite the advances in the development of computer-based quality control techniques, there remains no real substitute for the close inspection of processed data and, in particular, the flow hydrograph, by hydrometric staff with   field knowledge of the flow measurement station concerned. Many rivers are subject to artificial controls which can produce short but abrupt hydrograph changes. In such circumstances local expertise is imperative if the unwarranted rejection of genuine data is to be avoided.

11.8   THE DATA BASE APPROACH

It is now generally recognized in the water industry that information is as much of a resource as the water itself. It follows that the investment in data acquisition and processing will only be fully realized if all the users' needs are satisfactorily accommodated. The creation of a comprehensive flow archive cannot be an end in itself; provision must be made to relate this data to other relevant information and to form a data base capable of interrogation by all *bona fide* users. River flow processing can be seen as but a subset of hydrometric data processing which itself nests into the technical information system of the water management agencies (Figure 11.12).

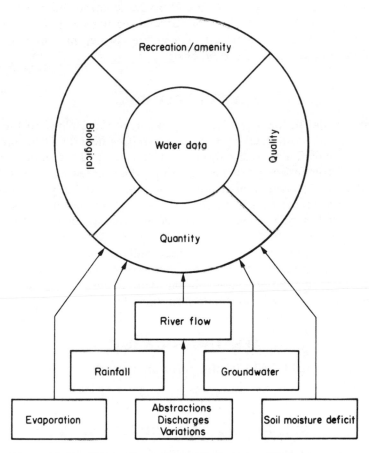

Figure 11.12   Schematic outline of the relationship between river flow and other categories of water data

The development of a dialogue between user and data base is essential and much can be learned from the relatively short history of electronic data processing. Systems have tended to develop to adapt the requirements of the user to the capabilities of the chosen computer. Such compromises have, all to often, resulted in management having to take decisions based on information 'up to the end of last month'. Water problems arise, and require solution, in a real-time world where the requirements for data can arise at random intervals. The development of data base management systems now offer the opportunity to reap the full benefits from a volume of assembled data and enable immediate and pertinent information to be provided to the decision-maker.

A data base approach implies not simply the computer storage of all relevant, flow related data but also the provision of standard methods of handling this data, to update or amend, to retrieve information for formal reports or publication, or *ad hoc* enquiry and, perhaps most importantly, to associate and compare the stored data in ways which may or may not be determinable at the stage of system design.

11.9 CONCLUSION

Advances in the technology of sensing and recording, allied to the use of microprocessors to undertake remote processing of field data and control the onward transmission of information, promise to extend the options open to the hydrometric network designer at a time when, at least in the developed world, the cost of manual monitoring is becoming prohibitive. At the other end of the processing system, computers are being harnessed to assist in the assembly of data sets in a manner which facilitates easy data manipulation and data dissemination. If the water industry is to fully benefit from these developments, then great emphasis must still be placed on vigilant field maintenance procedures, the standardization of processing techniques, and the recognition and correction of spurious data, to ensure that data users can make confident use of the information supplied to them.

ACKNOWLEDGEMENT

This material is published with the permission of the Director, Water Data Unit, Reading, UK.

REFERENCES AND BIBLIOGRAPHY

Benson, M. A. and Matalas, N. C., 1967. Synthetic hydrology based on regional statistical parameters, *Water Resources Res.*, **3** (4), 931–935.
Black, P. E., 1970, Runoff from watershed models, *Water Resources Res.*, **6**, 465–477.
Brandon, D. H., 1970. *Management Planning for Data Processing*, Brandon/Systems Press, Princeton, USA.
Briggs, R., Batt, L. S. and Bussell, R. B., 1975. Measurements of ground storage, surface flow and water quality, in *Engineering Hydrology Today*, Institution of Civil Engineers, UK.
Carter, R. W., 1965. Streamflow and water levels—effects of new instruments and new techniques on network planning, Joint WHO–IAHS Symposium on Design of Hydrologic Networks, Quebec, IAHS Pub. No. 67, 86–98.

Clarke, R. T., Mandeville, A. N. and O'Donnell, T., 1975. Catchment modelling to estimate flows, in *Engineering Hydrology Today*, Institution of Civil Engineers, UK.

Davey, P. G. *et al.*, 1974. Technical developments in Oxford PEPR system, Oxford Conference on Computer Scanning, UK.

Dumas, A. J. and Morel-Seytoux, H. J., 1969. Statistical discrimination of change in daily runoff, Hydrology Paper No. 34, Colorado State University, Fort Collins, Colorado, USA.

Eagleson, P. S. and Goodspeed, M. J., 1973. Linear systems applied to hydrologic data analysis and instrument evaluation, Division of Land Use Research, Technical Paper No. 34, Commonwealth Scientific and Industrial Research Organisation, Australia.

Gregory, K. J. and Walling, D. E., 1973. *Drainage Basin, Form and Process*, Edward Arnold, London, 456 pp.

Herbert, S. I., 1973. A user's guide to the river flow system, Devon River Authority, UK.

Herbertson, P. W., Douglas, J. W. and Hill, A., 1971. River level sampling periods, NERC, Institute of Hydrology Report No. 9.

HMSO, 1974. *The Surface Water Year Book of Great Britain, 1966–70, London* 160 pp.

Ibbett, R. P., 1975. Compression of Time Series Data, *J. Hydrology (N.Z.)*, **14**(1).

Institution of Water Engineers, 1969. Proceedings of the Symposium on River Flow Measurement, Loughborough University, UK.

ISO : 748. Liquid flow measurement in open channels—velocity area methods, International Organization for Standardization, Geneva.

Kitson, T. and Poodle, T., 1971. River hydrograph study as an aid to river pollution control, *Effluent and Water Treatment Journal* (September).

Littlewood, I. G., 1977. An introduction to data quality assurance, internal report, Water Data Unit, Reading, UK.

Marsh, T. J. and Stephenson, P. M., 1976. Surface water data processing—guide to practice, DoE, Water Data Unit, Technical Memorandum No. 5.

Martin, J., 1973. *The Design of Man-Computer Dialogues*, Prentice-Hall series in Automatic Computation, New York.

McGammon, R. B., 1975. Towards a computer based information and retrieval system for groundwater data, *J. Hydrology (N.Z.)*, **14**(1).

Moss, M. E., 1976. Design of surface water data networks for regional information, *Hydrological Science Bulletin*, **XXI**(1).

Nemec, J., 1973. Transnational transfer experiences, in Proceedings of the 1st International Conference on the Transfer of Water Resources Knowledge, Fort Collins, Colorado, USA.

Plinston, D. T. and Hill, A., 1974. A system for the quality control and processing of stream flow, rainfall and evaporation data, NERC, Institute of Hydrology Report No. 15.

Reardon, T. J., 1968. The system for the automatic processing of streamflow data in Queensland, Irrigation and Water Supply Commission, Queensland, Australia.

Rodda, J. C., 1969, Hydrological network design—needs, problems and approaches, WMO Reports on WMO/IHD Projects, Report No. 12, WMO, Geneva.

Thomas, D. M. and Benson, M. A., 1970. Generalization of streamflow characteristics from drainage basin characteristics, United States Geological Survey, Water Supply Paper 1975.

Toebes, D. and Ouryvaev, V. (Eds.), 1970. Representative and experimental basins, and international guide for research and practice, Unesco Studies and Reports in Hydrology, No. 4.

Trestman, A. G., 1964. New aspects of river runoff calculations, Israeli Programme for Scientific Translation, pp. 163.

United States Department of the Interior, 1971. Design characteristics for a national system to store, retrieve and disseminate water data, US Geological Survey, Office of Water Data Co-ordination.

Water Data Unit (DoE), 1976. Punched tape river level recorders, Technical Memorandum No. 6, Reading, UK.

Whetstone, G. W. and Grigoriev, V. J., 1972. Hydrologic information systems, Unesco/WMO Studies and Reports in Hydrology, No. 14, 72 pp.

Wiley, R. L., Hopkins, J. S. and Walker, L. R., 1974. The automatic digitisation of hyetograms using PEPR, Oxford Conference on Computer Scanning, UK.

WMO, 1971. Machine processing of hydrometeorological data, WMO Technical Note No. 115, Publication No. 324, WMO, Geneva.

WMO, 1972, National and regional computerised hydrological data banks and requirements upon related WMO systems, CHy-IV/Doc.33, WMO Commission (CHy), 4th Session, Buenos Aires.

WMO, 1976. Hydrological network design and information transfer, WMO Operational Hydrology Report No. 8, Publication No. 433, WMO, Geneva.

CHAPTER 12

The Use of Satellites in Hydrometry

ROBERT A. HALLIDAY

12.1　INTRODUCTION

Much of the water resources data routinely collected at a gauging station is used for archival purposes, for example in conducting water resources studies or in project design. However, there is an increasing need for data on an immediate or 'real-time' basis for water management purposes. Such data may be used for flow or flood forecasting, operation of structures, etc.

Under most circumstances real-time data requirements can be met only through the use of automated telemetry systems. These systems usually depend on conventional telephone or radio links but, in the 1970s, water resources agencies in North America started using earth-orbiting satellites to relay data from water resources stations to the station operators. Data then were disseminated to users on a near real-time basis.

It was subsequently demonstrated that relaying hydrometric data by satellite was reliable and cost-effective. Telemetry service could be extended to any site from which real-time data were needed. Dependence on on-site recorders could be reduced by using satellite telemetry as the principal data collection tool.

This chapter reviews the general principles and methods of telemetry, discusses the main components in a satellite data collection system, draws comparisons with conventional telemetry, and gives some North American examples. As one of the negative aspects of the space age has been the proliferation of acronyms and initials, Section 12.9 lists some frequently used ones.

12.2　PRINCIPLES OF TELEMETRY

12.2.1　General

Any telemetry system consists of four major elements. These are sensors that measure or detect changes in a parameter, encoders that convert the sensor output to forms suitable for transmission, a transmisson system that provides the link from a sensor to another location, and a data reception and distribution facility that sorts, decodes, checks and distributes the incoming data. Many alternatives exist in selecting sensors, encoders, transmission media, and data reception/distribution systems. Therefore, regional or national requirements for water resources telemetry can be met if funds are available. Figure 12.1 illustrates a typical satellite data collection system.

12.2.2　Sensors

The hydrometric parameter most often telemetered is water level. These levels are detected by means of a float operating in a stilling well, or by a

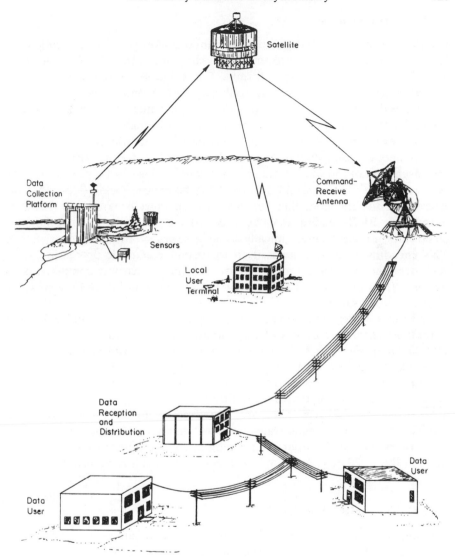

Figure 12.1   Satellite data collection system

pressure sensing unit which measures the water pressure over a fixed point in the stream bed—this pressure being proportional to the water level.

Other parameters may include: water velocity, say, from an acoustic flowmeter or from an *in situ* point velocity meter; water quality data such as water temperature, pH, specific conductance, and dissolved oxygen; and meteorological data such as precipitation, air temperature, wind speed, and wind direction. Other chapters in this text discuss the operation of some of these sensors in detail.

12.2.3   Data encoding

Prior to transmission, all data from hydrometric sensors must be encoded in a format compatible with the transmitter. This means that the data are converted to digital or analogue format. For example, a water level sensor consisting of a float in a stilling well may turn a pulley and shaft, whose rotation is directly proportional to the change in water level. The rotations of the pulley shaft may be converted to digital data by means of a shaft position encoder, or to an analogue voltage by means of a potentiometer.

A frequently used digital format is binary coded decimal (BCD). In this case each digit of the water level reading is encoded as a 4-bit binary number. A reading of 9.832 m would become 1001 1000 0011 0010. (Note that the decimal is not encoded, therefore its location must be known to the data user.) This BCD reading may be encoded electronically using solid-state electronics, or electromechanically using a series of two-position switches. This encoding technique lends itself to further electronic processing as all such data can be represented easily by two-state electronic components, in which a '0' is represented by zero (or low) voltage and a '1' by a specified higher voltage (say 5 V).

The same water level reading may be converted to a voltage using a potentiometer. If the output of the potentiometer is such that a reading of 10.000 corresponds to 5 V, then the reading 9.832 m will become 4.916 V. Note that in this case 1 mV corresponds to 2 mm of water level change, so the accuracy of the potentiometer can be a critical factor in determining the accuracy of the reading that is transmitted.

Most water resources data are encoded in digital format, the only significant exceptions being water quality data and air temperature data. In these cases the usual output is an analogue voltage, although analogue currents in the 4–20 mA range and analogue frequencies have also been used.

12.2.4   Transmission media

Transmission media may be divided into two broad categories, namely, line and space. When the distance from the sensor to the data user is short, an electrical conductor may be used to connect the two points, thus enabling reliable transmission of data. More often it is necessary to share lines with other users by connecting the sensor network to the public telephone system. It has been demonstrated that data can be collected using telephone systems over distances of several thousand kilometres. Such systems are widely used.

Because of terrain difficulties line transmission is not always feasible, thus it becomes necessary to collect data by means of transmission through space. As the electromagnetic spectrum is a finite resource all uses of the spectrum are regulated internationally by the International Telecommunications Union, and on a national basis by licensing authorities within each country. Figure

12.2 shows some ITU frequency assignments and some of the terminology used in referring to radio frequencies. Several methods have been used to collect water resources data by radio. These are: terrestrial line of sight, meteor burst, and satellite. Terrestrial line of sight techniques using v.h.f. (say 150 MHz) and s.h.f. (say 4 GHz) radio have been used for several years and are well developed. The equipment required for terrestrial radio transmisson is available from several manufacturers.

The meteor burst technique is a passive relay system in which radio trans-missons are bounced off the ionized trails left by meteorites as they enter the earth's atmosphere. Experiments have shown that these trails are so numerous that it is almost always possible to relay data at hourly or more frequent intervals. Implementation of a system using about 160 interrogat-able transmitters is now under way in the USA.

In contrast to meteor burst, the satellite telemetry technique is an active relay system which depends on the use of a spacecraft-carried repeater known

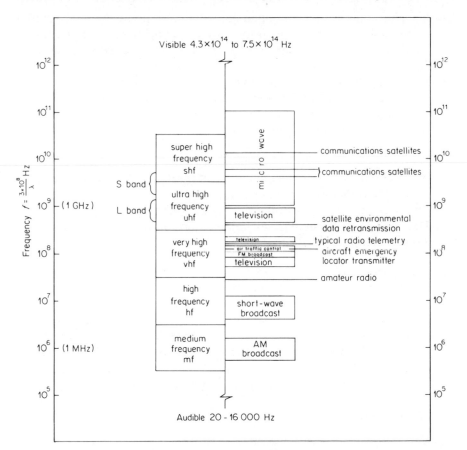

Figure 12.2 Radio frequencies in the electromagnetic spectrum

as a transponder. Provided that the satellite is in line of sight of the transmitter, the message will be received and either retransmitted immediately or tape recorded for playback when the spacecraft is in view of a ground station. Several experiments in which satellites are used for the relay of hydrologic data are under way in North America.

12.2.5 Data reception and distribution

All telemetry systems, no matter how rudimentary, must have a data reception and distribution system to receive data, to perform conversions to usable formats, to verify accuracy, and to disseminate the data to the ultimate user. When small numbers of sites are involved, these functions may be performed manually; however, as systems become larger and more complex it is almost mandatory that computers be used. Some of the operations that may be performed under computer control are:

(1) Gauging stations may be polled according to a predetermined schedule, or on the basis of the value of the parameters being monitored.
(2) Satellite tracking antennae may be operated or searches for meteor trails conducted.
(3) Data may be converted to engineering units from BCD readings or analogue voltages using predetermined calibration curves.
(4) Accuracy of data may be verified if the communications system uses parity checks or convolutional encoding routines for error detection and correction.
(5) Searches for new extreme values may be carried out and checks for possible sensor failure conducted by flagging improbable events or performing correlations with adjacent sites.
(6) Statistics on system performance may be generated.
(7) Data may be distributed to several users on various schedules, perhaps on a direct computer-to-computer basis.

The degree of sophistication built into a data reception and distribution centre is a matter of what the user needs, is williing to pay for, and is technically possible. In the case of some telemetry systems, such as those associated with satellite relay of data, a sophisticated data reception facility is mandatory. However, these facilities may be continental in coverage; therefore the cost to individual users can remain reasonable.

12.3 SATELLITE TELEMETRY

12.3.1 Background

The first experiments in relaying small quantities of data from a given location to a central receiving station were conducted by the United States National

Aeronautics and Space Administration (NASA) using the Omega Position Location Equipment (OPLE) system and the Applications Technology Satellite. Further tests (for example, the Eole experiment) all showed that such systems were feasible. Generally, small numbers of sites were involved and it was not until the Landsat (formerly ERTS) programme was developed that persons in the hydrologic community became aware of the possibilities of satellite telemetry of data.

12.3.2 Landsat

The main purpose of the Landsat mission is to provide repetitive global high resolution imagery of the earth's surface using a multispectral scanner and return beam vidicon. In addition to these sensors the satellites also carry a data collection system (Figure 12.3) that can relay messages containing 64 bits of sensor data from a small transmitter known as a Data Collection Platform (DCP) to a centrally located receive site. This is possible only when the satellite is in mutual view of a DCP and the receive site, a situation that occurs several times each day.

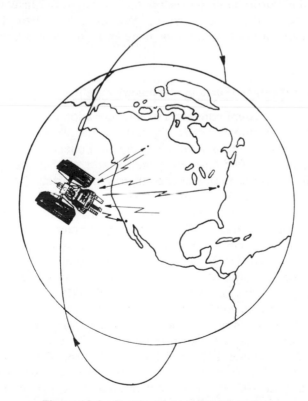

Figure 12.3   Landsat data collection system

Landsat-1 and Landsat-2, launched by NASA on 23 July, 1972, and 22 January, 1975, respectively, are in near-polar, sun-synchronous orbits. The two spacecraft orbit the earth at an altitude of 900 km every 103 minutes, crossing the equator southbound at 0942 hours local sun time, and are exactly nine days out of phase. These precise orbits ensure the best performance of the imaging systems but are not necessary for operation of the data collection system.

The Landsat data collection system operates on a self-timed random access basis. That is, each DCP transmits at approximately 180 s intervals and, provided that the DCP and a centrally located receiving antenna are in mutual view of the satellite, the message is relayed to the receive site. Since all DCPs compete for a single satellite channel, it is possible for two DCPs to occasionally interfere with each other's transmissions. The probability of this happening is small, even when 1000 DCPs are in view of the satellite, since each DCP message is transmitted as a 38 ms burst.

Experience has shown that data can be obtained from DCPs at mid-latitudes on three or four orbits of the satellite each day. This occurs during southbound orbital passes in the morning, when imagery is obtained, and during the northbound passes in the evening. Because the satellite orbits converge in the polar regions, data from high latitudes, such as northern Canada, are obtained on as many as 11 of the satellite's 14 orbits each day.

12.3.3 World Weather Watch geostationary satellites

A global observing system consisting of five operational geostationary satellites is now being implemented as part of the World Meteorological Organization's World Weather Watch. (A geostationary satellite is one having a circular orbit 36 000 km above the equator so that the speed of the spacecraft on its orbital path matches that of the earth's rotation. Such a satellite will appear to be stationary with respect to the earth and will have a fixed coverage zone.) All satellites in the system should be in operation in 1978, in time for the First GARP Global Experiment (FGGE), an atmospheric research programme. The system will consist of two USA GOES satellites stationed at 75° W. and 135° W. longitude, a European Space Agency Meteosat Satellite at 0°, a Soviet GOMS satellite at 70° E. and a Japanese GMS satellite at 140° E. longitude (GMS and Meteosat were launched in July and November 1977, respectively).

The two NASA prototypes of the GOES spacecraft, SMS-1 and SMS-2, were launched on 17 May 1974 and 6 February, 1975. GOES-1 and GOES-2 were launched in October 1975 and June 1977, respectively. GOES-2 is in use at 75° W. and SMS-2 is in use at 135° W. with GOES-1, and SMS-1 is acting as spares at 105° W. This configuration is shown in Figure 12.4. Additional GOES launches will be made as the need arises.

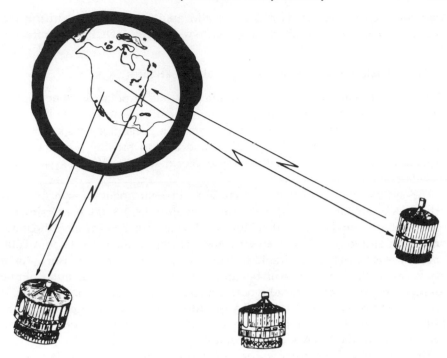

Figure 12.4   GOES data collection system

The GOES spacecraft are multi-purpose in nature, carrying a Visible and Infrared Spin-Scan Radiometer, a Space Environmental Monitoring System, a weather facsimile (WEFAX) retransmission system, and a data collection system. The other geostationary World Weather Watch satellites are similarly equipped.

The GOES data collection system consists of 183 channels on each satellite; 100 of these are for use with interrogated DCPs, 50 with ordered (as opposed to random access) self-timed DCPs and 33 for interrogated international DCPs. Each satellite can retransmit data from over 10 000 DCPs. The 33 international channels on GOES are identical with those carried by Meteosat, GOMS and GMS. Meteosat has also 33 domestic channels and GMS has 100 domestic channels.

All users of the GOES data collection system are assigned channels and precise time slots during which data may be transmitted. The time slots are typically 60 s in length and occur at 1, 3, or 6 h intervals depending on user needs. Provision is also made for emergency channels. In the case of the interrogated DCPs, a command is generated by the ground station and retransmitted to the DCP by the satellite. On receipt of this command, the DCP transmits. The ordered self-timed DCPs depend on an accurate clock to maintain their transmissions within the allocated time slot. Any drift in the timing of a DCP could result in interference with another DCP's transmission.

12.3.4   World Weather Watch polar orbiting satellites

One problem with geostationary satellites is that in polar or some mountainous regions it is not possible to obtain a line of sight to the satellites. For example, at 60° latitude the elevation angle to a geostationary spacecraft is 22° or less. To overcome this problem and also to provide position fixing capability, say for buoys or balloons, the World Weather Watch system will include polar orbiting satellites.

The system will consist of two satellites in near-polar sun-synchronous orbits. One satellite will orbit the earth at an altitude of 830 km, crossing the equator southbound at about 0800 hours, while the other will be an orthogonal orbit at an altitude of 870 km, crossing the equator northbound at about 1600 hours. These orbits are desirable for some of the spacecraft's sensors. As in other cases, the satellites will be multi-purpose, carrying an atmospheric sounder, a high resolution radiometer, a space environment monitor, and a data collection system. The NASA prototype of the satellite, known as Tiros-N, is scheduled for launch in early 1978. Subsequent satellites will become part of the NOAA series of satellites.

The data collection system carried by Tiros-N has been developed by the French Centre National d'Etudes Spatiales (Project Argos) and, like the Landsat system, operates on a random access self-timed basis. Unlike Landsat, however, the Argos system will handle messages that can vary in length from 32 to 256 bits; also, the transmission interval can vary from 40 to 200 s. (The shorter transmit intervals are used for platform location calculations based on the doppler shift in the carrier frequency of the incoming data.) Up to four DCP messages can be handled simultaneously. The messages are recorded on the satellite for playback when the spacecraft is in view of a receive site. The DCP data can also be relayed immediately to users by means of a v.h.f. beacon (about 137 MHz).

The number of orbits each day during which messages can be relayed in an operational Tiros-N system will depend on the latitude of the DCP. The average value will vary from seven orbits at the equator to 11 at 45° latitude to 28 at the poles.

12.3.5   Commercial satellites

Up to this point the satellites discussed all operate or will operate at u.h.f. frequencies. However, there are a number of commercial geostationary communications satellites, both international and domestic, that operate at much higher frequencies, i.e. in the s.h.f. band. It has been proposed that these satellites could be used for hydrometric telemetry. In 1977, a demonstration to prove the technical feasibility of such a system was begun jointly by the United States Geological Survey, Comsat General Corporation, and Telesat Canada.

The DCPs transmit to a satellite on a random self-timed basis at the same interval at which parameters are measured, i.e. a new reading may be transmitted as frequently as every 15 min. It has been calculated that one transponder on a commercial satellite could handle as many as 3000 000 DCPs.

12.3.6 The Datasat concept

The satellites described so far have been multi-purpose in one sense or another. The situation could arise whereby the data collection system on a satellite fails thus exposing the user of a telemetry system to the risk of losing record. As the data collection system is a relatively inexpensive part of a satellite, the satellite operator may be reluctant to launch a spare satellite or to activate an 'in-orbit' spare.

For this reason, some studies of a dedicated data retransmission satellite have been conducted. An operational 'Datasat' system could consist of two or three relatively inexpensive polar orbiting satellites. The satellites would operate on a random access basis and would handle variable length messages from several thousand DCPs. No programs that would authorize construction of a Datasat system have been funded.

12.4 DATA COLLECTION PLATFORMS

12.4.1 Description

The associated interface electronics, radio transmitter, and antenna used to transmit data from a hydrometric station to a satellite is known as a Data Collection Platform (DCP). Many companies have manufactured DCPs that operate at u.h.f. frequencies of 400–403 MHz; therefore, there arc differences in each manufacturer's product. A DCP electronics unit is small, usually about 0.01 to 0.02 m^3 in volume and 5 kg or less in mass (the DCPs used with commercial satellites are somewhat larger). Antennae that are as small as 0.35 m square and 5 mm thick have been used (the antenna used with commercial satellites are 1.2 m diameter dishes). All DCPs are manufactured to demanding performance specifications using modern electronic technology and must be certified by the operator of the satellite system.

The DCP collects data, say water level, from sensors, encodes the data in the required format, then transmits the data and a unique identification code to a satellite. The rate at which the message is transmitted and its length is determined by the satellite system. Message rates vary from 100 to 5000 bits a second and message lengths range from 32 to over 100 bits. DCPs that may be used with either of two satellite systems have been manufactured.

12.4.2 Interfacing

Data collection platforms are easy to interface with existing hydrometric sensors. Typically, sensors having serial digital, parallel digital, or analogue voltage outputs may be wired directly to a DCP. In other cases it may be necessary to use a relatively simple interface between the sensor and the DCP. These interfaces may consist, for example, of a voltage amplifier to scale an analogue voltage output to the correct level.

12.4.3 DCP memories

In those instances where data are collected more frequently than the allowed transmission interval, it becomes necessary to use a DCP memory to enable all required data to be transmitted. Typically, a DCP memory will store several hundred bits of data in a shift register, such as that shown in Figure 12.5.

Three important factors concerning memory are the total length of memory, the maximum message length that may be transmitted at any one

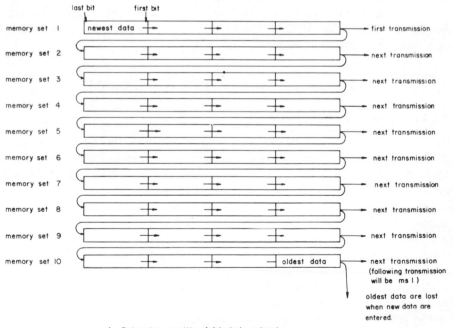

Figure 12.5   Schematic of DCP shift-register memory

time, and the quantity of data word that is entered into the memory at each collection interval. With some satellite systems, such as GOES, entire contents of the memory can be transmitted at once. In other cases, such as Landsat or Tiros-N, the total memory contents have to be transmitted as a series of memory subsets.

In either case the use of DCP memory enables the more efficient use of the data collection system and the reconstruction, for example, of an entire stage hydrograph, based on sampling intervals of 15 min. It becomes possible then to consider satellite telemetry as a possible primary data collection tool, thus eliminating the need for on-site recording.

12.4.4 Power sources

All DCPs are battery operated and are generally designed to operate with a d.c. power supply of about 12.5 ± 1.5 V. Because up-do-date electronic components are used in the construction of DCPs, power consumption is low. Batteries will often last longer than one year.

A DCP battery should have sufficient capacity to operate the unit for extended periods of time and to provide the peak current draw that takes place during a transmission. These conditions should be met under all temperatures that will be encountered at the site. Battery manufacturers can provide specifications for their products.

Batteries that have been used include lead–acid, alkaline and nickel–cadmium rechargeable cells and alkaline or air-depolarized potash non-rechargeable cells. Rechargeable batteries tend to be cheaper in the long run provided that the logistics problems of transporting batteries in and out of a site for recharging can be overcome. One frequently used solution to this problem is that of connecting a small (say 3.5 W) solar panel to the battery. The size of battery and solar panel combinations will depend on the power consumption of the DCP and the incoming solar radiation at the DCP site. The DCP manufacturer can supply power consumption figures and national meteorological agencies can supply solar radiation data.

12.5 GROUND STATIONS

12.5.1 Introduction

The ground stations required for reception of hydrometric data relayed by satellite tend to be more complex than those associated with most other telemetry systems. However, the number of data reception facilities required tends to be low since coverage from one site may include an entire continent.

The operator of a satellite system must have a ground station that can command the satellite and monitor its health. This station usually can also receive telemetered data and distribute the data to users. In some cases a large user of data may wish to operate a passive receive facility, i.e. one that cannot command the satellite but can only receive data. In some cases, operation of such a 'local user terminal' can reduce data distribution costs, decrease data distribution time, or reduce data loss due to line problems between the command station and the user.

12.5.2 Command and data acquisition station

The command and data acquisition stations operated by NASA for the Landsat program and by NESS for the GOES program are typical of the types of station associated with satellite telemetry systems. These sites are in operation whenever the satellite is in view of the station and have considerable built-in redundancy so that the risk of data loss is minimized.

The Landsat command and data acquisition system consists of three tracking stations located in Alaska, California and Maryland. All data received are formatted and sent immediately by land line to a ground data handling centre at Greenbelt, Maryland where the data are sorted into user files and distributed via teletype on a near real-time basis, and by punch cards and computer listings on a delayed basis.

The GOES data are received by one of several antennae at Wallops Island, Virginia (Figure 12.6) then sent by land line to Suitland, Maryland, where the data are sorted into user files and sent to users by high speed teletype lines or

Figure 12.6   Command and data acquisition station

placed in disc storage for interrogation by users of dial-up low speed teletype terminals. The Wallops Island facility will also handle the Iiros-N satellites with data being distributed from Suitland or Toulouse, France.

12.5.3   Local user terminal

The expression 'local user terminal' can be misleading, as the only implication is that the terminal cannot command a spacecraft. The range of a Landsat ground station is about 3200 km, while that of a DCP is about 2500 km. Therefore it is possible for Landsat data to be relayed 5000 km or more to a local user terminal. A GOES local user terminal equipped to receive data

Figure 12.7   Landsat local user terminal

from certain GOES channels would receive all data transmitted on those channels in the western hemisphere. In actual practice the mini-computers used in local user terminals are programmed so that they compare the platform identification number transmitted by the DCP to those numbers stored in the computer. Only data from matching numbers are stored for further processing.

Since the local user terminal can depend on the command centre for support in the event of equipment failures, redundant items can be reduced significantly thereby reducing the complexity and cost of the terminal. Several such terminals are in operation in the USA (Figure 12.7), one is in operation in Canada, and one is in operation in Chile. The stations that can receive Landsat imagery that are in operation or planned for Argentina, Brazil, Iran, Italy and Zaire could be upgraded to receive DCS data at relatively low cost.

The v.h.f. beacon that will be carried by Tiros-N also lends itself to construction of simple, inexpensive local user terminals. In this case the cost of a suitable tracking antenna becomes very small.

12.6 COMPARISON WITH CONVENTIONAL TELEMETRY

12.6.1 Introduction

Although use of satellite telemetry systems for water resources data collection is a relatively new phenomenon, it is possible to make comparisons with other telemetry systems and to draw conclusions concerning the potential for more widespread use of such systems. These comparisons are by no means rigorous since satellite data collection systems and DCPs are still evolving. The trend, however, is to more versatile and less expensive systems. In addition, the spacecraft now used in satellite telemetry are multi-purpose in nature. This means that, while the data collection system may not exactly meet all of a user's needs, the user is not asked to contribute to the cost of the satellite. In the case of a conventional telemetry system, all of a user's technical requirements may be met but the user has to pay all costs. Both technical and cost aspects should be considered in selecting a telemetry system.

12.6.2 Technical comparison

The first consideration in developing a hydrometric telemetry system is the question of site selection and the system design. In the cases of conventional telemetry, the availability of telephone lines or perhaps use of radio links must be considered. Compromises between the best sites from a hydrologic standpoint and the sites that are economically feasible may have to be made. Satellite telemetry stations (and meteor burst ones), on the other hand, may usually be installed at any site from which data are required. The only technical aspect of site selection is that if a geostationary satellite is considered, there must be a clear line of sight to the satellite. Tables of antenna headings, both in elevation and in azimuth, are readily available.

As standard satellite DCPs are generally designed for multi-parameter inputs of serial digital, parallel digital and analogue data, the future expansion of the system to include additional parameters in the data message is ensured. This is not always the case for conventional telemetry systems.

Data collection platforms may be installed easily by technicians that install conventional hydrometric instrumentation. A knowledge of electronics is not required; all that is necessary is an ability to follow a simple checklist. Normally two persons can install a DCP at an existing gauging station in a few hours. Because installation is simple, DCPs can be deployed quickly in response to critical situations, for example during floods. Also, once installed the DCP is not subject to interruptions in service such as those caused by destruction of telephone lines, radio towers or by power failures. Unfortunately, in the case of conventional systems, such failures often occur just when the demand for hydrometric data is most critical.

Maintenance of DCPs in the event of failure is by field substitution. Actual repairs can be performed at a depot operated by the manufacturer or by in-house electronics personnel.

Since site selection, system expansion, and installation and maintenance of DCPs are relatively straightforward when satellite telemetry is used, the need for elaborate system design studies prior to establishing a telemetry system is eliminated. However, it should be realized that, where feasible, it may be a better decision to add another remote unit to conventional telemetry system than to start to embark on a satellite data collection program.

All operators of satellite data collection systems provide a data distribution service at a reasonable price. However, to increase the timeliness of the data or to reduce reliance on land lines, a user may wish to establish a local user terminal. Some users with as few as 25 DCPs have done this while others having 100 DCPs have not. Using a terminal, data can be made available to a user within seconds of retransmission by a satellite.

Satellite telemetry terminals are more complex than the terminals associated with conventional telemetry systems. The main factor in this increase in complexity is the requirement for large antennae (say 5 m dishes), either fixed for geostationary satellites or tracking for polar orbiting ones, to acquire data from the satellite. As stated earlier, however, a terminal can provide coverage for a whole country or even a continent.

12.6.3 Cost comparison

The costs of satellite telemetry versus other systems should be examined on the basis of capital costs and costs of data reception and distribution and/or maintenance. For ease of conversion to various currencies, the following approximate capital costs are in United States dollars:

landline	$1400 (water level only, assumes no-cost connection to telephone systems)
v.h.f. radio (terrestrial)	$15 000–20 000 (water level only)
microwave (terrestrial)	$30 000 (water level only)
meteor burst	$6000 (multi-parameter)
satellite (u.h.f.)	$3000–5000 (multi-parameter)
satellite (s.h.f.)	no initial cost (multiparameter)

Often it is simply not economically feasible to install land lines and terrestrial radio systems because of rugged terrain and long distances.

If the data reception and distribution facilities provided by the satellite operator are satisfactory then related capital expenditures, depending on the degree of automation required, may consist either of £3000 for purchase of a teletype compatible terminal or £10 000–15 000 for purchase of a mini-computer with telephone modem. On the other hand, if a local user terminal is established, this could cost $50–150 000 or more depending on the satellite system used. The cost in excess of that associated with a sophisticated data reception and distribution centre for a conventional telemetery system would be in the order of $20 000–50 000.

Estimated annual costs for data reception, distribution and maintenance of the equipment installed at the gauging station are:

landline	$600
v.h.f. radio	$1000
microwave	$2000
meteor burst	$2000
satellite (u.h.f.)	$300
satellite (s.h.f.)	$3000–5000

The operators of commercial satellites propose to provide a service on a lease basis; therefore, the user pays no capital costs but instead pays a relatively high rental fee. These rentals will probably be determined more precisely in 1978.

The annual cost of maintaining a local user terminal could be as much as $10 000; a relatively minor part of this sum would be devoted to antenna maintenance.

In comparing costs of satellite telemetry against those of other systems, it can be seen that landline systems are very competitive if one or two parameters are required and if connection to the telephone system is feasible. Since DCPs consist almost entirely of high quality electronic components, it is fair to say that there is considerable scope for cost reductions in the future.

This is not the case for other telemetry systems, all of which have a larger percentage of fabricated metal components.

12.7 EXAMPLES OF HYDROMETRIC TELEMETRY

12.7.1 Background

Operation of hydrometric data collection networks requires that small quantities of data be collected from widely dispersed, often remote sites. The very nature of the task is ideally suited to the use of satellite data collection systems. This is especially true when there is a need to obtain data on a near real-time basis. Examples of applications in Canada and the USA are described in the following sections.

12.7.2 Canadian examples

Canada is a large (10 000 000 km^2), sparsely populated (23 000 000) country having a severe winter climate. Collection of hydrometric data is difficult and expensive, and provision of real time data is economically impossible under most circumstances. Real time data are available by telephone from fewer than 5 per cent of the Canadian hydrometric stations.

Therefore in 1972, when the Landsat data collection system became available, ten DCPs were installed in northern and western Canada in the hope that at least one water level reading would be telemetered successfully each day from each site. These data would then be used for flow and flood forecasting and for operation of gauging stations. It was soon demonstrated that many more messages than originally expected were telemetered and system capability surpassed all expectations. The network of DCPs was then expanded to 30, some of which were operated on the GOES system. Data uses have been as diverse as providing a water level forecast for barge operators on a major river system, to supplying data for division of flow on a river flowing from Canada into the USA to providing data for regulation of an anadromous fisheries agreement.

A Landsat DCP site on the Trent–Severn waterway is shown in Figure 12.8. The power line visible in the figure is required for operation of an acoustic flowmeter and not for the DCP.

Studies for Canadian data collection satellites have also been conducted, although construction of such systems have not been funded. In 1977, Landsat and GOES data collection system receive capability was installed and tests conducted using commercial satellite DCPs. Tests on the suitability of using satellite data collection as a prime data collection tool are also under way.

Figure 12.8   Landsat DCP on the Severn river

12.7.3   US examples

Since 1972 the US Geological Survey, Water Resources Division, has been conducting experiments with 100 Landsat and, more recently, about the same number of GOES DCPs. The Survey has also partially funded a programme which calls for the use of about a dozen DCPs with a commercial satellite. All data that the Survey transmits by satellite are sent directly to the Survey's computing centre in Reston, Va. where the data are reformatted to simulate data from field water level and water quality recorders.

In this way it is possible to evaluate the use of satellite data collection as the primary means of collecting data, thus dispensing with the need for on-site recorders or, at least, reducing travel costs in servicing instruments by reducing the number of visits to a site. Some of the data collected via satellite is also made available to several co-operating agencies on a near real-time basis.

The New England Division of the US Army Corps of Engineers has, since 1972, operated a network of 25–30 DCPs in the northeastern United States. These DCPs have transmitted water level, meteorological and water quality data via Landsat to Waltham, Mass. The data are used mainly for river regulation and flood control in New England.

In 1975 an inexpensive semi-automatic tracking antenna was established, thus enabling the Corps to receive data directly. This local user terminal has proved nearly 100 per cent reliable.

The results obtained from the Landsat system are being compared with those from an Automatic Hydrologic Radio Reporting Network that has been in operation since 1970. Major points of comparison will be cost, reliability and operational effectiveness. Up to this time the Landsat system compares favourably.

12.8 CONCLUSIONS

During the 1970s, the concept of relaying small quantities of hydrologic data from national networks of water resources stations to data collection agencies using polar orbiting or geostationary satellites has become increasingly attractive. Experiments with various satellite systems have shown that data can be retransmitted reliably and distributed to users on a near real-time basis.

The DCPs that are installed at the gauging stations have proved to be reasonably priced, easy to install and interface with sensors, trouble free and able to withstand extreme environmental conditions. Data reception and distribution presents no major problems, operators of small numbers of DCPs may use the facilities of the satellite operator.

In addition to meeting real time data needs, it appears that satellite telemetry can also be used in basic data collection. Gauging stations may then be visited only as required since the operator is aware of conditions at the gauging station at all times.

12.9 ACRONYMS AND INITIALS

BCD	Binary Coded Decimal
CNES	Centre National d'Etudes Spatiales
CDA	Command and Data Acquisition
DCP	Data Collection Platform
DCS	Data Collection System
ERTS	Earth Resources Technology Satellite
ESA	European Space Agency
ESSA	Environmental Survey Satellite
FGGE	First GARP Global Experiment
GARP	Global Atmospheric Research Program

GOES	Geostationary Operational Environmental Satellite
GOMS	Geostationary Operational Meteorological Satellite
GMS	Geostationary Meteorological Satellite
ITU	International Telecommunications Union
MSS	Multispectral Scanner
NASA	National Aeronautics and Space Administration
NESS	National Environmental Satellite Service
NOAA	National Oceanic and Atmospheric Administration
OPLE	Omega Position Location Equipment
RAMS	Random Access Measurement System
RBV	Return Beam Vidicon
SMS	Synchronous Meteorological Satellite
SSARR	Streamflow Synthesis and River Regulation
TIROS	Television Infrared Observation Satellite
TWERLE	Tropical Wind, Energy Conservation, and Reference Level Experiment
VHRR	Very High Resolution Radiometer
VISSR	Visible and Infrared Spin-Scan Radiometer
WEFAX	Weather Facsimile
WMO	World Meteorological Organization
WWW	World Weather Watch

ACKNOWLEDGEMENTS

Figures 12.1 to 12.4 originally appeared in a paper by the author in the WMO *Casebook on Hydrological Network Design Practise* (WMO No. 324) and appears in this text through the courtesy of the World Meteorological Organization. Figures 12.6 and 12.7 are presented through the courtesy of the National Environmental Satellite Service and the New England Division, U.S. Army Corps of Engineers, respectively.

BIBLIOGRAPHY

Araya, Maurisio, 1976. Programs de empleo de sistemas de collection de detos por satellite en Chile, Division NASA–Universidea de Chile, Santiago, Chile.

Buckelow, Timothy D., 1976. Operation of Landsat automatic tracking system, New England Division, US Army Corps of Engineers, Waltham, Mass., USA.

Castruccio, Peter A. and Loats, Harry L., 1977. Data collection system requirements correlation study, Type III final report to NASA, ECO systems International Inc., Gambrills, Md., USA.

Cooper, Saul and Ryan, Philip T. (Eds.), 1973. Data collection system, Earth Resources Technology Satellite-1, Proceedings of Workshop (30–31 May, NASA Wallops Flight Center), NASA SP-364, 132 pp. National Aeronautics and Space Administration, Washington, DC, USA.

Delmas, G., 1975. Présentation du project ARGOS, Centre Nationale d'études spatiales, le 31 mai, Toulouse, France.

Edelson, Burton, I, 1977. Global satellite communications, *Scientific American*, February, New York, NY, USA.

Exner, Michael L., 1977. Use of the GOES/DCS for data collection in remote areas, Report to US Bureau of Land Management, Synergetics, Boulder, Colo., USA.

Halliday, R. A., 1975. Data retransmission by satellite for operational purposes, Proceedings of International Seminar on Modern Developments in Hydrometry (8–13 September, Padova, Italy), vol. II, pp. 490–499, WMO No. 427, World Meteorological Organization, Geneva, Switzerland.

Halliday, R. A., 1977. Hydrologic relay by satellite from remote areas, Proceedings of World Water Conference, Technical and Scientific Sessions on Water Resources (14–25 March, Mal del Plata, Argentina), in press.

Heaslip, George B., 1976. Satellites viewing our world: the NASA Landsat and the NOAA SMS/GOES, *Environmental Management*, **1**(1), Springer-Verlag, New York, NY, USA.

Higer, A. L., Coker, A. E. and Cordes, E. H., 1973. Water-management models in Florida from ERTS-1 data, Proceedings of Third Earth Resources Technology Satellite-1 Symposium (December 10–14, Washington, DC, NASA SP-351), National Aeronautics and Space Administration, Washington, DC, USA.

Kallio, Nicholas A., 1976. Operational hydrometeorological data collection system for the Columbia river, US Geological Survey, Portland, Oregon, USA.

Kite, G. W. and Reid, I. A., 1976. Discrete sampling of hydrologic data, Environment Canada, Ottawa, Ontario, Canada.

Leader, Ray E., 1974. Meteor burst communications, Proceedings of 42nd Annual Meeting of the Western Snow Conference (16–20 April, Anchorage, Alaska, USA).

Ludwig, H., 1974. The NOAA Operational Environmental Satellite System–status and plans, Proceedings of 6th Conference on Aerospace and Aeronautical Meteorology (12–15 November, El Paso, Texas), Amer. Meteor. Soc., Boston, Massachusetts, USA.

Lystrom, David J., 1972. Analysis of potential errors in real-time streamflow data and methods of data verification by digital computer, US Geological Survey Open-File Report, Portland, Oregon, USA.

Mathematica, Inc., 1974. The potential market for satellite data collection and localization services in the United States (27 September, 94 pp., Princeton, New Jersey, USA.

Moody, D. W. and Preble, D. M., 1975. The potential impact of satellite data relay systems on the operation of hydrologic data programs, Proceedings of World Congress on Water Resources (12–17 December, New Delhi, India), vol. V, pp. 261–273, Indian Committee for IWRA, New Delhi 110001, India.

NASA, 1976. *Earth Resources Technology Satellite Data Users Handbook*, Document No. 76SD4258, Goddard Space Flight Center, Greenbelt, Maryland, USA.

Nelson, Merle L., 1975. Data collection system geostationary operational environmental satellite—preliminary report, NOAA Technical Memorandum NESS 67, 48 pp. National Ocean and Atmospheric Administration, National Environmental Satellite Service, Washington, DC, USA.

Painter, J. Earle and Seitner, J., 1973. Environmental monitoring via the ERTS-1 data collection system, Proceedings of American Astronautic Society Conference (19–21 June).

Painter, R. B., 1973. The potential application of satellites in river regulation, *Water and Water Engineering*, **77**(934), 487–491.

Paulson, Richard W., 1973. An evaluation of the ERTS data collection system as a potential operational tool, Proceedings of 3rd Earth Resources Technology Satellite-1 Symposium, vol. I, Section B (10–14 December, Washington, DC, NASA SP-351, Paper W-8, pp. 1099–1111, National Aeronautics and Space Administration, Washington, DC, USA.

Paulson, Richard W., 1975. Use of earth satellite technology for telemetering meteorological station data, Proceedings of International Seminar on Modern Developments in Hydrometry (8–13 September, Padova, Italy), vol. II, pp. 476–489, WMO 427 World Meteorological Organization, Geneva, Switzerland.

Paulson, Richard W., 1976. Use of earth satellites for automation of hydrologic data collection, US Geological Survey, Reston, Va., USA.

Persoons, E. *et al.*, 1974. Telemeasurement in hydrology, Proceedings of International Seminar on Water Resources Instrumentation, 4–6 June, vol. 1, pp. 243–253, International Water Resources Association, Champaign, Illinois, USA.

Reid, I. A., Terroux, A. C. D. and Halliday, R. A., 1976. Preliminary results using GOES data collection system, Environment Canada, Ottawa, Ontario, Canada.

Robinove, Charles, J., 1969. Space technology in hydrologic applications, Proceedings of 1st International Seminar for Hydrology Professors 13–25 July, vol. 1, pp. 88–107, Dept. of Civil Engineering, University of Illinois, Urbana, Illinois, USA.

Santeford, Henry S., 1976. Meteor burst communication system—Alaska winter field test program, NOAA Technical Memorandum NWS HYDRO-30, National Weather Service, Silver Spring, Md., USA.

Schumann, H. H., 1973. Hydrologic applications of ERTS-1 Data Collection System in Central Arizona, Proceedings of 3rd Earth Resources Technology Satellite-1 Symposium, vol. I, Section B (10–14 December, Washington, DC, NASA SP-351, pp. 1213–1223, Paper W-14. National Aeronautics and Space Administration, Washington, DC, USA.

Sharp, Warren L., 1976. Water control management using earth satellites, US Army Corps of Engineers, Vicksburg, Miss., USA.

Sitborn, P., 1975. Platform location and data collection by satellite systems—the EOLE experiment, *IEEE Trans. on Geoscience Electronics*, **GE-13**(1), 2–17, IEEE, New York, NY, USA.

Spohn, C. A. and Puerner, J. H., 1976. Role of geostationary satellites in data collection and relay during the first GARP global experiment, National Environmental Satellite Service, Washington, DC, USA.

Strangeways, Ian, 1975. Telemetering river level from a large, remote tropical area, Proceedings of International Seminar on Modern Developments in Hydrometry (8–13 September, Padova, Italy), vol. II, pp. 304–310, WMO 427, World Meteorological Organization, Geneva, Switzerland.

Vannostrand, G. C. and Meyerson, G., 1973. Data collection platform/field test set program (April 23), NTIS No. E73-11008, NASA No. CR-133801, 42 pp., General Electric Company, Beltsville, Maryland, USA.

Walker, B. A., Card, M. L. and Roscoe, O. S., 1973. A planning study for a multi-purpose communications satellite serving northern Canada, Proceedings of Satellite Systems for Mobile Communications and Surveillance (13–15 March, London, England).

Ward, P. L. *et al.*, 1974. Development and evaluation of a prototype global volcano surveillance system utilizing the ERTS-1 satellite data collection system (February), NTIS No. E74-10689, NASA No. CR-139222, 168 pp., National Center for Earthquake Research, US Geological Survey, Menlo Park, California, USA.

Water Survey of Canada, 1971. Namakan Lake at Kettle Falls, Radio system for transmission of water level data, Internal Report (November), Water Survey of Canada, Dept. of the Environment, Winnipeg, Manitoba, Canada.

Wolff, Edward A. *et al.*, 1975. Satellite data collection user requirements workshop (21 May), Draft Final Report, National Aeronautics and Space Administration, Greenbelt, Maryland, USA.

World Meteorological Organization, 1974. Guide to hydrological practices, pp. 4.5–4.9, WMO-No. 168, Geneva, Switzerland.

World Meteorological Organization, 1975. Information on meteorological satellite programmes operated by members and organizations, WMO-No. 411, Geneva, Switzerland.

CHAPTER 13

River Flow in Arid Regions

J. RODIER AND M. ROCHE

13.1 INTRODUCTION

Often little is known about the surface hydrology of arid regions. Flow measurement presents a particular problem in such areas, and even when carefully made the irregular character of runoff in both time and space constitutes a serious obstacle in the determination of the parameters of the hydrological regime. These regions, however, can only be satisfactorily developed with exceptional effort to obtain the optimum utilization of existing resources as a whole, including surface water resources. A rational study of surface runoff in arid zones is often effected by means of a series of co-ordinated measurements, of which the measurement of flow constitutes an important part.

Some details concerning these measurements, however, can only be properly understood if they are placed in their general context, as defined by the natural conditions of the environment and the strategy to be implemented in making them.

13.2 GENERAL CHARACTERISTICS OF THE HYDROLOGICAL REGIME IN ARID ZONES

Despite the establishment of some quantitative indices suggested by Martonne, Koppens and Thornthwaite it will not be attempted here to offer an exact and quantitative definition of arid and semi-arid regions. It is not always a simple matter to classify all arid regions of the world since aridity, although related to low values of annual rainfall in relation to potential evaporation, depends on many other factors such as temperature, concentration of rainfall in time, seasonal rainfall, wind and geomorphological data.

Generally, however, it may be stated that in countries with a mean annual temperature of more than 18°C, arid regions are those which have an annual rainfall below a vaguely defined limit of between 150 and 200 mm, and semi-arid regions are those with a rainfall limit of 500–600 mm. In countries with a mean annual temperature lower than 18°C, those zones which receive less than 100 mm of rainfall per year may be considered arid zones and those receiving less than 300 mm per year may be considered semi-arid zones.

The limits given above, however, may be appreciably modified in a certain number of special cases, notably when the water resources are concentrated in narrow valleys (palm groves and river oases).

13.3 CHARACTERISTICS OF SURFACE RUNOFF

In arid regions, runoff only occurs when the conditions are favourable, such as on soils of low permeability and on steep slopes, and on a few days per annum. Runoff is made up of one or more flood flows between which events the river bed is dry with the possible exception of instances of significant subterranean resources. Appreciable base flow may occur in some cases which may last all the year round. Flood flows, on the other hand, may occur over a short season, for example during two months in summer in some tropical arid zones, or in autumn or spring with low probability of occurrence in winter and summer, as in the north of the Sahara. Figure 13.1 shows three typical rivers in arid regions.

Irregularity of runoff is extreme from one year to the next, as in the case of the Enneri Bardague at Bardai (Tibesti massif, Saharra, mean annual rainfall

(a)

(b)

Figure 13.1 (a) Mountain torrent (Ennedi, Chad); (b) Wadi Guir (Morocco) from the air; (c) Rude Bahn (Baluchistan) from the air

(c)

15–20 mm), where maximum values for flood flows over a nine-year period were as follows:

1954	$425 \, \text{m}^3 \, \text{s}^{-1}$	1959	$0 \, \text{m}^3 \, \text{s}^{-1}$
1955	0	1960	≥ 5
1956	0	1961	5
1957	0	1962	3 flood flows:
1958	≥ 5		$4 \, \text{m}^3 \, \text{s}^{-1}, 9 \, \text{m}^3 \, \text{s}^{-1}$
			and $32 \, \text{m}^3 \, \text{s}^{-1}$

Although the above example may illustrate an extreme case, nevertheless variations such as these are not uncommon.

Floods resulting from intense rainfall are often of short duration and torrential. The lack of vegetation may lead to considerable soil erosion, especially when the floods are the result of convective storms and the flows are sediment laden. In addition, hydrographic degeneration may occur, a phenomenon in arid regions which has considerable importance in the evaluation of flood flows.

13.4 HYDROGRAPHIC DEGENERATION

For a river in a normal river system the bed is well defined and each tributary contributes to the main river as far as the estuary or the delta. There is minimal loss of water if the river regime is not interfered with by man. However, when degradation is present the continuity of runoff is no longer obvious any more than the continuity of the bed itself. Minor floods sometimes produce no runoff and simply moisten the permeable sedimentary deposits of the bed downstream of the basin.

As soon as the slope reduces, the river overtops the banks and is then lost in the flood plains when only a small fraction returns to the river as runoff. Effluent arms leave the bed and are generally lost in marshy depressions. The river may reach inner delta zones from whence only slight runoff occurs towards the downstream area.

As a general rule, when there is significant slope at the head of the basin, and the soil is not too permeable, fine channels are formed with well-defined beds close to the watershed. These channels rapidly increase in size with distinct low banks as the flow becomes concentrated.

On reaching a zone with slight slope the bed is generally unstable, producing a lowering of the banks with one or more depressions becoming detached from the main bed. The runoff generally ends in a short delta which takes the remaining solid and liquid deposits into a clay-bottomed depression. Often, however, before reaching this stage the principal bed may join a major depression and, if the slope of the latter is steep enough, regular and possibly fairly low runoff may be observed, although possibly fairly small, especially if

Figure 13.2   Rapid hydrographic degeneration near the Piedmont, Archei pond (Ennedi, Chad)

the depression is covered with trees and bushes in their natural state. Runoff in the principal depressions is fed by successive tributaries and terminates at the point where their flows become inadequate.

If these tributaries are torrential they may direct their flows both upstream and downstream on reaching the main depression. As soon as all the swampy hollows in the depression have been filled, runoff occurs from the group as a whole towards the downstream area. In a desert regime the greater the annual runoff the more its downstream limits enter the main depression. At the downstream end of this main collecting area, runoff sometimes occurs only every 50 or 100 years.

Figure 13.3   Hydrographic degeneration—Rude Bahn downstream of Bahn Kalat (Baluchistan) from the air

It is this phenomenon that is known as hydrographic degeneration. The common notion of a drainage basin becomes less marked, and discharge in $m^3 s^{-1}$ per square kilometre no longer has much meaning. Careful account must be taken of these phenomena in setting up gauging stations and selecting sites for dams. An extreme case of hydrographic degeneration occurs when, during a flood, runoff occurs upstream at the beginning of the flood and downstream at the end of it. Figures 13.2 and 13.3 show typical examples of hydrographic degeneration.

13.5 LOGISTIC AND MATERIAL PROBLEMS PECULIAR TO ARID ZONES

Hydrological studies, especially flow measurement, are made particularly difficult in arid zones because of the sparseness of the population and because reliable observers are difficult to find.

The few who make up the permanent population are often nomadic and frequently their only interest in an instrument is shown in damaging it or making off with it. Even in developed arid countries it is sometimes necessary to set up apparatus which is bullet proof.

Access to gauging stations often poses serious problems, as very often there is no road available. Sand and sand-laden winds present problems outside mountain areas and in the latter it is even sometimes necessary to travel along the river bed when possible. However, clay zones which may be accessible whey dry become impassable when wet and the roads become dangerous for transportation.

Obviously the above problems may be substantially alleviated by spending a goodly sum of money but this is not usually available in arid countries other than in those which possess oil resources.

13.6 STRATEGIC PRINCIPLES FOR THE STUDY OF RUNOFF IN ARID REGIONS

It is not possible to install proper networks of gauging stations in arid zones, except in zones small in area and in rich countries. In semi-arid zones, however, adequate networks are possible but these of necessity are less dense than in humid regions.

In the arid and semi-arid zones the greater part of the basic data is obtained from temporary expeditions carried out over a period of two or three years. As a result of hydrographic degeneration the selection of gauging sites has to be studied with the greatest care, and it is necessary to carry out systematic measurements in order to follow the decrease in flows from upstream to downstream.

The scarcity of runoff data, even of a qualitative nature, has led hydrologists, in the course of expeditions, to try to locate traces of the latest floods or the most violent flows and to subsequently make use of these data for the calculation of discharge.

13.7 GENERAL ORGANIZATION FOR RUNOFF MEASUREMENTS IN ARID ZONES

It is possible to remedy the lack of hydrometric networks and the low raingauge density by means of a combination of the following operations:

(1) Extensive studies of the terrain of rainfall and flow over a period of two or, better still, three years.
(2) Studies on representative catchments on the structure of the rain and of the rainfall–runoff relation for the given geomorphological conditions.
(3) The installation of reference gauging stations.

13.7.1   Extensive studies of the terrain

These studies are carried out during the seasons when the probability of floods is greatest. Having set up a network of rain gauges, water level recorders and staff or ramp gauges the hydrologist proceeds around his base by the means of transport at his disposal. An example of transport by caravan is shown in Figure 13.4. He studies the geomorphological characteristics of the terrain (relief, geology, pedography, hydrogeology, the aspect of the hydrometric network) and the vegetation cover of the catchments including the river banks. During this reconnaissance the opportunity is taken to measure floods as they occur or, if the flood has recently passed, he is required to determine traces of the flood and its peak, with a view to computing the corresponding discharge from the slope area method (ISO 1070). He also gathers information on historical floods if at all possible.

Brief studies are also made on the variation in rainfall with altitude and exposure, and in very arid regions he is required to demarcate zones receiving no rain.

Figure 13.4   Caravan for extensive hydrological survey (Ennedi, Chad)

Rapid manoeuvrability is obviously essential if the best results of such a study are to be obtained. However, even if the necessary financial resources are available to utilize air transport (plane or helicopter), it is not always possible to employ this form of transport after major floods due to poor runway conditions, in the case of planes, or because helicopters are reserved for more urgent flood relief tasks. For the most part, therefore, cross-country vehicles are used for reconnaissance work. Figure 13.5 shows a typical organizational map for an extensive hydrological study in an arid zone.

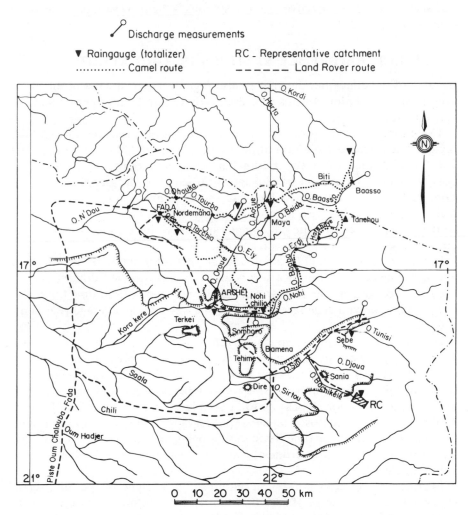

Figure 13.5   Example of field work organization for an extensive hydrological study in arid zone

13.7.2 Representative catchments

These catchments are set up to obtain rainfall–runoff relations, and generally within the framework of the main hydrometric studies. These catchments are usually small, rarely exceeding 10 km^2, and hydrographic degeneration is absent.

13.7.3 Reference stations

These stations are installed at locations where a permanent resident population exists and consist essentially of a rain gauge and a water level recorder and staff gauge. The objective of these stations is the provision of a statistical series of long-term data.

13.8 REFERENCE STATIONS IN ARID REGIONS

Because of the difficulties peculiar to arid regions, especially the problems of access, gauging stations of this type are costly to install and operate and are generally few in number.

The station should be installed on a river which is fully representative of the region. The catchment should not be too small and should extend to at least $200–300 \text{ km}^2$, an area of about 1000 km^2 being preferable. In so far as it is possible, runoff should not have been disturbed by hydrographic degeneration. These two conditions, which may seem contradictory, are aimed at facilitating correlation with other catchments. If the important resources are in the upper reaches of the catchments it may be necessary to modify the above rules and select smaller catchments.

In almost all cases it will be possible to acquire information of the effects of hydrographic degeneration if a secondary station is installed downstream of the reference station. The former will only be referred to from time to time so as to be familiar with degeneration effects in a wet year, an average year and a dry year.

Discharge measurements made at reference stations in arid zones are generally carried out by the velocity-area method, the techniques for which are described in Chapter 1. Natural controls are, however, rare although it should be pointed out that rocky sills are sometimes buried under deposits but exposed during major floods, as for example at Wadi Zeroud in Tunisia. Similar sills may be disclosed on aerial photographs and require to be verified by sounding, since at the time the station is being installed the sill is generally buried beneath the sand. Artificial controls are sometimes installed if funds are available.

In practically all cases the station requires to be designed to permit precise measurement in the range of zero flow to major flood flows. It is necessary to anticipate that flood flows may have velocities of the order of at least 5–6 m s^{-1}, and measurement by current meter is then impracticable. Discharge measurements then have to be made by floats (ISO 748) or by the slope-area method.

Peak flows of some $1000 \text{ m}^3 \text{ s}^{-1}$ should be anticipated for a catchment of 1000 km^2 and up to $10\,000 \text{ m}^3 \text{ s}^{-1}$ in extreme cases (Texas).

13.9   REPRESENTATIVE BASIN STATIONS IN ARID REGIONS

The same methodology is used for discharge measurements at representative gauging stations as for reference stations, but with more rudimentary means. In all cases it is necessary to avoid bank overflow by some remedial works such as dikes or a curtain of stakes. Bed stabilization may be carried out by the installation of gabions but these are usually limited to about 1 m^3 (see Section 13.10). Portable cableways may be used with a 50 kg sounding weight and in many cases it is usually possible to install a catwalk, as shown in Figure 13.6. Generally, therefore, although representative stations demand a measure of improvization the quality of the calibration curves is usually high.

Figure 13.6   Footbridge and water level recorder at a representative basin station: kori Iberkoum (Air, Niger Republic)

Representative stations used for special studies, however, are installed on a more permanent basis and may include an artificial control or a measuring structure.

13.10   BED CONTROLS

It is an important condition, even in arid regions, that a velocity-area station should have a stable stage–discharge relation. This involves either a natural control or a section control (see Chapter 1). In arid zones, gabions are frequently used as bed controls. These consist of baskets of thick wire netting, each with a capacity of about 3 m^3 and filled with blocks and boulders, which are anchored in the alluvia by iron bars. The top of these bars should finish level with the bed. Solid mass constructions, such as continuous concrete, should be avoided in arid zones because they are too sensitive under pressure.

Figure 13.7   Reference station of the kori Teloua—carrying out a
locking by 'gabions' (Air, Niger Republic)

Figure 13.7 shows the installation of a gabion bed control at a reference station on the kori Teloua.

13.11 WATER LEVEL RECORDER

An autographic recorder is virtually indispensable in arid regions. However, the river flow may sometimes contain more than 100 kg m^{-3} ($10\,000 \text{ mg l}^{-1}$) of sediment and intake pipes become frequently blocked even if made as large as 300 mm in diameter. Additional intakes are therefore desirable in order to

Figure 13.8   Water level recorder at Adoubdoub on Tamgak (representative basin), (Air, Niger Republic)

record higher water levels as the lower intake becomes blocked on a rising river. Measures require to be taken, therefore, to mitigate this situation and a porous filtering mass of sandy concrete (50 kg cement to 1 m^3 sand) is sometimes incorporated in the base of the stilling well to record low water levels.

The water level recorder requires to be set up so that it is protected from flood flows and trees carried down by the current. Pressure-actuated recorders, although sometimes used, are susceptible to silting up of the sensor and blockage of the conduit. Figure 13.8 shows a typical water level recorder installation in a representative catchment.

13.12 MEASUREMENT FROM BRIDGES

When a bridge is available it is sometimes convenient to use it to carry out discharge measurements. However it should be noted that:

(a) flow is usually irregular upstream of the bridge due to the high velocities;
(b) during flood flows in arid zones, bridges are especially vulnerable and it is then dangerous and sometimes prohibited to use them.

13.13 MEASUREMENT FROM CABLEWAYS

It is generally preferable to install cableways, since it is seldom practical to carry out measurements from boats in arid zones. Manoeuvrability is dangerous in velocities over 3.5 m s^{-1}, even for the expert.

Cableways used are similar to those described in Chapters 1 and 8. Due to the inexperienced local staff available, however, the remotely controlled type operated by a winch from the bank is generally preferred.

13.14 MEASUREMENT WITH FLOATS

At certain stations it is sometimes necessary to abandon measurements by current meter and to proceed to float measurements.

For float measurements a measuring reach should be selected where the flow is as uniform as possible and close to the cableway. Six reference points are chosen, three on each bank at the upstream end, downstream end, and at the centre of the reach (ISO:748). It is desirable for the length of the measuring reach to be about 3–5 times as long as the width of the river. A temporary telephone line or walkie-talkie set is useful for communication between the upstream and downstream reference points.

Various types of float are described in Chapters 2 and 8 but in arid zones it is sometimes convenient to use floating half-submerged trees descending with the current (on Wadi Zeround, 600 were counted on 27 September, 1969). This method is only appropriate, however, for deep rivers about 100 m wide.

Other floats include lemonade bottles and wood blocks. However, even in good conditions these floats are difficult to observe.

Float measurements are generally carried out in velocities of the order of 4–7 m s^{-1} and if carefully performed can yield acceptable results. If the float is adequately immersed and wind influence is negligible then the float speed is taken as the surface velocity of the water. For rivers with sandy beds and depths exceeding 3 m or 4 m on average, the coefficient to reduce surface velocity to average velocity is about 0.92 to 0.95. For greater reliability, however, the relation between mean velocity in the vertical and surface velocity is found from current meter gauging.

Floats are observed over periods of 10–15 min and during the measurement the corresponding stage is noted. A curve of surface velocities is established from which a mean surface velocity is deduced. The mean surface velocity is then multiplied by the reduction coefficient, corresponding to the appropriate stage, and by the cross-section area to provide discharge. The position of a float in the river may be found by employing two stopwatches. One measures the time taken for the float to pass from the upstream cross-section to the downstream cross-section (t_1) and the second measures the time taken (t_2) for the float to cross the diagonal between the upstream reference point on the left bank and the downstream reference on the right bank. If the upstream and downstream cross-sections are parallel then from Figure 13.9

$$l_1 = L_1 \frac{(t_1 - t_2)}{t_1}$$

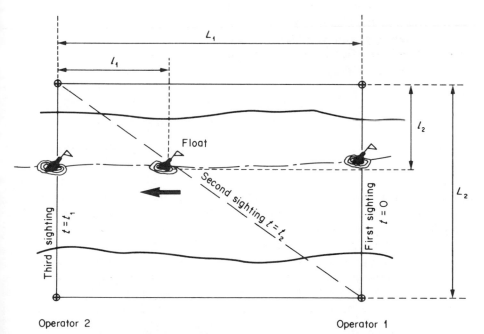

13.9   Velocity measurement by floats (position fixing)

and

$$l_2 = L_2 \frac{t_1 - t_2}{t_1}$$

Alternatively, and more conveniently, the position of the float may be found by using the cableway and lowering the sounding weight close to the water surface. The position of the float may then be estimated by sighting the float in relation to the sounding weight.

13.15 MEASUREMENT OF LOW FLOWS

In arid zones it is advisable to determine the reading of the water level recorder or staff gauge at zero flow. This will probably be one measurement in arid zones which may be determined with the greatest precision.

However, it should be noted that in beds of alluvium, runoff may cease at the measuring section and may reappear further downstream. This sort of resurgence occurs when the mass of alluvium no longer offers a sufficient depth for bed flow.

Measurement of very low flow (less than $0.2 \text{ m}^3 \text{ s}^{-1}$, for example) in a stream which yields more than $1000 \text{ m}^3 \text{ s}^{-1}$ in flood often demands special improvization. The usual gauging station is seldom sensitive enough, even if an artificial control has been installed. A small permanent canal may be constructed on which a portable weir may be used to gauge the low flows. Alternatively a small temporary canal may be used and a miniature current meter employed to measure velocities.

Sometimes a second water level recorder is required with a natural scale to record the lowest water levels, or discharge measurements are made at regular intervals and by interpolation between them a complete recession is established.

13.16 MEASUREMENT OF FLOOD FLOWS

Flood flows too great to be measured by current meter may also be measured by the slope-area method (see Chapter 5). To determine the flow by this method it is necessary to approach as closely as possible the hydraulic conditions necessary for the application of the Manning formula. The measurement reach should be as rectilinear as possible, the cross-section within the reach should vary as little as possible and the bed slope should be fairly regular. The reach should be of sufficient length for the slope to be significant. High water staff gauges require to be installed upstream and downstream of the measuring reach at a distance apart equal to at least 75 times the mean depth of the anticipated flood (see Chapter 5). It is usually wise to duplicate these staff gauges with crest-stage gauges, all gauges being levelled-in to convenient bench marks.

The uncertainty in the result of a slope-area measurement depends to a significant extent on the uncertainty in the measurement of slope and the uncertainty in the value used for Manning's *n*. A suitable precision level should be used to install staff gauges and preferably one with an automatic level setting device, especially if, as is often the case in hydrology, the operator is not a professional surveyor. The slope measured should be that of the surface water slope at the flood peak. If the staff gauges are at the same datum, the difference in readings divided by the distance between them is equal to the slope. The advantage of the slope-area method is its ability to obtain a measurement of discharge of past floods by using flood marks or traces. This is an important hydrological facility in arid zones.

Flow is calculated from the Manning formula as follows:

$$Q = \frac{1}{n} R_h^{2/3} S^{1/2} A$$

where Q is the discharge $(m^3 s^{-1})$; S the slope (in fraction form); R_h the hydraulic radius (m); A the wetted area (m^2); and n is the coefficient of roughness. For values of n see ISO 1070 and Benson (1968).

13.17 INVESTIGATION OF FLOOD LEVELS

There are many arid zones or even desert zones where permanent villages exist. Rural populations are often greatly motivated by flood flows as a result of the damage they cause to their fields and homes. Life goes on close to watercourses the more arid the region and 'valley civilization' is one form of life based on the oasis. These valleys are not always majestic rivers like the Nile, but more frequently ill-fated wadis. As far as possible, the investigation should be followed on a regional plan at least within the framework of a drainage basin, which permits comparisons and verifications which are always fruitful (regional enquiry). When possible it is a good idea to return to the same village several times and to question the largest number of witnesses possible, using different interpreters where this is feasible (local enquiry).

The local enquiry may centre on just the greatest known flood, but it may go further and seek to gather information about all the most significant flood flows. Attempts are made therefore to establish chronology. The memory of river dwellers is often excellent concerning this type of event, but is typically that of the countryman. It is always very difficult to discern chronology, even for relatively recent events (over a period of about 10 years, for example). The date of a flood (the year at least) may only be learned accurately if it coincides with some event which stands out in the life of a witness, the birth of a child for instance, departure on military service, or if it marks something special in village life or even some event of political interest for the country (war especially).

The regional enquiry is not always solely a series of local enquiries, it may appeal to other sources of information, administrative archives, newspaper articles, travellers' accounts, and so on, and naturally data from regularly observed stations, if sufficiently old-established ones exist.

The local enquiry should be concerned with both chronology and levels of the flood flows. For this aspect of the enquiry it is necessary to consider diverse evidence and places of observation. For the greatest flood in living memory (providing that it is not too far back, say not more than 20 or 30 years) it frequently happens that good agreement of evidence is obtained. It is customary to have the flood levels pointed out on both banks of the watercourse, but with exceptions the most coherent indications provided are for the one on which the village is sited. For each level indication it is necessary to make a careful examination of its location and surroundings and more particularly the local conditions of flow.

At the time of the local enquiry a topographical survey is made on the terrain and the checking of bed variation, particularly of cross-section profiles, takes on greater importance the more historical the flood. During one enquiry, it was observed at one site that all the cultivated terraces had disappeared increasing the width of the apparent bed from a few metres to more than 100 m.

In uninhabited regions it is possible to refer to ancient traces, but one obviously needs to be cautious. The best traces may be observed in gorges, in the shape of continuous lines over vertical rocky walls; these traces are extremely varied in nature. Some information may occasionally be extracted from considering levels of erosion of banks which are relatively free from crumbling. Alluvium deposits may also furnish some indications. Frequent mention is made of cases of clay deposits or silt in the rock fissures and hollow trees. These elements must be carefully examined and their origin duly criticized. This involves flair and common sense, as scarcely any general methodology exists. Fairly frequently one observes deposits of floating debris which is relatively and sometimes even extremely coarse in texture. The location of some of this, in caves or holes hollowed out in the rocky walls of gorges, is at such a depth that there are difficulties of interpretation. Flood marks of this kind are rarely able to serve as a basis for serious evaluation.

13.18 DILUTION GAUGING

If a chemical is injected into a river with a flow rate q and concentration c it will become diluted in the river water if sufficient mixing is present. If C is the concentration of the chemical in the river then, from Chapter 4,

$$Q = q\frac{c}{C}$$

where Q is the discharge.

This method is used for measuring relatively low to medium discharges. However, faced with the considerable difficulties in current meter gauging, which are so often encountered in arid or semi-arid regions, the dilution method has been applied to the measurement of floods.

The high sediment load, however, the rapid variation in flow during a flood and occasionally the quality of the water, are technical obstacles to the use of this method. Nevertheless it has been possible in certain instances to overcome these problems and in Madagascar a discharge of some 2200 m^3 s^{-1} was measured with sodium dichromate to an accceptable degree of uncertainty. However, the method requires specialized staff which are rarely available in arid regions. For this reason tests carried out in the north and south of the Sahara scarcely reached the operational stage. In addition the cost of the large amount of dichromate required was most discouraging. It is possible, however, in a country which has both semi-arid and arid regions to form a specialized team which could be sent in emergency to a gauging station where there could be stored a permanent stock of chemical for dilution gauging. Even so, the major technical difficulty is still the rapid change in discharge during a flood.

13.19 MEASUREMENT OF CHANGE IN BED LEVEL DURING FLOODS

If the natural bed is unstable or if it has not been artificially stabilized, it is necessary to be able to determine its actual level at the time of the flood peak. This may be performed by burying a sounding weight of about 25 kg in the alluvium, to a depth greater than it is expected the bed will attain at the time of maximum scouring. To this weight a chain is fastened which will just reach the bed level prior to the flood. In order that the chain can be recovered again, it is lengthened by means of a cord to which is attached a float. Up to a dozen of these devices are installed in the cross-section. The section of chain which is cleared by the flood by removal of the surrounding alluvium follows the direction of the current and takes up a position approximately at a right angle to the remaining buried portion of the chain. The entire chain is buried once again by alluvium as the flow subsides. The chain may finally be exposed from each of the devices by digging and the minimum bed level ascertained. The same result may be obtained by burying a pile of bricks in the bed, the layers of top bricks being washed away by the current.

The measurement of depths in arid regions, however, generally pose special problems where beds are changing and are unstable. Frequently if the bed configuration is altered with each flood flow the mean depth remains the same and, except for low flows, the calibration curve remains unchanged. However, in some cases a violent flood may gouge out the bed by several metres and it may then be necessary to wait some years in order to establish the same cross-section profile which existed prior to the flood. This phenomenon is experienced in the semi-arid region of Madagascar and on many rivers in the arid and semi-arid regions of North Africa. An example of such a flood on the Wadi Zeroud occurred during a period of four days in September 1969, where the surface velocities exceeded $8 \, \text{m s}^{-1}$ and the bed levels during the flood varied from about $+1.7$ m to about -10 m.

13.20 CONCLUSIONS

Every endeavour is made in arid regions to apply standard methods of hydrometric principles and practices. The problems and difficulties in applying these methods, however, are considerable. Physical problems such as abnormal site conditions, hydrographic degeneration, mobile beds and rapid flow variation are severe, while staffing problems, security of installations and mobility in difficult terrain present formidable obstacles for hydrometric measurements. Nevertheless the economic development of these regions depends to a large extent on their investment in hydrometry, so that their water resources may be exploited for the benefit of the community.

BIBLIOGRAPHY

Andre, H., Richer, C. and Douillet, G., 1973. Les jaugeages par la méthode de dilution en 1970, Colloque sur l'Hydrométrie (Koblenz), Publ. IAHS 99, AISH–UNESCO–OMM.

Barnes, H. H., Jr., 1974. Measurement of flash floods in the United States—State of the art, Symposium on Flash Foods (Paris), Publ. IAHS 112.

Benson, M. A., 1968. Measurement of peak discharge by indirect methods. WMO Technical note 90, Publ. WMO 225, TP. 119, Geneva.

Benson, M. A., 1973. Measurement and estimation of flood discharges, Symposium on Hydrometry (Koblenz, 1970), Publ. IAHS 99, IAHS–UNESCO–WMO.

Cruette, J., 1974. Méthodologie pour la mesure des crues brutales, Symposium sur les Crues brutales (Paris), Publ. AISH 112.

Cruette, J. and Rodier, J. A. (1971). Mesures de débits de l'Oued ZEROUD pendant les crues exceptionneles de l'Automne 1969, Cahiers ORSTOM, Série Hydrologie, vol. VIII, No. 1, Paris.

ISO: 1070, 1973. Liquid flow measurement in open channels—slope-area method. International Organization for Standardization, Geneva.

Kornitz, D., 1973. Hydrometric stations in arid zones, Symposium on Hydrometry (Koblenz, 1970), Publ. IAHS 99, IAHS–UNESCO–WMO.

Riggs, H. C., Flash flood potential from channel measurements, Symposium on Flash Floods (Paris), Publ. IAHS 112.

Smith, R. E. and Chery, D. L., Jr., 1974. Hydraulic performance of flumes for measurement of sediment laden flash floods, Symposium on Flash Floods (Paris), Publ. IAHS 112.

Zebidi, H. and Kallel, R., Exploitation d'un réseau d'alerte: son utilisation pour l'étude des crues brutales (Paris), Publ. AISH 112.

CHAPTER 14

Aerial Methods of Measuring River Flow

V. V. KUPRIANOV

14.1 INTRODUCTION

Methods of measuring discharge by means of aeroplanes have been under development in the USSR since 1965. Some 4000 discharge measurements have been carried out in rivers where existing methods are unsuitable due mainly to the difficulty of access. Two methods of aerial measurement are employed, one which measures the surface velocity by floats and the other which uses the method of vertical velocity integration. In the first method, floats are dropped from an aeroplane and their paths traced by aerial photography; in the second method, a hydrobomb is dropped from the aircraft and oil, released when the bomb hits the bed, traces the distribution of velocity.

14.2 GENERAL

Aerial methods have become more and more popular in hydrometric research and indeed have become recognized practice in the USSR for measuring floods and observing the drifts in isolated areas. Aircraft are also used in surveying and in the photography of glaciers and swamps. Special instrumentation is now available and the field work has become routine, the major bulk of the work being transporting the data to the laboratory and processing. The method is particularly applicable in the field of river flow measurement where 10–15 gauging stations can be covered by one aeroplane thereby offering a considerable saving in staff. The method is not restricted by river width and has the distinct advantage of making gaugings in isolated areas which otherwise would be practically impossible. Aerial hydrometric methods of measuring river discharge depend on the successive photographing of special devices dropped from the plane into the stream. The discharge is then computed from the photographic data.

14.3 SURFACE-VELOCITY METHOD

In practice, surface floats containing special dyes, for better differentiation on photographs, are most widely used. The floats are photographed twice during travel and, as the photograph scale is known, the travel distance L during time Δt is determined. The numerical value of the surface velocity v_n is calculated from

$$v_n = \frac{L}{\Delta t}$$

Discharge is calculated by summing the partial discharges $v_{ni}h_i$ (where h_i is the depth of flow) in the usual way and then multiplying by a reduction coefficient, K, to reduce the surface velocity to mean velocity. If Q is the total discharge, n the number of segments, and b the width of each segment, then

$$Q = K \sum_{i=1}^{i=n} 0.5(q_i + q_{i+1})b_i \qquad (14.1)$$

where $q_i = v_{ni}h_i$.

The coefficient K may be obtained from current meter measurements or from empirical formulae. If the latter method is used preference is given in the USSR to the following formula derived by G. V. Zheleznyakov:

$$K = \frac{(2.3\sqrt{g} + 0.3C)C}{(2.3 + \beta_*)g + 0.3C)C\beta_* g} \qquad (14.2)$$

where C is the Chezy coefficient, and β_* is the stream shape factor determined from the ratio of the mean depth of flow to the maximum depth of flow at the given gauging station. Values of K from equation (14.2) have been tabulated for convenience. Alternatively, values of K may be obtained from Table 14.1, also suggested by Zheleznyakov.

In recent years the State Hydrological Institute has gathered a large amount of field data and from this information I. F. Karasseff has suggested the following empirical formula for K:

$$K = 0.77 + 0.043\,(\tilde{C} - 3.8)^{1/2} \qquad (14.3)$$

in which $\tilde{C}$ is a dimensionless expression for the Chezy coefficient, where $\tilde{C} = C/\sqrt{g}$. Equation (14.3) is valid for $C > 12^{1/2}$ when it gives a minimum value of $K = 0.77$.

Table 14.1   Values of K for different channel characteristics

Channel (floodplain) characteristic	Mean depth (m)		
	<1	1–5	>5
Channels straight, clear, earthen (clay, sand), shingle, gravel	0.80	0.84	0.86
Channels meandering, partially overgrown by grass, stony. Floodplains—relatively well developed, with vegetation (grass, sparse undergrowth)	0.75	0.80	0.83
Channels and floodplains—considerably overgrown, with deep scours. Channels—meandering, consisting of large cobblestones	0.65	0.74	0.80
Floodplains—completely covered by forest of the taiga type	0.57	0.69	0.75

The complexities involved in estimating a value of K for particular field conditions are such that the uncertainty in K is of the order of ±10 per cent. The choice of formula is therefore largely irrelevant. More accurate values can, however, be obtained from field data and an empirical relation so established for individual rivers. Figure 14.1 demonstrates the application of

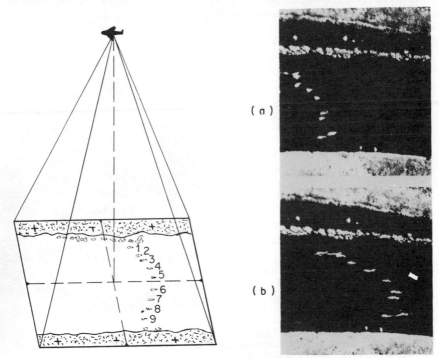

Figure 14.1   Aerial determination of the surface velocity using repetitive aerial photography: (a) first position of floats; (b) second position of floats

the surface velocity method from floats dropped from an aircraft. This method requires a preliminary survey of the measuring sections to have been made. On small and medium-sized rivers, air survey may be substituted by photography from the bank taken from a high level or from a specially built tower. This method is generally applicable under conditions of intensive ice flows or drifting timber, both of which may serve as floats. The floats are photographed by a special camera installation, the optical axis of the camera being at a fixed angle of inclination above the horizon.

14.4 INTEGRATED-VELOCITY METHOD

In this method several devices, known as 'hydrobomb' floats, are dropped from an aeroplane. These hydrobombs in coming in contact with the bed eject a special oil which follows the velocity distribution both in the horizontal and the vertical. Aerial photography fixes the point the hydrobomb float hits the surface and photographs the path which the oil traces. Figure 14.2 demonstrates the procedure and is self-explanatory. The integration method does not require a preliminary survey of the measuring section although ground markers are required to determine the scale of the photographs. The choice of site is made by direct flights over any given area.

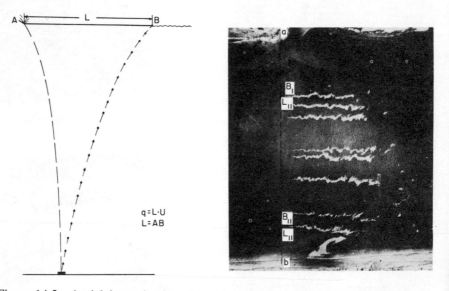

$$q = L \cdot U$$
$$L = AB$$

Figure 14.2   Aerial determination of discharge using the integration method of computation: ab is the hydrobombs' dropping line; U the speed of rising oil; q the elementary discharge; B the point where oil surfaces; and AB $= L$ is the distance from the point of plunge to the point of surfacing

14.5 ACCURACY

The uncertainty in a single determination of discharge for both the surface-velocity method and the integration method is estimated at between 10 and 15 per cent (95 per cent confidence level). In many cases, however, aerial methods compare favourably with the velocity-area method performed by current meter and sometimes they may even be better. This is especially the case in rivers having wide flood plains, where routine current meter measurements may lead to significant error. Other conditions being equal the percentage uncertainty in aerial measurements increases as velocities decrease. Discharge measurements in flood plains are therefore of less accuracy than in channel measurements.

Investigations have shown that under average conditions of an air survey scale of $1:5000$, wind speed of $2-3\,\mathrm{m\,s^{-1}}$ and symmetrical distribution of kinematic and morphological stream elements about the centre of the photograph, the uncertainty in discharge measurement in the main channel is of the order of 5–6 per cent for stream velocities of $1-2\,\mathrm{m\,s^{-1}}$. Under the same conditions in flood plains, but with stream velocities of $0.1-0.5\,\mathrm{m\,s^{-1}}$, the uncertainty increases to about 10–15 per cent.

The use of aerial methods is limited under unstable stage–discharge conditions, especially if channel morphology changes between the period when a bed survey is made and when the discharge measurement is made.

14.6 CONCLUSIONS

The aerial methods of discharge measurements employed in the USSR have been carried out on over 150 rivers under varying flow conditions. The range of flow of these measurements varied from about $100\,\mathrm{m^3\,s^{-1}}$ to $100\,000\,\mathrm{m^3\,s^{-1}}$. Analysis of the data so far obtained has shown that the use of aerial methods for the measurement of discharge of medium and large rivers in the northern region and in Siberia is economically justified. Aerial methods are almost indispensable during catastrophic floods either from snow-melt or from heavy rainfall or both.

The main item of cost is of course the flight cost which amounts to some 90 per cent of the total, ground surveys making up about 5–10 per cent. For small rivers with good mixing it is expected that dyes may be used, the discharge being computed by determining the concentration of the dyes from an aeroplane by telephotometer at the measuring section where complete mixing has taken place.

It will be expedient to choose the aerial method appropriate to the site conditions or a combination of different methods.

BIBLIOGRAPHY

Anon, 1974. *Metodicheskie Rekomendatsii po izmereniu raskhodov vody rek aerometodami* (Methodological recommendations on water discharge measurements in rivers by airborne methods), Gidrometeoizdat, Leningrad, 133 pp.

Cameron, H. L., 1952. The measurement of water current velocities by parallax methods, *Photog. Eng.*, **18**(1), 99–104.

Dyer, A. I., 1970. River discharge measurement by the rising float technique, *J. Hydrology*, **XI**(2), Amsterdam.

Karasseff, I. F., 1973. Raspredelenie prodolnikh skorostey techenia v poimakh i ruslakh rek (Distribution of longitudinal current velocities in flood plains and river beds), *Trans. GGI*, **202** (Gidrometeoizdat), 3–39.

Maliavskiy, B. K., 1965. *Metodi opredelenia gidrologicheskykh kharakteristik rek s samoleta* (Methods of determination of river hydrometrical characteristics from plane). Publishing House 'Transport', Moscow.

Shumkov, I. G., 1973. Opredelenie s samoleta skorosti i napravlenia vetra nad vodnoi poverhnostyu. (Determination of wind direction and velocity above water surface from an aeroplane), *Trans. GGI*, **203**, 223–232.

Shumkov, I. G., 1973. Pogreshnosti operatsii letno-syemochnogo i kameralnogo protsessov pri izmerenii raskhodov vodi s samoleta (Errors of the aerial photographing and data processing procedure while measuring water discharges from aeroplanes), *Trans. GGI*, **202**, 39–63.

Shumkov, I. G., 1975. Opredelenie coefficienta perekhoda ot fictivnogo k deystvitelnomy raskhodu vodi s pomostchyu aerophotosyemki (Determination of transition coefficient from apparent to actual water discharge with the help of air survey), *Meterologia i gidrologia*, **N8**, 100–103.

Uida Kazimierz, 1972. Wykozzystanie zdjec fotogrametrycznych w hydrometrii, *Gaz. Obserw. PIHM*, **25**(8), 10–13.

Zdanovich, V. G. and Sharikov, Y. D., 1970. Perspectivi primenenia aerometodov dlia izuchenia raskhodov vodi (The perspective use of aerial methods for water discharge study), *Meteorologia i Hydrologia*, **N10**, 56–62.

CHAPTER 15

The Work of the World Meteorological Organization

J. NĚMEC

15.1 INTRODUCTION

Hydrometry, by definition, is in its wider sense (Chebotarev, 1964) that part of hydrology concerned with measuring and observing the elements of the hydrological cycle. In the more restricted sense adopted for this book, it is the activity concerned with the measuring of the regimes of rivers, lakes and reservoirs, including techniques and instrumentation (WMO, 1974a). It is therefore not obvious, at least at first glance, what international implications such an activity can have. Nevertheless, as also indicated in Chapter 16, there are at least two international organizations, namely the World Meteorological Organization (WMO) and the International Organization for Standardization (ISO), concerned with several aspects of hydrometry, both in its more general and its restricted sense.

The activities of the ISO, as its name indicates, consist exclusively of standardization. The WMO, however, conducts a larger variety of international activities in the field of hydrometry. Here again, the need for international co-operation has two facets (Němec, 1976). One is the need for intellectual and technical co-operation, for the transfer of technology and exchange of scientific knowledge; the second is specific to hydrology, to the character of a river basin. The watershed dividing line crosses many frontiers in international basins. Without international co-operation, promoted in the first place by international governmental organizations such as the World Meteorological Organization, the measures of flow across the border, often of much-coveted water, would be and, despite these endeavours often are, a source of international conflict.

The Operational Hydrology Programme (OHP) of the WMO was established by the Seventh WMO Congress as one of the main programmes of this intergovernmental organization. In addition to international co-operation in meteorology the WMO, early in the 1940s, accepted responsibility for international co-operation in hydrology, but the full impact of this was felt only at the beginning of the 1950s and particularly in the 1960s, with the impetus of the International Hydrological Decade launched by Unesco[1]. Thus, in 1965, the first edition of the WMO *Guide to Hydrological Practices* (then *Hydrometeorological Practices*) was published, containing the very first attempt towards standardization of hydrometric instruments and methods of measurement at the international level.

15.2   WMO'S ACTIVITIES IN HYDROMETRY

Since the publication of this Guide, the WMO's activities in the field of hydrometry may be summarized under three headings:

(1) Preparation of internationally agreed guidance material, including efforts towards standardization.
(2) Co-ordination of international projects, in particular on intercomparison of instruments and methods of measurement.
(3) Hydrometric field operations within technical co-operation projects in developing countries.

Some of these activities are briefly described below. For details, the reader is referred to the referenced WMO publications and documents. As also indicated in Chapter 16, all these activities are co-ordinated with all other governmental and non-governmental organizations concerned, in particular ISO, UNESCO, IAHS[2], FAO[3], and WHO[4] (Němec, 1976).

15.3   WMO GUIDANCE MATERIAL AND STANDARDIZATION IN HYDROMETRY

Published WMO guidance material which applies to standardization in hydrometry can be divided into three groups, as indicated in Table 15.1

[1] United Nations Educational, Scientific and Cultural Organization.
[2] International Association of Hydrological Sciences.
[3] Food and Agriculture Organization of the United Nations.
[4] World Health Organization.

Table 15.1 WMO publication series as applied to standardization in the field of hydrology

Type of WMO publication	User of publication	Level of standardization of publication; adopted by:
1. Technical Regulations Content: Definitions of technical terms, recommendations concerning hydrological observing networks and stations; hydrological observations, warnings and forecasts as well as meteorological observations and forecasts for hydrological purposes	Generally applicable to all Hydrological Services	Highest level of standardization. It is desirable that given rules be implemented by all national Hydrological Services Adopted by: Congress
1(a). Annexes to Technical Regulations* Content: All necessary detailed standard and recommended practices and procedures to support the appropriate Technical Regulations	As (1) above	Adopted by: Executive Committee if authorized by Congress
1(b). Appendices to Technical Regulations* Content: Texts which are appended to the Technical Regulations and have the same status as the Technical Regulations to which they refer	As (1) above	As 1(a) above
2. Guide to Hydrological Practices Content: General information about practices, procedures and instrumentation for carrying out hydrological work. Not a manual but rather a supplement to Technical Regulations	Same as above but less needed by Services having national guides	Next to highest level of standardization. The practices, procedures and instrumentation described are recommended as reliable for implementation by all national Hydrological Services Adopted by: Commission for Hydrology

[continued overleaf

Table 15.1 (cont.)

Type of WMO publication	User of publication	Level of standardization of publication; adopted by:
2(a). Annexes to the Guide to Hydrological Practices Content: Now used for material not sufficiently standardized for main text of the Guide	Applicable mainly to all Services without national operational	Presently at a level just below that of the main text of the Guide Adopted by: Commission for Hydrology
3. Manuals Content: Describes in greater detail the practices and procedures contained in the Guide	Same as (2)	As (2) above, but covering a specific topic in detail Adopted by: President of Commission for Hydrology

* Not yet existing in the field of hydrology.

(WMO, 1976, p. 91). The table shows that the highest level of standardization is provided by the WMO *Technical Regulations*, vol. III—Hydrology (WMO, 1975a), the following parts of which concern hydrometry:

Hydrological observing networks and stations:
  Classification of hydrological observing stations
  Networks of hydrological observing stations
  Location of hydrological observing stations
  Identification of hydrological observing stations
  Information relating to hydrological observing stations
  Supervision of hydrological observing stations

Hydrological observations:
  Composition of observations

Table 15.2   Parts of the WMO *Guide to Hydrological Practices* concerning hydrometry

---

Chapter 1   General
   1.4   Functions of hydrological services
      1.4.1   Basic data functions
   1.5   Organization of hydrological and meteorological services
      1.5.1   Existing patterns
      1.5.2   Technical considerations in organizational planning
      1.5.3   Recommended organization of hydrological and meteorological services
Chapter 2   Instruments and methods of observation
   2.4   Water levels of rivers, lakes and reservoirs
      2.4.1   General
      2.4.2   Gauges for measurement of stage
      2.4.3   Procedures for measurement of stage
      2.4.4   Frequency of stage measurements
   2.5   Discharge measurements
      2.5.1   General
      2.5.2   Measurement of discharge by current meters
      2.5.3   Measurement of discharge by float method
      2.5.4   Measurement of discharge by dilution method
      2.5.5   Measurement of corresponding stage
      2.5.6   Measurement of discharge by indirect methods
   2.6   Stream-gauging stations
   2.7   Sediment discharge
   2.8   Water temperature
   2.9   Ice on rivers, lakes and reservoirs
Chapter 3   Design of networks
   3.1   General principles for design of networks
   3.2   Density of observation stations for a minimum network
Chapter 4   Collection, processing and publication of data
   4.3   Streamflow computation
   4.5   Processing
   4.6   Publication

Observing and reporting programme for hydrological obsèrving stations
Equipment and methods of observation
Collection, processing and publication of hydrological data

However, the standardization and guidance effort is most comprehensive in the WMO *Guide to Hydrological Practices*. Its third edition (WMO, 1974b) contains exhaustive guidance on internationally recommended procedures and techniques in hydrometry. The parts of the Guide concerning hydrometry are indicated in Table 15.2.

15.4 MANUAL ON STREAM GAUGING

The most significant WMO contribution to hydrometric guidance is its *Manual on Stream Gauging* (WMO, in press), which provides detailed and extensive coverage of modern stream gauging procedures and can be used directly as an operator's (hydrometrist's) handbook in any country where detailed national manuals are missing.

15.5 TECHNICAL NOTES

The above basic guidance and standardization publications of WMO are supplemented in different fields of hydrometry by WMO Technical Notes. Such Notes exist on 'Measurement of peak discharge by indirect methods' (WMO, 1968), 'Machine processing of hydrometeorological data' (WMO, 1971a), and 'Use of weirs and flumes in stream gauging' (WMO, 1971b). Several problems inherent in hydrometry are also discussed in the WMO *Casebook on Hydrological Network Design Practice* (*WMO*, 1972). A review of hydrometric networks in all WMO member countries, together with names and addresses of national agencies in charge of the operation of hydrometric stations, is included in 'Statistical information on activities in operational hydrology' (WMO, 1977).

15.6 TECHNICAL COMMISSION FOR HYDROLOGY

All the above guidance material and publications have been basically prepared by working groups and rapporteurs of the WMO Technical Commission for Hydrology (CHy), assisted by the officers of the Secretariat in the WMO headquarters in Geneva. The CHy, and intergovernmental body of experts, meets at four-year intervals and examines new material, approves it or recommends its approval by other bodies, and decides on future work required. The last session of the CHy (in Ottawa, 1976) appointed a Working Group on Improvement and Standardization of Instruments and Methods of Observation for Hydrological Purposes, composed of Rapporteurs on Sediment Transport (measurement and estimation), New Methods of Discharge

Measurement, Levels and Discharge Measurements under Difficult Conditions, Intercomparison of Principal Hydrometric Instruments, Groundwater (measurement), and on Remote Sensing of Hydrological Elements.

Ways of preparing guidance material may vary from case to case, but the basic approach remains the collection of information on national instruments and practices, collation and discussion of the material by all members of a working group, and its final approval by an intergovernmental body. Table 15.1 indicates the bodies authorized to approve different levels of guidance and standardization material. The reader will note that this method ensures an international consensus and acceptance of the published material; it does not, however, provide for the co-ordination of guidance issued by different international organisations. To provide for such co-ordination, an Inter-Agency Panel on Standardization of Instruments and Techniques in Hydrology was established by WMO, UNESCO, FAO, WHO, IAEA, ISO and IAHS. WMO has provided the secretariat for this panel and its activities are described in WMO/IHD Report No. 18 (WMO, 1973).

15.7 INTERNATIONAL PROJECT ON INTERCOMPARISON OF PRINCIPAL HYDROMETRIC INSTRUMENTS

Projects requiring co-operation among a large number of countries and national institutions, and involving rather complex and sometimes difficult agreements on specifications and procedures for project implementation, cannot be conducted by other than an international governmental organization. WMO has wide experience in such projects; in the field of hydrology, the project on intercomparison of conceptual models used in operational hydrological forecasting (WMO, 1975b) has generated world-wide interest, some hydrologists (Chapman, 1975) considering that 'the project must go down as a most significant event in the history of hydrology'.

With such experience to hand, an international project for comparative tests of principal hydrometric instruments was launched by the CHy in 1972. In launching this project it was considered that instrument manufacturers are developing and marketing new types of instrument at short intervals, the parameters of which are, with very rare exceptions, superior to their forerunners. But, as a rule, the hydrological services make their instrument purchases in accordance with their annual budgets, the result of which may be a highly miscellaneous assortment of instruments, together with their inherent drawbacks, such as difficulties in obtaining spare parts, rapid replacement, training, etc. Within each major regional unit, e.g. within the same project, the number of types of instrument serving the same purpose should be limited to a few. It will be noted that, although efforts at modernization are undeniably important, a great variety of types is undesirable in a network. To achieve optimum operation any change in type should be preceded by a careful analysis of intercomparison tests. It is advisable to decide in advance which identical type of instrument will be introduced in fairly large numbers during the reconstruction of a network. Considerable benefits accrue from large orders. However, large orders involve greater risks of poor choice, thus emphasizing the need for careful intercomparison before a particular type is decided upon.

It is generally true that instruments of growing complexity are marketed at correspondingly higher prices. The instruments require less frequent attendance, but often better-trained operators. Whereas the simple, mechanical, clockwork-driven recording gauges called for no special skill, electric/electronic telemetering instruments can be repaired by skilled technicians only. Thus not only are the purchasing costs higher, but higher expenditure must be envisaged for the operation and maintenance of the already expensive network.

In many instances the factors dominating the comparison are the purchase costs, although the costs of operation and maintenance are also of interest. Cases are known in which funds could be mobilized to establish an expensive network but, owing to financial difficulties, operation and maintenance were unsatisfactory and the network deteriorated within a short time.

Besides balancing technical and economic parameters, consideration must also be given to environmental influences. Reliability of operation may in some instances become the consideration of paramount importance and the purchase of a smaller number of more reliable and more expensive instrument may be advisable, instead of a larger number of less reliable and cheaper types. Hydrologists prefer less reliable and continuous time series to a great number of unreliable, inaccurate and incomplete ones. It may thus be noted that economics must not be allowed to play a decisive role in the choice, but should be considered only when the cheaper instruments have identical technical (operational) parameters.

Other factors, which are difficult to include in a comparative study of a general character, include the import policies of a particular country, the terms of payment, of delivery, other services of the manufacturer, such as co-operation in installation, training of personnel, guaranteed supply of spares, a local pool of spares, servicing, etc.

All these considerations demonstrate strikingly that the results of any intercomparison furnish no more than the basis for the choice of a type of instrument, without relieving the professional responsibility of the technician in charge of the purchase.

It is obvious that not all hydrological services, and least of all those in the developing countries, can afford the funds, time and skill required for intercomparison tests, and thus the WMO initiative was most welcome. The project is implemented in two phases, the first consisting of testing current meters and water level recorders owned (but not necessarily produced) by each participating country (its hydrometric agencies). In the first phase, the technical, operational and economic characteristics of the instruments were noted and tests conducted in laboratory and field according to detailed

Table 15.3   List of water level recorders tested

Country (where tests performed)	Instrument offered for testing	Implementation* (%)	Additional instruments tested	Implementation (%)
France	Ott, XV	100	LAG 2	100
	Ott, X	100	Seba Omega	100
	Ott, XX	100	Nivotran	100
	Neyrpic, Telimnip	100		
	Ott, R16. (20.250)	100		
Germany, FR	Ott, X		Brackish water gauge	45
	Ott, XX			
	Fuess			
	Alpina			
	Seba Delta	65		
	Hagenuk			
	Killi			
	Ott, 20.095	65		
	Metrawatt			
Hungary	VR-1	100	VRD-1	100
	Ott, 20.061	100		
	Fischer-Porter	100		
	Metra 501	100		
Japan	OKO			
	Kyowa-Syoko LFT-3	65		
	Uijin IS-30	60		
	Uijin Fw-25	60		
	Ogasawara FI-460B	60		
	Ogasawara FL-200	50		
	Nakaasa Suiken 62	75		
	Ogasawara FL-710	35		
	Ogasawara WL-100	40		
	Takuwa VM-10	60		
	Sokkisha ND-7	90		
	Sokkisha NA-7	90		
	Sokkisha NA-R	90		
	Sokkisha NA-WF	75		
	Sokkisha NDC-30	90		
	Sokkisha NAC-30	90		
Netherlands	Ott, X		Plessey 9610 Wave	95
	Ott, XX	80	Datawell Wave	80
	Ott, 20.061			
	Fisher-Porter	85		
	Pneumaticlevel, Prikkler	10		
	Negretti G			
	Van Essen 9102 (Golfmeter)	30		
	Joens			
	Helios	90		

Table 15.3 (cont.)

Country (where tests performed)	Instrument offered for testing	Implementation* (%)	Additional instruments tested	Implementation (%)
Spain	Ott, X	50		
	Seba, X	50		
	SIAP, Normal	50		
	SIAP, Two-pen	50		
	Ott, XX	50		
	Neyrpic, Telimnip	50		
UK of Great Britain and N. Ireland	Stevens 7000 (punched tape)	100		
	Leupold & Stevens A71	100		
	EHY (Bubble-type)	90		
	Muntro IH95	90		
	Ott, R16. (20.250)	100	Ott, X	100
USA	Fisher-Porter	100		
	Leupold & Stevens A35	100		
	Friez FW			
	Leupold & Stevens F			
USSR	LPU-10	40	Metra 501	40
	KB-2	40		
	GR-38	40		
	Walday	40		

* Implementation = per cent executed of the full programme of tests suggested by WMO specifications.

Table 15.4   List of current meters tested

Country (where tests performed)	Instrument offered for testing	Implementation* (%)	Additional instruments tested	Implementation (%)
France	Ott, Arkansas	80	Traceurs radioactifs	50
	Ott, C-31	80	BEN CM01	65
	Ott, C-1	40	Ott, C-2	40
	Neyrpic, Dumas	30	Neyrpic, Neyrflux	80
Germany, FR	Ott, C-31			
	Ott, Texas	25		
	Killi			
	Seba			
Guyana	Ott, Arkansas			

Table 15.4 (cont.)

Country (where tests performed)	Instrument offered for testing	Implementation* (%)	Additional instruments tested	Implementation (%)
Hungary	Ott, C-31	110		
	Ott, C-1	110		
	VITUKI, M-1	120		
	Zhestovski, Zs-3	110		
India	Watts	80		
Japan	OKI	35		
	Takuwa	20		
	Takuwa CM200	30		
	Kyowa-Syoko N-II.	50		
Netherlands	Ott, Arkansas	25	Plachsee	35
	Ott, C-31	60		
	Ott, C-1			
	Colnabrock Zdim			
	Hydrowerkstätten Shallow			
	Plessey No. 21	25		
	Ott, Current			
	Pen, Current			
	TPD/TNO			
Spain	Ott, Arkansas	30		
	Ott, C-31	20		
	Ott, Torpedo Weights	15		
	Ott, C-1	20		
	Ott, Permanente	5		
	Gurley	5		
	SIAP	15		
	Neyrpic, Dumas	15		
UK	Braystoke	90		
USA	Price	65		
	Price Pygmy	50		
	Ice meter	50		
	Gurley 665			
	Hydro Products			
	Brain Com 316			
USSR	GR-21 M	40		
	GR-55	40		
	GR-42	20		
	Metra 560	20		

* Implementation = per cent executed of the full programme of tests suggested by WMO specifications. If above 100 per cent, additional tests performed.

specifications, to ascertain the accuracy, sensitivity, reliability and durability of the instrument under different conditons. The instruments tested are indicated in Tables 15.3 and 15.4 (Starosolszky and Muszkalay, 1977).

Forty-eight different types of water level recorders were tested, the full test programme being performed on 16 of these. Twenty-seven types of current meters were tested, only four of which completed the full test programme. For the results of the tests the reader is directed to the relevant WMO publication. Their evaluation indicates that, for water level recorders, large errors are inherent in some instruments under specific conditions. For float-operated recorders, the steel tape movement transmission instruments are the most accurate, and the bubble gauges the least precise. For current meters, the vertical axis instruments (cup or 'Price' type) are suitable only for steady flow conditions but are very accurate, the horizontal axis instruments being more universal. The uncertainties observed (0–16 per cent) are conditioned more by the flow conditions than by the instrument itself.

For the second phase of the project it is planned to exchange instruments so as to achieve a maximum objectivity. The plan envisages the testing of a selected few of the instruments exchanged, at a limited number of places representing as many field conditions as possible. It has been suggested that, in addition to water level recorders and current meters, ultrasonic flow meters, sediment samplers and snow density gauges be included in the inter-comparison. Preparations for the second phase are under way, under the auspices of CHy.

15.8 HYDROMETRIC FIELD OPERATIONS WITHIN THE WMO TECHNICAL CO-OPERATION PROJECTS IN DEVELOPING COUNTRIES

The WMO's technical co-operation activities include many projects of assistance to national hydrological services in developing countries. The majority of these projects, for which external contributions are provided under multilateral assistance programmes—be it the UNDP (United Nations Development Programme) or the WMO VAP (Voluntary Assistance Programme) or under bilaterally provided funds-in-trust—are aimed at the establishment of hydrometric networks. During the past few years the WMO has undertaken field projects comprising hydrometric activities in the following countries: Afghanistan, Bolivia, Brazil, Colombia, Guinea, Malawi, Mongolia, Pakistan, Paraguay, and the Philippines. The largest projects undertaken are, however, of a regional nature, involving co-operation among several countries, most often in large international river basins. Such projects have been undertaken in Africa in the basins of lakes Victoria, Kyoga and Mobutu Sese Seko—the Upper Nile (supporting co-operative endeavours of the governments of Burundi, Egypt, Kenya, Rwanda, Sudan, Tanzania, Uganda and Zaïre), in the upper basin of the River Niger (supporting co-operative endeavours of the governments of Guinea and Mali), agrometeorology and hydrology of the Sahelian countries (Benin, Cameroon, Cap Verde, Gambia, Mali, Mauritania, Niger, Upper Volta), and in Central America (Costa Rica, El Salvador, Guatemala, Honduras, Nicaragua, Panama). WMO-supported programmes for warning and prevention of the disastrous effects of tropical cyclones (also called hurricanes or typhoons), involving flood forecasting and thus hydrometric activities, are centred in three regions of the world: South-West Asia (the Typhoon Committee), Bay of Bengal (the Tropical Cyclone Committee) and the Caribbean (the Hurricane Committee), integrating respectively actions by almost all the governments of these regions.

To give the reader some idea of the extent of the operations in the above projects, two examples are briefly described. In the hydrometeorological Survey of the Catchments of Lakes Victoria, Kyoga and Mobutu Sese Seko, 45 stream gauging, 5 staff gauge and 10 lake level stations were established with the assistance of the WMO/UNDP project. At these stations, 1 self-propelled, 5 stationary and 6 portable cableways were installed. Sediment and water quality sampling and processing was initiated at 24 locations. Seven small index catchments were instrumented for intensive study of the rainfall runoff relationship.

In the project for the Extension and Improvement of the Hydrometerorological and Hydrological Services of the Central American Isthmus, 284 stage recording/stream gauging stations and 53 staff gauges were established. At 136 of these stations, sediment samples are taken and analysed. A current meter calibration tank was installed in Nicaragua for the region, with a repair shop servicing, in addition to the Central American countries, parts of the Caribbean and of South America. Computerized data processing (including stage–discharge evaluation) has been initiated in all the countries of the project and a regional streamflow yearbook is published regularly. The project has also published over 100 guidance and training publications in Spanish, including a stream gauging manual which is very popular in many Spanish-speaking countries of Latin America.

15.9 WMO TRAINING PROGRAMME IN HYDROMETRY

One of the most important activities of the WMO field projects is the training of qualified personnel from the level of observers, hydrometrists and hydrometric technicians to that of professional hydrologist. Indeed, both inside and outside its projects, WMO has organized perhaps a larger number of training courses in hydrometry than any other national and/or international institution involved in this field. The latest is a two-year course organized in co-operation with the French institution ORSTOM for training hydrometrists and hydrological technicians in Niamey (Niger) for hydrological services of the French-speaking Sahelian countries. These courses are acquiring a permanent, biennial character and they may well become the basic training centre of hydrometrists for the whole of Africa using French as the first foreign language. The courses have, among other subjects, 280 hours of theoretical and 640 hours of practical training in hydrometry, over two years, and are equipped with modern instrumentation including cableways and boats—although adapted to the conditions of Africa.

By far the largest training activity is, of course, performed on the job, by WMO staff and specialized instructors, on an individual basis.

15.10   CONCLUSIONS

It is obvious that, in such large-scale projects in a number of countries and variety of climatological and hydrological conditions, the activities of WMO staff members yield a large amount of practical experience and provide solutions to many hydrometric problems. To start with, this experience is gained in gauging the largest rivers of the world, be it the Amazon or the Nile. Problems of current meters have had to be solved in mountain rivers in Colombia, where suspended sediments destroyed all the Ott Arkansas meters, which had to be replaced by the Ott Universal. Damage to hydrometric installations by wildlife (in particular hippopotami and elephants) had to be prevented in the Lake Victoria project. In areas visited by nomadic tribes in the Sahel, special arrangements had to be made to prevent damage to recorders and staff gauges. These examples are only a few of the many problems which WMO experts have had to solve in hydrometric activities in assisting technicians of developing countries to collect data on flow in rivers all over the world.

It can therefore be confidently concluded that, with respect to instruments and methods of measurement, and all other practices in hydrometry, be it in the provision of guidance, intercomparison and testing, or field activities, WMO's experience ranks alongside that of the major hydrometric services of the world. With respect to variety of conditions and world-wide experience in hydrometry, WMO is probably second to none. It is therefore only natural that, at the International Seminar on Modern Developments in Hydrometry (WMO, 1975c), convened in 1975 by WMO in Padua, Italy, some 70 world-renowned specialists confirmed the leading role of the World Meteorological Organization in international co-operation in this basic field of hydrology, without which no rational water resources development can be conceived.

ACKNOWLEDGEMENTS

The author wishes to express his gratitude to the World Meteorological Organization and its Secretary-General, Dr D. A. Davies, for the kind permission to use WMO material and documents for this publication. The views expressed herein are, however, those of the author as an individual and should not be interpreted as necessarily representing the official views of the WMO.

REFERENCES

Chapman, T. G., 1975. Trends in catchment modelling, in *Prediction in Catchment Hydrology*, Australian Academy of Science, Canberra.

Chebotarev, A. T., 1964. *Hydrologic Glossary* (in Russian), Gidrometeoizdat, Leningrad.

Němec, J., 1976. International espects of hydrology, in *Facets of Hydrology*, Wiley, London.

Starosolszky, Ö. and Muszkalay, L., 1977. Report on intercomparison of principal hydrometric instruments (first phase, 1972–75), WMO, Geneva.

WMO, 1968. Measurement of peak discharge by indirect methods, by M. A. Benson, TN No. 90 (WMO—No. 225.TP.119), WMO, Geneva.

WMO, 1971a. Machine processing of hydrometeorological data, TN No. 115 (WMO—275), WMO, Geneva.

WMO, 1971b. Use of weirs and flumes in stream gauging, TN No. 117 (WMO—No. 280), WMO, Geneva.

WMO, 1972. *Casebook on Hydrological Network Design Practice* (WMO—No. 324), WMO, Geneva.

WMO, 1973. Standardization in hydrology and related fields—activities of FAO, Unesco, WHO, WMO, IAEA, ISO, IAHS; WMO/IHD Report No. 18 (WMO—No. 351), WMO, Geneva.

WMO, 1974a. *International Glossary of Hydrology*, 1st edn (WMO/Unesco) (WMO—No. 385), WMO, Geneva.

WMO, 1974b. *Guide to Hydrological Practices*, 3rd edn (WMO—No. 168), WMO, Geneva.

WMO, 1975a. *Technical Regulations*, vol. III—Hydrology (WMO—No. 49), WMO, Geneva.

WMO, 1975b. Intercomparison of conceptual models used in operational hydrological forecasting, Operational Hydrology Report No. 7 (WMO—No. 433), WMO, Geneva.

WMO, 1975c. *Modern Developments in Hydrometry*, vols. I and II, Proc. WMO/Unesco/IAHS/Univ. Padua Seminar (WMO—No. 427), WMO, Geneva.

WMO, 1976. Abridged final report of the fifth session of the Commission for Hydrology (WMO—No. 453), WMO, Geneva.

WMO, 1977. Statistical information on activities in operational hydrology, Operational Hydrology Report No. 10 (WMO—No. 464), WMO, Geneva.

WMO (in press). *Manual on Stream Gauging*, prepared by CHy Working Group on Hydrological Instruments and Methods of Observation, WMO, Geneva.

CHAPTER 16

International Standardization in River Flow Measurement

K. K. FRAMJI

16.1 INTRODUCTION

Although the origin of hydrology is somewhat arbitrarily traced to the year 1674—when Pierre Perrault published his book *De l'origine des fontaines*— the first recognition of the need to develop international co-operation in this field dates back to only 50 years ago, when the International Association of Scientific Hydrology (IASH) was established. Several circumstances have been said to bring about this development amongst which the most important are:

(1) Most developing countries were faced with the need of making better use of natural resources on the one hand and with water resources development being hindered by lack of hydrological data on the other hand.

(2) International co-operation offered the promise of overcoming the difficulties of the inadequacy of hydrological data, of the lack of institutional infrastructure and of the need for specialists.
(3) The development of measurement equipment and techniques, due to advances in the related modern sciences of electronics and nuclear techniques, and of computers, models and modern materials, have made the exchange of information between countries better and indeed necessary.

16.2 INTERNATIONAL ORGANIZATIONS ENGAGED IN HYDROMETRY

As well as the International Organization for Standardization (ISO), there are a number of international organizations which are active in the field of hydrometry. In a key paper 'Prospects of hydrometry in the light of modern technology', presented at the celebration of the Three Centuries of Scientific Hydrology, Paris, 9–12 September, 1974, the author endeavoured to deal with the achievements and aims of the international organizations in the search for new and improved instruments and techniques. In regard to the activities of various international organizations working in this field it was said:

> In the field of hydrometry there are several international organizations which are active but with different aims. The total interest of ISO lies in the development of standards on established methods and equipment for measurements and observations; others (e.g. FAO[1], ICID[2], WHO[3]) have the interest of a 'user' of the standards; others still (e.g. WMO[4]) are active both as promoters of standardization and in using the standards for making observations, while the IAHS[5] works and aims at promoting progress on new measuring methods and instruments and the IAEA[6] for the employment of special nuclear techniques in the measurements.
>
> For advancing the trends towards improvement in the accuracy or the performance of conventional instruments, the WMO project on the Intercomparison of Hydrometric Instruments is of special significance. Under this project the instruments most frequently used in surface water hydrometry, namely water level recorders for stage measurements and current meters for measuring velocities, are proposed to be scientifically intercompared by independent laboratory and field tests.

[1] Food and Agriculture Organization of the United Nations.
[2] International Commission on Irrigation and Drainage.
[3] World Health Organization.
[4] World Meteorological Organization.
[5] International Association of Hydrological Sciences.
[6] International Atomic Energy Agency.

At the intergovernmental level, international co-operative actions have been fostered in the International Hydrological Decade sponsored by Unesco[7] (followed by the Unesco long-term International Hydrological Programme started in 1975) and in the Operational Hydrology Programme developed by the WMO. The latter concentrates on the development of hydrological networks, the improvement of data measurements, collection and processing and in ensuring the supply of hydrological data for design purposes and the development of forecasting services.

16.2.1 ISO–WMO liaison

The WMO assumes the technical secretariat to the Inter-Agency Panel of Standardization of Instruments, Methods and Techniques, which held its first session in Geneva in 1971, and its second session in Padova in 1975. At these sessions a review was made of the standardization activities of each participating organization[8].

A representative of WMO is usually present at the ISO/TC 113 meetings. It has been mutually accepted that there is a need for close co-operation at the working level between ISO/TC 113 and the WMO Commission for Hydrology in order to minimize duplication of efforts and to ensure that the ISO Standards are acceptable to the WMO Commission for Hydrology and the WMO *Guide to Hydrological Practices* and its Technical Regulations. The Inter-Agency Panel on Standardization of Instruments, Methods and Techniques in Hydrology also recommended, in general, that any publication of particular interest to one agency such as ISO should also be circulated for review to the other agency before it is published. At the appropriate stage of the evolvement of the ISO Standards, therefore, these are sent to the WMO for information and comment.

At the fifth session of the WMO Commission for Hydrology (CHy-1V) in Ottawa, in 1976, the Commission noted with appreciation the information on ISO activities in the field of hydrology provided by the ISO Central Secretariat in Geneva. It noted that the activities of Technical Committee ISO/TC 113, 'Measurement of liquid flow in open channels', and ISO/TC 147, 'Water quality', were of direct concern to the Commission for Hydrology and that close co-operation between these two committees and the WMO Secretariat had been established during the CHy-1V inter-sessional period. The Commission realized that several of its activities were closely related with the current activities of these committees. In order to obviate any possible duplication of effort and to ensure that ISO standards are also acceptable to it, the Commission requested the WMO Secretariat to develop, in co-operation with the ISO Central Secretariat, mutual procedures which would ensure the avoidance of duplication and the acceptability of the standards. In this

[7] United Nations Education, Scientific and Cultural Organization.
[8] FAO, IAEA, Unesco, WHO, WMO, ISO, IAHS.

Table 16.1 International liaison

International organization/ committee	Secretariat
ISO/TC 30 and ISO/TC 115	AFNOR[1], Paris
ISO/TC 147	American National Standards Institute
IEC/TC 4[2]	American Society of Mechanical Engineers
ISO/TC 59/SC 103	British Standards Institution
Liaison organization—Category A:	
ICID[3]	Secretary-General, ICID, India
WMO	Secretary-General, WMO, Geneva
Liaison organization—Category B:	
1. ESCAP[4]	Executive Secretary, ESCAP, Sala Santitham, Bangkok, Thailand
2. IAHR[5]	Secretary, IAHR, c/o Hydrological Laboratory, 61 Raam, Delft, Netherlands
3. IAHS	Secretary-General, Assoc. Internationale d'Hydrologic Scientifique 61, Avenue des Rones, Gentbroglee, Belgium
4. PIANC	Secretary-General, Association Internationale Permanente des Congres de Navigation, 60 rue Juste Lipse, Bruxelles, Belgium
5. ICMG (International Current Meter Group)	Chief of the Special Service Section, Hydrological Department, Swiss Federal Water Resources Bureau, 27, Bollwork, CH-3011 Berne, Switzerland

[1] Association Francaise de Normalization.
[2] International Electrotechnical Commission.
[3] International Commission in Irrigation and Drainage.
[4] Economic and Social Commission for Asia and the Pacific (UN).
[5] International Association of Hydraulic Research.

connection, the Commission noted with appreciation that specific action had been taken by WMO and other international organizations to co-ordinate standardization activities of common interest through the efforts of the Inter-Agency Panel on Standardization of Instruments, Methods and Techniques in Hydrology, whose activities, as well as the standardization activities pursued by the WMO in co-operation with other organizations, are described in WMO/IHD Report No. 18.

In view of the parallel activities of the two organizations it has been appreciated that there should be strong liaison between the WMO and the ISO. It is also important that WMO and ISO should continue to be the organizations operating in the field of international hydrometry, the former in respect of Technical Regulations and recommended practices, and the latter in the field of standardization of methods and equipment. It is, therefore, necessary that the areas in which each organization are involved are therefore clearly defined as follows:

Technical Regulations, Recommended Practices	WMO Technical Regulations, WMO *Guide to Hydrological Practices*
Suggested Practices	Technical Notes, Operational Hydrology Reports, Casebooks, Glossaries, Manuals and Proceedings
Standardization of methods and equipment	ISO standards

16.2.2 Liaison with other international organizations

Liaison or co-ordination has been established with other committees of ISO and other international organizations as shown in Table 161.

16.3 THE WORK OF THE ISO

Prior to 1954, international standardization of the ISO in river flow measurements formed a part of its work of standardization of measurements of fluid flows in general. On the suggestion of India, in 1954 the Technical Committee ISO/TC 30 of the International Organization for Standardization, which was then dealing with the standardization of fluid flow measurement, took upon itself the task of standardization of flow measurement in open channels and set up a separate Sub-Committee (SC 1) to deal with the subject. On account of its long experience in the field, India was requested to assume the Secretariat of this Sub-Committee—ISO/TC 30/SC 1— Measurement of Liquid Flow in Open Channels. The Sub-Committee comprising nine Participating (P) members—Belgium, France, Germany, Italy, Roumania, UK, USA, USSR and India (Secretariat) and 3 Observer (O) members, held its first meeting in Munich in 1956 under the Chairmanship of Shri Kanwar Sain of India, the then Chairman of the Central Water and Power Commission, Government of India. The Sub-Committee had before it for consideration four draft proposals from India relating to: glossary of terms used in measurement of flow of water in open channels; measurement of flow of water in open channels by the velocity area method; measurement of flow of water through open channels using notches, weirs and flumes; and standard forms of recording measurement of flow of water in open channels.

At the first meeting, the Sub-Committee constituted three Working Groups (WG) namely:

WG 1—Liquid flow measurement by velocity area methods.
WG 2—Liquid flow measurement by notches, weirs and flumes.
WG 3—Glossary of terms relating to liquid flow measurement in open channels.

Dr Wittmann (Germany), Prof. Schlag (Belgium) and Mr Griffiths (UK) were elected as the Chairmen of the Working Groups WG 1, WG 2 and WG 3, respectively.

At its second meeting in 1958, in London, under the Chairmanship of Shri Kanwar Sain, the Sub-Committee decided to constitute another Working Group (WG 4) to deal with measurement of liquid flow in open channels by dilution methods with Mr Remenieras of France as its Chairman.

Taking into account the large amount and specialized nature of the work to be processed in the field of liquid flow measurement in open channels and its importance for measuring river and canal discharges uniformly and accurately, the ISO Council at its meeting held in New Delhi in 1964 agreed to elevate this Sub-Committee to a Technical Committee—ISO/TC 113— and prescribed the following scope:

SCOPE—Standardization of rules and methods relating to different techniques for the measurement of liquid flow with particular reference to sediment transport in open channels including:
(1) terminology and definitions,
(2) rules for inspection, installation, operation,
(3) instruments and equipment required,
(4) conditions under which measurements are to be made, and
(5) rules for collection, evaluation, analysis and interpretation of measurement data including errors.

The Indian Standards Institution (ISI) was asked to hold the Secretariat for the Committee.

ISO/TC 113 held its first plenary meeting in London in 1965 under the Chairmanship of Shri K. K. Framji, formerly Chief Engineer and Joint Secretary, Ganga Basin Water Resources Organization, Ministry of Irrigation and Power, Government of India. At this meeting the Technical Committee set up two additional Working Groups—WG 5 Flow measuring instruments and equipment, and WG 6 Sediment flow (renamed Sediment transport).

The second plenary meeting of TC 113 was held in Paris in 1968. After an interval of five years due to the unforeseen circumstances the Secretariat had to face, the third plenary meeting of ISO/TC 113 was held in New Delhi in 1973.

In the recent past the back-log of the years 1968 to 1973 has been successfully cleared by meetings held annually, in The Hague (1974), Zurich

(1975) and London (1976). The seventh meeting of the Technical Committee and all its Working Groups was held in 1977 in Washington (USA). At this meeting the Working Groups were elevated to the status of Sub-Committees and a Seventh Sub-Committee added (SC 7 Measurement of flow under difficult conditions).

At present, the Technical Committee comprises 13 P and 18 O member countries from various parts of the world, but efforts are afoot to give this work wider coverage by encouraging member countries wherever water flow in open channels is being measured to become active participants and give the Committee the benefit of their experience and knowledge.

16.4 PROGRAMME

The Technical Committee has been very active in the recent past, thanks to the active participation of many countries and the Chairmen of the Sub-Committees for whom no details are too small and no trouble too much for dedicating themselves to the systematic and efficient analysis of methods and equipment for these measurements. The assistance of a competent Secretariat is available to all seven Sub-Committees and to the Technical Committee and contributes greatly to the progress of the work. The Chairmanship of the Sub-Working Committees has been distributed amongst the various countries most interested in the field and currently are:

SC 1—R. W. Herschy (UK).
SC 2—J. P. Th. Kalkwijk (Netherlands).
SC 3—R. A. Halliday (Canada).
SC 4—R. Roche (France).
SC 5—F. Smoot (USA).
SC 6—K. K. Framji (India).
SC 7—Y. G. Filippov (USSR).

The progress so far achieved by, and the programme of, the individual Working Groups (now Sub-Committees) are indicated in the following section. A list of Standards is given in the Appendix.

16.4.1 Sub-Committee 1—Velocity-area methods

Systematic revision of standards which were prepared over five years ago, are being undertaken with dynamism in keeping with the changes that are taking place in scientific technology and the experience gained in the use of the standards in practice in different countries. Accordingly, the Standards—ISO: 748 'Liquid flow measurement in open channels by velocity-area method'; ISO: 1070 'Liquid flow measurement in open channels by the slope-area method; ISO: 1100 'Establishment and operation of a gauging station and determination of stage–discharge relation'; and ISO: 2425 'Measurement of flow in tidal channels'—are all currently under revision.

In the current revision of ISO: 748, provisions for measurements under ice cover have been added to guide the practices in countries such as Canada or countries in Europe or America, where the streams and rivers are frozen during winter. New and fast methods of measurement, such as the moving boat method and the ultrasonic method of measurement have been or are also being added. In fact, the moving boat method has already been established as the most useful method for measurements of large discharge rivers in a short measuring time and promises to be the method of the future for such conditions.

In ISO: 1100, the methods continue to be distinguished, as they should be, between those appropriate for stable channels and those suitable for unstable channels, since there is indeed a sharp distinction in the behaviour of the two kinds of open channel. As in the original ISO 1100, simple statistical tests are proposed to be provided in the revised version to establish the objectivity and freedom of bias of the stage–discharge curves drawn by eye from very different values of the points under conditions of instability in the unstable channels. Two or more separate curves at each change in control for the stage–discharge curves have been advocated and the method to determine the changes in control are specified. The section on *errors*—or as more correctly now termed *uncertainties*—has been amplified as a result of an in-depth study made for this purpose. Originally, examples of errors in measurements by velocity-area methods computed from component uncertainties (widths, depths, velocities) had been specified for a wide range of rivers or open channels (3000–8000 $m^3 s^{-1}$). Preceding the revision of this error section, the Working Group had, after much deliberation, formulated ISO 1088 'collection of data for the determination of errors in measurement of liquid flow by the velocity area method'. Based on the information received from different parts of the world in accordance with the requirement of this standard, elaborate research was done, first by the national standards counterpart of ISO/TC 113 from the Netherlands, and then by the UK and Dutch members, and the foundation was laid for the revision of the *errors* section in so far as the velocity-area method of measurement for flow in open channels was concerned. As stated subsequently, this chapter on *uncertainties* was then jointly finalized in co-operative collaboration with the concurrent work being done on the same subject by Technical Committee TC 30 (Measurement of fluid flow in closed conduits).

16.4.2 Sub-Committee 2—Notches, weirs and flumes

Standards on 'Liquid flow measurement in open channels by the use of thin plate weirs and flumes' (ISO 1438), by *broad crested weirs* (ISO: 3846), and by *the brink depth method* required for measurement of waste water and sullage from a free overfall in rectangular channels (ISO: 3847), have already been published, while other standards for measurement by *triangular profile weirs* (ISO: 4360), by *flat V-weirs* (ISO: 4377) and by *round nosed weirs* (ISO: 4361) are under printing. Revision has been taken up of ISO: 1438 for measurement by *thin plate weirs*, while a separate standard on measurements by *standing wave flumes* (ISO:4359) has been finalized for different throat sections, namely rectangular, trapezoidal and U-shape. To cover the measurements of flows for varying sections and varying discharges of the river, a proposal on measurement by compound weirs is currently under consideration.

A useful contribution of Sub-Committee 2 is the Appendix, 'Guide for the selection of weirs and flumes for the measurement of the discharge in open channels'. This guide lists the criteria for selection of the appropriate structure and the accuracy of measurements by different types of weirs and flumes under steady, uniform flow conditions.

16.4.3   Sub-Committee 3—Glossary of terms

The original (1st edition Recommendation 1968; Standard 1973) ISO: 772 'Vocabulary of terms and symbols used in connection with the measurement of liquid flow with a free surface' has been revised and supplemented according to the principle that there should be a complete revision after every four years and an intermediate supplement should be provided in between every two years. The Sub-Committee is now engaged on the revision for unification of the symbols, having regard to its own and other ISO standards and the work of other international organizations such as WMO, etc.

16.4.4   Sub-Committee 4—Dilution methods

The Sub-Committee's existing standards—ISO: 555/I 'Constant rate injection method' and ISO: 555/II 'Sudden injection method'—are currently being revised with particular reference to the controversial views on the appropriate 'mixing length' and the corrections for unsteady flow. A draft proposal is also under review relating to measurement of flow in open channels using radioactive isotopes, having particular regard to the health hazards to human and fish life and the pollution problems entailed.

16.4.5   Sub-Committee 5—Flow measuring instruments and equipment

Recent important contributions by this Working Group have been the Standards, (completed but under voting for approval) on *cableway systems* (ISO: 4375), *water level measuring equipment* (ISO: 4373) and *echo sounders* (N 129). On the anvil is a draft proposal on *acoustic velocity meters* (N 139). With the rapid advances in electronics communication systems and the growing need for data retrieval from distant observation stations, there has been the need for development in many countries of hydrometric telemetry methods and a standard on the subject is currently under active consideration. It is the consensus of view of ISO/TC 113 that there should be no duplication or overlapping with the WMO in this area, but on the other hand there will be useful co-operative action by the separate efforts of the WMO and ISO, because the approaches, needs and contents will be different but of mutual interest.

16.4.6 Sub-Committee 6—Sediment transport

In this relatively new field of flow measurement, extreme care has had to be exercised to ensure that standards are not ventured upon in advance of crystallization of knowledge and experience in an admittedly uncertain field. Therefore, with the utmost caution standards have been formulated only on the *functional requirements and characteristics of suspended sediment load samplers* (ISO: 3716), on *methods of measurement of suspended sediment* (ISO: 4363) and of *bed material samping* (ISO: 4364). A proposal on the *methods for analysis of sediment concentration, size and specific gravity* (N 81) is near completion. The programme of work has an impressive array of hopeful and helpful items, but the way is difficult, especially where bed load measurement and equipment are concerned. The *uncertainties* in measurement of sediment transport are even far more complex to assess than those of measurement of liquid flow in open channels.

16.4.7 Sub-Committee 7—The measurement of flow under difficult conditions

The objectives of this Sub-Committee are in the field of flow measurement under extreme natural conditions. In particular it will be concerned with arid and semi-arid regions, remote regions and regions of the world where ice conditions are present. The Chairman was appointed at the eighth meeting of TC 113 in Leningrad in 1978.

16.4.8 Technical Committee TC 113

Conscious of the great responsibilities and the *uncertainties* in the complex subject of standardization of methods and equipment for the measurement of flow in open channels, Technical Committee TC 113 has established liaison with the other international bodies interested in the same field, such as other Technical Committees of the ISO and the IEC or other international organizations, for example WMO, IASH, IAHR, and ICID. In particular, TC113 functions in close co-ordination with TC 30 and TC147, the former on the subjects of errors and uncertainties in the measurements of flow in closed conduits, and the latter in measurements relating to quality of water.

APPENDIX INTERNATIONAL STANDARDS FOR FLOW
MEASUREMENT IN OPEN CHANNELS

ISO: 555/1	1973	Liquid flow measurement in open channels—dilution methods, Part I Constant rate injection method.
ISO: 555/11	1974	Liquid flow measurement in open channels—dilution methods, Part II Sudden injection method.
ISO: 748	1973	Liquid flow measurement in open channels by velocity-area methods. (Revision expected to be published in 1978.)
ISO: 772	1973	Vocabulary and symbols. (Revision expected to be published in 1978.)
ISO: 1100	1973	Liquid flow measurement in open channels—establishment and operation of a gauging station and determination of the stage–discharge relation. (Under revision.)
ISO: 1070	1973	Liquid flow measurement in open channels—slope-area method. (Under revision.)
ISO: 1088	1973	Collection of data for determination of errors in measurement by velocity-area methods. (Under revision.)
ISO: 2425	1974	Measurement of flow in tidal channels. (Under revision.)
ISO: 2537	1974	Liquid flow measurement in open channels—cup-type and propeller-type current meters.
ISO: 1438	1975	Thin plate weirs and flumes.
ISO: 1438	1979*	Thin plate weirs. (Revision of ISO: 1438, 1975).
ISO: 3454	1975	Liquid flow measurement in open channels—sounding and suspension equipment.
ISO: 3455	1976	Liquid flow measurement in open channels—calibration of current meters in straight open tanks.
ISO: 4363	1977	Methods of measurement of suspended sediment in open channels.
ISO: 4364	1977	Bed material sampling.
ISO: 3716	1977	Functional requirements and characteristics of suspended sediment load samplers.
ISO: 3846	1977	Liquid flow measurement in open channels by weirs and flumes—rectangular broad crested weirs.
ISO: 3847	1977	Liquid flow measurement in open channels by weirs and flumes—end depth method.
ISO: 4359	1978*	Liquid flow measurement in open channels—flumes.
ISO: 4360	1978*	Liquid flow measurement in open channels by weirs and flumes—triangular profile weirs.
ISO: 4361	1978*	Liquid flow measurement in open channels by weirs and flumes—round nosed broad crested weirs.

ISO: 4373 1978* Water level measuring devices.
ISO: 4369 1978* The moving boat method.
ISO: 5168 1978   Calculation of the uncertainty of a measurement of flow rate.
ISO: 4377 1978* Flat V-weirs.
ISO: 4375 1978* Cableway system.
DP: 6418 1978* Ultrasonic (acoustic) velocity meters.
ISO: 4366 1978* Echo Sounders.

Reports

ISO: Data 1978* Investigation of the total error in measurement of velocity-area methods

* Expected publication date.
Note: In 1973 all existing Recommendations became Standards.

Index